Lecture Notes in Mechanical Engineering

Lecture Notes in Mechanical Engineering (LNME) publishes the latest developments in Mechanical Engineering—quickly, informally and with high quality. Original research reported in proceedings and post-proceedings represents the core of LNME. Volumes published in LNME embrace all aspects, subfields and new challenges of mechanical engineering. Topics in the series include:

- Engineering Design
- Machinery and Machine Elements
- Mechanical Structures and Stress Analysis
- Automotive Engineering
- Engine Technology
- Aerospace Technology and Astronautics
- Nanotechnology and Microengineering
- Control, Robotics, Mechatronics
- MEMS
- Theoretical and Applied Mechanics
- Dynamical Systems, Control
- Fluid Mechanics
- Engineering Thermodynamics, Heat and Mass Transfer
- Manufacturing
- Precision Engineering, Instrumentation, Measurement
- Materials Engineering
- Tribology and Surface Technology

To submit a proposal or request further information, please contact the Springer Editor of your location:

China: Ms. Ella Zhang at ella.zhang@springer.com
India: Priya Vyas at priya.vyas@springer.com
Rest of Asia, Australia, New Zealand: Swati Meherishi at swati.meherishi@springer.com
All other countries: Dr. Leontina Di Cecco at Leontina.dicecco@springer.com

To submit a proposal for a monograph, please check our Springer Tracts in Mechanical Engineering at https://link.springer.com/bookseries/11693 or contact Leontina.dicecco@springer.com

Indexed by SCOPUS. All books published in the series are submitted for consideration in Web of Science.

More information about this series at https://link.springer.com/bookseries/11236

Ram P. Bharti · Krunal M. Gangawane
Editors

Recent Trends in Fluid Dynamics Research

Select Proceedings of RTFDR 2021

Springer

Editors
Ram P. Bharti
Department of Chemical Engineering
Indian Institute of Technology Roorkee
Roorkee, Uttarakhand, India

Krunal M. Gangawane
Department of Chemical Engineering
National Institute of Technology Rourkela
Rourkela, Odisha, India

ISSN 2195-4356 ISSN 2195-4364 (electronic)
Lecture Notes in Mechanical Engineering
ISBN 978-981-16-6927-9 ISBN 978-981-16-6928-6 (eBook)
https://doi.org/10.1007/978-981-16-6928-6

This Springer imprint is published by the registered company Springer Nature Singapore Pte Ltd.
The registered company address is: 152 Beach Road, #21-01/04 Gateway East, Singapore 189721, Singapore

Foreword

A National e-Conference on "Recent Trends in Fluid Dynamics Research (RTFDR-21)" was organized by fluid dynamics and heat transfer research groups of the Department of Chemical Engineering, National Institute of Technology Rourkela, during April 2–4, 2021, while the world was sailing through a roaring disaster of the pandemic. Fluid is an intriguing part of our lives; hence, fluids engineering is an excellent old discipline that evolved as an interdisciplinary forum with time. Researchers, practitioners, and educators across the globe congregated here to meet practical challenges the world encountered today, starting from water metering to irrigation to space technology with innovative solutions adopted. Microfluidics and nanofluidics, computational fluid dynamics, multiphase flow modeling, and its practical realization are some of the essential milestones that accentuated the phenomenal growth of fluids engineering witnessed in the twenty-first century. Cryosurgery and controlled drug delivery are the new paradigms aspired by fluid engineers deciphering the biological flow rheology.

Professor Animesh Biswas (Director, NIT Rourkela) has always encouraged this kind of vigorous academic activity. The faculty and research scholars of the chemical engineering department made it a successful event on e-format. I want to take the opportunity to congratulate my young faculty colleagues and conveners of RTFDR-21, Prof. Krunal Madhukar Gangawane, Prof. Akhilesh Kumar Sahu, Prof. Hara Mohan Jena, and Prof. Basudeb Munshi, for their excellent efforts in the successful organization of the conference.

The conference sought conceptual, analytical, empirical, experimental, or theoretical contributions in the following areas of fluid dynamics. Five brilliant keynote lectures were delivered by Prof. Rajendra P. Chhabra (IIT Ropar), Prof. S. Pushpavanam (IIT Madras), Prof. Suman Chakraborty (IIT Kharagpur), Prof. Subrata K. Mujumdar (IIT Guwahati), and Prof. Ram Prakash Bharti (IIT Roorkee). We extend our gratitude for accepting our invitation as keynote speakers. It is an honor to have stalwarts of their stature amongst us and sharing their thoughts and experiences through various models.

There were five technical sessions, where 55 papers were presented. Out of technical contributions received, 23 papers have been selected and peer-reviewed for publishing in the Springer conference proceedings.

I hope that this proceedings becomes a primer for teachers, researchers, and professional developers.

Prof. Madhusree Kundu
Chairman, RTFDR-21
National Institute of Technology Rourkela
Rourkela, India

Preface

A National Conference on "Recent Trends in Fluid Dynamics Research (RTFDR-21)" was organized by the Department of Chemical Engineering at the National Institute of Technology Rourkela (India) during April 2–4, 2021. The conference aimed to bring together scientists, university professors, and graduate researchers to present the research trends of fluid dynamics research and their applications.

The conference witnessed 55 research papers presented in 5 technical sessions covering the broader scope of fundamental and applied fluid dynamics explored theoretically, computationally, and experimentally. In addition, five keynote lectures were also delivered by eminent researchers in fluid dynamics: *Hidden treasures of fluid mechanics and heat transfer in the Dittus-Boelter equation* (Prof. Raj. P. Chhabra, IIT Ropar); *Nonlinear dynamics, flow visualization* (Prof. S. Pushpavanam, IIT Madras); *Microfluidics, Nanofluidics and CFD* (Prof. Suman Chakraborty, IIT Kharagpur); *Hydrodynamics and transport in Millichannel-based fixed-bed device* (Prof. Subrata K. Majumder, IIT Guwahati); and *Droplet dynamics in microfluidic systems* (Prof. Ram Prakash Bharti, IIT Roorkee). The highly rated 27 (out of 55) contributed papers were selected and peer-reviewed for publication in this conference proceedings.

This book presents select proceedings of the Conference on "Recent Trends in Fluid Dynamics Research (RTFDR-21)." It signifies the current research trends in fluid dynamics and convection heat transfer for both laminar and turbulent flow structures. The topics covered include fluid mechanics and applications, microfluidics and nanofluidics, numerical methods for multiphase flows, cavitation, combustion, fluid–solid interactions in turbulence, biological flows, CFD, experimental fluid mechanics, experimental heat transfer, fluid power, non-Newtonian rheology, and boundary layer theory. The book also comprises various fundamental and application-based researches of fluid dynamics and heat transfer by theoretical and experimental approaches. The book can be a valuable reference for beginners, researchers, and professionals interested in fluid dynamics research and allied fields.

Editors

Rourkela, India Prof. Krunal M. Gangawane
Roorkee, India Prof. Ram P. Bharti

Acknowledgments

The success of any conference depends mainly on the encouragement and guidelines of many others. We take this opportunity to express our gratitude to the people who have been instrumental in the successful organization of the Conference "Recent Trends in Fluid Dynamics Research (RTFDR-21)." For many reasons, we want to acknowledge our indebtedness and warmest thanks to Professor Animesh Biswas (Director, National Institute of Technology Rourkela, and *Chief Patron*, RTFDR-21). His proclivity for excellence and unbridled support for organizing national symposiums, conferences, and workshops on relevant subject areas is well known, and we all thank him for that. We sincerely thank Professor Madhusree Kundu (*Chairman*, RTFDR-21) for showing great interest in organizing this conference since its inception. We are also grateful to all the *local and national advisory committee members* of RTFDR-21 for their continued interest and support.

We convey our token of appreciation to the *esteemed speakers* of the conference for serving the audiences with their mesmerizing talks. We are also grateful to the *technical session chairs* (Prof. Anoop Kumar Gupta, IIT Patna; Prof. Avinash Chandra, TIET Patiala; Prof. Chandi Sasmal, IIT Ropar; Prof. Deepak Sahu, NIT Jalandhar; Prof. Ravikant Gupta, Banasthali Vidyapith; Prof. Naresh Thota, NIT Warangal; Prof. Neelkanth Nirmalkar, IIT Ropar; and Prof. Swati Patel, IIT Ropar) for accepting our invitation and for their mesmerizing conduct of the session and evaluations of the presentations.

We wish to express our most profound appreciation to the *contributing authors* of the conference papers and their patience in undergoing the review process for a couple of months. We thank all the *contributing authors* and *audiences/participants* for their great interest in RTFDR-21. We also would like to express our gratitude to the *anonymous reviewers* for their dedicated time and efforts to bring out the best form of the manuscript.

The efforts put in by the *students* (Debabrata Panda, Abhishek Kumar, and Sudhanshu Kumar) were phenomenal. They tried their level best by taking time from their busy academic schedule to help complete this conference proceeding, and we thank you all.

The guidance and support received from all the *faculty members* and *local and national advisory committee members* who contributed to the conference and contributing to this conference proceeding were vital for its success. I am grateful for their continued interest, constant support, and help.

Editors
Rourkela, India Prof. Krunal M. Gangawane
Roorkee, India Prof. Ram P. Bharti

Contents

Editors and Contributors

About the Editors

Prof. Ram P. Bharti is an *Associate professor* in the Department of Chemical Engineering at Indian Institute of Technology (IIT) Roorkee, India. He received his B.Tech. (2000), M.Tech. (2002), and Ph.D. (2006) degrees (all in chemical engineering) from SLIET Longowal, IIT Bombay, and IIT Kanpur, respectively. Subsequently, He worked as a postdoctoral fellow (2007–2009) in the Department of Chemical & Biomolecular Engineering at the University of Melbourne, Australia. He is working as a faculty member at IIT Roorkee since 2009. His research interests include computational fluid dynamics (CFD), convective hydrodynamics of non-Newtonian fluids and bluff bodies, microfluidics, electrokinetic flow in micro channels, and development of computational algorithms for complex flow simulations. Prof. Ram has co-authored over 30 refereed international journal papers and supervised large number of research (M.Tech. & Ph.D.) theses. He has reviewed large number of articles published in internationally recognized journals and research theses. In addition to various national and international research collaborations, he is also serving as a member of the editorial board of the International Journal of Aerospace Sciences, Scientific and Academic Publishing (SAP), California.

Dr. Krunal M. Gangawane is an Assistant Professor of Chemical Engineering at National Institute of Technology (NIT) Rourkela, India. He has done his graduation in Chemical Engineering (B.Tech.) from the University of Pune in 2007. Later, He received his M.Tech. and Ph.D. Degrees in Chemical Engineering from Indian Institute of Technology Roorkee in 2010 and 2015, respectively. His Ph.D. research was based upon the development of CFD code based on the lattice Boltzmann method for convective heat transfer problems. His current area of research is Magnetoconvection, Nanofluidics, enhanced oil recovery, aerogels, etc. He has more than 20 publications in the journals of repute. He joined UPES Dehradun as Assistant Professor in 2015 and, subsequently, moved to NIT Rourkela in March, 2018 as Assistant Professor in the Chemical Engineering department. Dr. Gangawane received "Best

Researcher-Individual Excellence- December 2017' during MANTHAN-2017 at UPES Dehradun. He was chosen as a Member of Academia-Industry interaction (2016) in UPES for conducting research at Reliance R&D. He had a research collaboration with the Firat University (Turkey), Ghent University (Belgium), King Saud University (Saudi Arabia) on the topic of 'Convection heat transfer in enclosed bodies for different fluids.' He is also the recipient of the ISRO-RESPOND sponsored project in March 2020. He has acted as a TOPIC EDITOR (Magnetohydrodynamics convection) in the journal of Frontiers in Mechanical Engineering During 2020–21. Dr. Gangawane is the reviewer of various peer-reviewed international journals.

Contributors

Richa Agarwal Department of Chemical Engineering, Banasthali University, Banasthali, India

Greeshma K. Anand Department of Chemical Engineering, Birla Institute of Technology and Science, Pilani, Pilani-Campus RJ 333031, India

M Arun Department of Mechanical Engineering, National Institute of Technology Karnataka, Surathkal, Mangalore, India

Ashok Kumar Baranwal Department of Chemical Engineering, BIT Mesra Ranchi, Ranchi, Jharkhand, India

Ram Prakash Bharti Complex Fluid Dynamics and Microfluidics (CFDM) Lab, Department of Chemical Engineering, Indian Institute of Technology Roorkee, Roorkee, India

Gautam Choubey Department of Mechanical and Aerospace Engineering, Institute of Infrastructure, Technology, Research and Management (IITRAM), Ahmedabad, India

Jitendra Dhakar Complex Fluid Dynamics and Microfluidics (CFDM) Lab, Department of Chemical Engineering, Indian Institute of Technology Roorkee, Roorkee, India

Pradeep Dhondi Complex Fluid Dynamics and Microfluidics (CFDM) Lab, Department of Chemical Engineering, Indian Institute of Technology Roorkee, Roorkee, India

Urlaganti Krishna Gopika Sreenidhi Institute of Science and Technology, Hyderabad, Hyderabad, India

Prateek Gupta Department of Chemical Engineering, National Institute of Technology Rourkela, Odisha, Rourkela, India

Ravikant R. Gupta Department of Chemical Engineering, Banasthali University, Banasthali, India

Mohammad Hussain Department of Chemical Engineering, Laboratory of Transport Phenomenon, NIT Rourkela, Odisha, India

Hara Mohan Jena Department of Chemical Engineering, National Institute of Technology Rourkela, Rourkela, Odisha, India

Sangmo Kang Department of Mechanical Engineering, Dong-A University, Saha-gu, Republic of Korea

Ravikumar Karmur Department of Mechanical and Aerospace Engineering, Institute of Infrastructure, Technology, Research and Management, Ahmedabad, India

Shrutika Kashetti Department of Chemical Engineering, Birla Institute of Technology and Science, Pilani, Pilani-Campus RJ 333031, India

Deepak Kumar Department of Chemical Engineering, National Institute of Technology Rourkela, Odisha, Rourkela, India

Roshan Kumar Department of Chemical Engineering, BIT Mesra Ranchi, Ranchi, Jharkhand, India

Vinayak Malhotra Department of Aerospace Engineering, SRM Institute of Science and Technology, Chennai, India

Monisha Mridha Mandal University School of Chemical Technology, GGSIPU, Dwarka, New Delhi, India

Ranjith Maniyeri Biophysics Laboratory, Department of Mechanical Engineering, National Institute of Technology Karnataka (NITK), Surathkal, Mangalore, Karnataka, India

K. L. Vidhu Manohar Biophysics Laboratory, Department of Mechanical Engineering, National Institute of Technology Karnataka (NITK), Surathkal, Mangalore, Karnataka, India

Aishwarya Mulampaka Department of Chemical Engineering, National Institute of Technology Warangal, Warangal, Telangana, India

Basudeb Munshi Department of Chemical Engineering, Laboratory of Transport Phenomenon, NIT Rourkela, Odisha, India

Atharav Naik Department of Mechanical and Aerospace Engineering, Institute of Infrastructure, Technology, Research and Management, Ahmedabad, India

Manjappatta Pazhiyottumana Neeraj Biophysics Laboratory, Department of Mechanical Engineering, National Institute of Technology Karnataka (NITK), Surathkal, India

Parthkumar Patel Department of Mechanical and Aerospace Engineering, Institute of Infrastructure, Technology, Research and Management, Ahmedabad, India

Tirth Patel Department of Mechanical and Aerospace Engineering, Institute of Infrastructure, Technology, Research and Management, Ahmedabad, India

Sankalp Patidar Department of Mechanical and Aerospace Engineering, Institute of Infrastructure, Technology, Research and Management, Ahmedabad, India

Yogendra Nath Prajapati Department of Chemical Engineering, BIT Mesra Ranchi, Ranchi, Jharkhand, India

Madasu Krishna Prasad Department of Mathematics, National Institute of Technology, Chhattisgarh, India

K. S. Rajmohan Department of Chemical Engineering, National Institute of Technology Warangal, Warangal, Telangana, India

Lomas Rishi University School of Chemical Technology, GGSIPU, Dwarka, New Delhi, India

Akhilesh K. Sahu Department of Chemical Engineering, National Institute of Technology Rourkela, Odisha, Rourkela, India

Shuvam Samadder Complex Fluid Dynamics and Microfluidics (CFDM) Lab, Department of Chemical Engineering, Indian Institute of Technology Roorkee, Roorkee, India

Saroj Kumar Samantaray Department of Chemical Engineering, Laboratory of Transport Phenomenon, NIT Rourkela, Odisha, India

Priya C. Sande Department of Chemical Engineering, Birla Institute of Technology and Science, Pilani, Pilani-Campus RJ 333031, India

Priya Sarkar Department of Mathematics, National Institute of Technology, Chhattisgarh, India

Nikhil Sharma Sreenidhi Institute of Science and Technology, Hyderabad, Hyderabad, India

Saumya Shekhar Department of Aerospace Engineering, SRM Institute of Science and Technology, Chennai, India

S. M. Shravankumar Sreenidhi Institute of Science and Technology, Hyderabad, Hyderabad, India

Pedina Sibakrishna Department of Chemical Engineering, National Institute of Technology Rourkela, Rourkela, Odisha, India

T. Sirichandana Sreenidhi Institute of Science and Technology, Hyderabad, Hyderabad, India

Srivatsa Thimmaiah Department of Mechanical Engineering, National Institute of Technology Karnataka, Surathkal, Mangalore, India

Bhushan Thombare Department of Aerospace Engineering, SRM Institute of Science and Technology, Chennai, India

Sumit Tripathi Department of Mechanical and Aerospace Engineering, Institute of Infrastructure, Technology, Research and Management (IITRAM), Ahmedabad, India

Teja Reddy Vakamalla Department of Chemical Engineering, NIT Calicut, Kattangal, Kozhikode, India

Mona Mary Varghese Department of Chemical Engineering, NIT Calicut, Kattangal, Kozhikode, India

Akepogu Venkateshwarlu Complex Fluid Dynamics and Microfluidics (CFDM) Lab, Department of Chemical Engineering, Indian Institute of Technology Roorkee, Roorkee, India

Tabish Wahidi Department of Mechanical Engineering, National Institute of Technology Karnataka, Surathkal, Mangalore, India

Ajay Kumar Yadav Department of Mechanical Engineering, National Institute of Technology Karnataka, Surathkal, Mangalore, India

Pabbala Monish Yadav Department of Mechanical and Aerospace Engineering, Institute of Infrastructure Technology Research and Management (IITRAM), Ahmedabad, India

Computational Study of Mixing of Shear Thinning Fluids with Modifications in Rushton Turbine Impeller

Aishwarya Mulampaka and K. S. Rajmohan

Abstract Mixing is one of the most common and vital operations in industries such as the chemical, cement, food, and polymer industry. It is necessary to understand the hydrodynamics of the mixing tank and the mixing behavior of the fluids to assess the quality of mixing and the energy consumption to optimize the mixing process. From the literature survey, it is observed that there are plenty of experimental and computational studies done on Newtonian fluids whereas only limited studies related to non-Newtonian fluids have been reported. Generally, the impeller geometry is selected based on the viscosity of a fluid. The turbulence effect around the region of the impeller, pumping capacity, and power consumption can be best studied with the help of computational fluid dynamics. The simulation of the mixing process is done using ANSYS Fluent, where the velocity vector plots, contours, and streamlines can be studied and analyzed, which in turn will help to optimize the design. In the present research work, study related to various turbulence models and rotating approaches in the CFD, prediction of hydrodynamic behavior of Newtonian and non-Newtonian fluids is made to validate the model. Modifications in the Rushton turbine impeller are made by introducing cuts in the blade of the Rushton turbine impeller. A comparison of the results regarding power consumption and mixing time will be made to identify the optimum design with less power consumption and shorter mixing time for shear thinning fluids.

Keywords Shear thinning fluids · Computational fluid dynamics · ANSYS Fluent · Power consumption

A. Mulampaka · K. S. Rajmohan (✉)
Department of Chemical Engineering, National Institute of Technology Warangal, Warangal 506004, Telangana, India
e-mail: rajmohan@nitw.ac.in

R. P. Bharti and K. M. Gangawane (eds.), *Recent Trends in Fluid Dynamics Research*, Lecture Notes in Mechanical Engineering,
https://doi.org/10.1007/978-981-16-6928-6_1

Nomenclature

K	Consistency index (kg s^{n-2} /m)
n	Flow behavior index
ρ	Density (kg/m^3)
ε	Dissipation due to turbulence kinetic energy (m^2/s^3)
μ	Viscosity (Pa s)
D_i	Impeller diameter (m)
N	Impeller rotational speed (rps)
η	Apparent viscosity (Pa s)
Np	Power number
k	Specific turbulent kinetic energy(m^2/sec)
P	Power required for running agitator (W)
N_q	Flow number
ϕ_v	Viscous dissipation function (s^2)

1 Introduction

Mixing is one of the oldest and extremely important operations with countless applications in almost every chemical, food processing, and polymer industry. It involves blending of two immiscible fluids, homogenization, dispersion of solids into liquids, and gases in liquids. Based on the type of operation, they can be classified as batch, continuous, and semi-batch. It is important to study the hydrodynamics of the mixing tank and the mixing behavior to assess the quality of mixing and the energy consumption to optimize the mixing process. From the literature survey, it has been found that there are many experimental and computational studies done on Newtonian fluids and few studies are related to non-Newtonian fluids. Non-Newtonian fluids are of vital importance in the food and pharmaceutical industries. Based on the fluid properties, the impeller is selected. The mixing operation greatly depends on the impeller geometry, presence of baffles, type of tank, and presence of draft tubes. The turbulence effects around the region of the impeller can be best studied with the help of computational fluid dynamics. The simulation of the mixing process is be done using ANSYS Fluent, where the velocity profiles, vector, contour, and streamline plots can be studied and analyzed, through which the optimization of the design can be done. In the present research work, study related to different turbulence models and rotating approaches in the CFD, prediction of hydrodynamic behavior of Newtonian and non-Newtonian fluids is made to validate the model. Energy saving is the most concerning aspect in mixing operations.

Many studies have been done to study the effect of impeller geometry on power consumption [1]. The power draw depends greatly on the impeller blades, blade spacing, clearance, disc thickness, shaft inclination, and eccentricity [2]. Modifications in the Rushton turbine impeller made by Rao [3] are limited to Newtonian fluids. Both experimental and simulation work have been reflected. Based on this research work, Houari Ameur has adapted the modification in Rushton turbine blades to viscoelastic fluids from the literature data mentioned [4]. The studies show that the cavern size (well mixed) region is great for standard Rushton turbine but with additional power consumption when compared to blades with cuts [5]. In the present research work, the idea of modified blades from [3] has been adapted to test the effect of power consumption when shear thinning fluids (pseudoplastic) are being used. The velocity flow field, power number, and mixing time will be compared and the best design with less power consumption and shorter mixing time is predicted with the help of computational fluid dynamics. The effect of fluid flow and impeller characteristics on the hydrodynamic behavior of a continuous stirred tank reactor (CSTR) has been studied using the computational fluid dynamics (CFD) method [12]. A detailed description of tank geometry, baffles, impeller, and draft tubes are given with a description of experimental and computational techniques are provided with Classification on types of discretization schemes and modeling approaches [13].

2 Problem Statement

The geometry of the mixing tank is created using SpaceClaim, with standard dimensions. Data validation with shear thinning fluids is done with reference to [6]. The dimensions of the standard Rushton blade turbine are mentioned in Table 1. Modification in the blade design is done with reference to [3], and the dimensions are taken from the same reference, mentioned in Table 1. The main idea is to predict how the power consumption and mixing time vary when the cuts in the blade are introduced, which are depicted in Fig. 1. This can help in selecting the best design suitable for the mixing operation with less power consumption and shorter mixing time.

Table 1 Tank and blade dimensions

Tank	Dimensions (mm)
Tank height (H)	270
Tank diameter (T)	270
Impeller diameter (D = T/3)	90
Blade height (D/5)	18
Blade length (D/4)	22.5
Baffle width (T/10)	27
Baffle thickness (T/100)	3

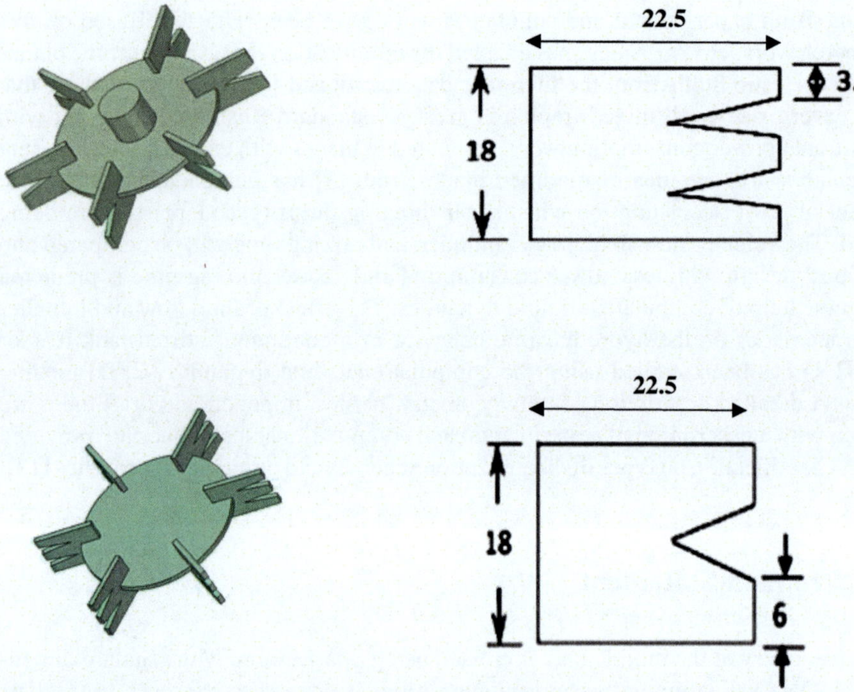

Fig. 1 V—cut turbine and W—cut turbine image with dimensions

2.1 Mathematical Model

A non-Newtonian fluid is a fluid that deviates from "Newton's law of viscosity." The viscosity of a non-Newtonian fluid is variable and may vary with stress, with time, or with a combination of both. Based on this behavior, the fluids are classified as pseudoplastic, thixotropic, rheopectic, and dilatant. The different non-Newtonian fluids as jam, butter, carboxymethyl cellulose, xanthan gum, sauces, yogurt, detergents, etc. Based on their viscous behavior, the fluids can be classified as follows [7]. The working fluids xanthan gum, carboxymethyl cellulose at different concentrations are taken with reference to [6]. The rheology of the working fluids is shown in Table 2.

Table 2 Rheological properties of working fluids

Working fluid	Wt%	Consistency index (K) [kg s^{n-2} /m]	Flow behavior index (n)
Carboxymethyl cellulose	0.1	13.2	0.85
Xanthan gum	0.045	9.5	0.8
Xanthan gum	0.08	34.0	0.64
Natrosol	1	10.8	0.59

2.2 Governing Equations

The commercial software, ANSYS Fluent, which uses a control volume technique to discretize the conservation equations, is used to solve the conservation of mass and momentum energy along with the other equations and to generate flow fields. The governing equations of continuity, momentum, and temperature are as follows [8].

$$\frac{\partial \rho}{\partial t} + \nabla \cdot (\rho \overline{v}) = 0 \tag{1}$$

$$\frac{\partial}{\partial t}(\rho \overline{v}) + \nabla \cdot (\rho \overline{v}\overline{v}) = -\nabla P + \nabla \cdot (\overline{\overline{\tau}}) + \rho \overrightarrow{g} \tag{2}$$

$$\rho_{C_P} \frac{\partial T}{\partial t} + \rho C_p \nabla(\overline{v}T) = ke_{ff} \cdot \nabla^2 T - \phi_v \tag{3}$$

where $\overline{v}$ is velocity vector, T is temperature, P is the static pressure, $\overline{\overline{\tau}}$ is the stress tensor, $\overrightarrow{g}$ is the gravitational body force, K_{eff} is an effective thermal conductivity, C_P is the heat capacity of the liquid at constant pressure, and Φ_v is the viscous dissipation function.

The stress tensor $\overline{\overline{\tau}}$ is expressed as

$$\overline{\overline{\tau}} = \eta\left[\nabla\overline{v} + \nabla\overline{v}^T\right] - \frac{2}{3}\nabla \cdot \overline{v} I \tag{4}$$

where η is the apparent viscosity, I is the unit tensor.

The turbulent flow, which is induced by the Rushton turbine, is modeled by realizable k-ε turbulence model. The governing equations of turbulence kinetic energy, k, and its rate of dissipation, ε are

$$\nabla.(\rho k \overline{v}) = \nabla\left[\left(\mu + \frac{\mu_t}{\sigma_k}\right)\nabla \cdot k\right] + G_k - \rho\varepsilon \tag{5}$$

$$\nabla \cdot (\rho \varepsilon \overline{v}) = \nabla \cdot \left[\left[\mu + \frac{\mu_t}{\sigma_\varepsilon}\right]\nabla \cdot \varepsilon\right] + C_2 \rho \frac{\varepsilon'}{k + \sqrt{\nu\varepsilon}} \tag{6}$$

in which C_2 is constant. σ_k and σ_ε are the turbulent Prandtl numbers for k and ε, respectively. The following values are used for the constants [8]

$C_2 = 1.92$, $\sigma_k = 1.0$, $\sigma_\varepsilon = 1.2$.

In the equation, G_k represents the generation of turbulence kinetic energy due to mean velocity gradients and calculated as

$$G_k = -\rho \overline{u_i^{'} u_j^{'}} \frac{\partial u_j}{\partial x_i} \tag{7}$$

The turbulent or eddy viscosity μ_t is computed by combining k and ε as follows [8]

$$\mu_t = \rho c \mu \tfrac{k^2}{\varepsilon} = \mu_t s^2 \tag{8}$$

where S is the modulus of the mean rate of the strain tensor

$$S = \sqrt{2 S_{ij} S_{ij}},\ S_{ij} = \frac{1}{2}\left(\frac{\partial u_j}{\partial x_i} + \frac{\partial u_i}{\partial x}\right)$$

The variable C_μ is calculated as

$$C_\mu = \frac{1}{A_0 + A_s \frac{kU^*}{\varepsilon}} \tag{9}$$

where $U^* = \sqrt{S_{ij} S_{ij} + \overline{\Omega}_{ij} \overline{\Omega}_{ij}};\ \overline{\Omega}_{ij} = \Omega_{ij} - 2\varepsilon_{ijk}\omega_k$

Ωij is the mean rate of rotation of tensor viewed in a moving reference frame with angular velocity ω_k. The model constants A_0 and A_s are given as

A0 = 4.04, $A_s = \sqrt{6} cos\varnothing$

$$\phi = \frac{1}{3} cos^{-1}(6W),\ W = \frac{S_{ij} S_{jk} S_{ik}}{\overline{S}^3},\ \overline{S} = \sqrt{s_{ij} s_{ij}}$$

For a Newtonian fluid, the impeller Reynolds number for the stirred tank is given by

$$Re = \frac{\rho N D_i^2}{\mu} \tag{10}$$

where N is the rotational speed of the impeller, and ρ is the density of the fluid. For non-Newtonian fluid, the power law is used to model viscosity and is given as

$$K = \dot{\gamma}^{n-1} \tag{11}$$

in which η is apparent viscosity, K is consistency index, and n is the flow behavior index. The impeller Reynolds number for pseudoplastic fluid is calculated using the Metzner-Otto method [9]

$$\mathrm{Re} = \frac{\rho N^{2-n} D_i^2}{K \cdot k_s^{n-1}} \tag{12}$$

where k_s is Metzner-Otto constant with $k_s = 11.5$.

The power number of impellers is calculated by

$$N_p = \frac{P}{\rho N^3 D_i^5} \tag{13}$$

where P is the power input, which is calculated from the torque, Γ applied on the impeller shaft. Mathematically, calculated as

$$P = 2\Pi N\Gamma$$

3 Numerical Method

The three-dimensional flow of the shear thinning fluids in the mixing tank with different impellers is simulated using ANSYS workbench 17.2. This computer tool uses the finite volume method to solve the momentum and energy equations. The geometry is created using SpaceClaim CAD tools, and the computational domain is discretized with tetrahedral mesh as shown in Fig. 2. A grid independence test is conducted to check if the number of nodes in the mesh affects the velocity magnitude. The results are listed in Table 3. The second order upwind scheme has been used to

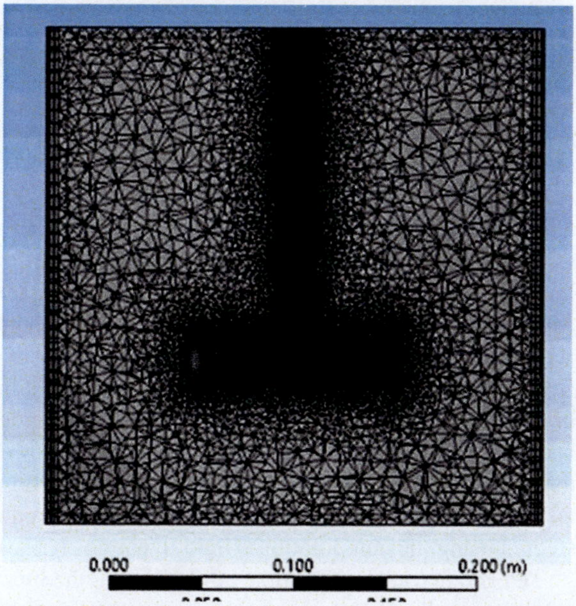

Fig. 2 Magnified view of mesh of the mixing tank

Table 3 Grid independence test

Element size (m)	No. of nodes	No. of Elements	Velocity (m/s)
4e-003 (fine)	475,647	2,325,578	0.960531
5e-003 (medium)	278,179	1,331,897	0.958653
6e-003 (coarse)	188,757	895,474	0.960149

discretize the convective terms in the momentum equations. A SIMPLE algorithm was used for solving pressure–velocity coupling.

In this study, we are using the standard k-ε model, it is the most commonly used model, it is robust and with less computational cost and has been useful in the engineering community for many years. It gives stable calculation and very suitable, especially for high Reynolds numbers.

Multiple reference frames (MRFs) model modified form of the rotating frame model uses several rotating and non-rotating frames. In the MRF approach, for the rotating frame, the impeller does not move. In a stationary frame with tank walls and baffles, the wall and baffles do not move. The rotating frame is under motion.

The non-Newtonian power-law model is activated by writing console code, and the materials are created with the corresponding rheological properties as mentioned in Table 2. Relative convergence criteria of 10–6 for the continuity and x- and y-components of velocity are defined.

4 Validation

Computational fluid dynamics is a multi-disciplinary subject that deals with fluid mechanics, numerical analysis, and data structures. It is a tool used to solve the conservation equations for mass, momentum, and energy. These equations are in the form of partial differential equations, which is an extremely tough task to solve analytically. CFD discretizes these equations from nonlinear termed equations to linear algebraic form, which are further solved to get accurate results for the corresponding fluid domain [10]. During the theoretical study of the subject, assumptions are made that the fluid is in the laminar region but most of the industrial processes deal with the turbulent conditions of fluid. Hence, it is necessary to study the effect of turbulence on the mixing phenomena for better understanding. Many studies have been done with various impellers to observe the power consumption and flow field inside the mixing tank system. The effect of fluid flow and impeller characteristics on the hydrodynamic behavior of a continuous stirred tank reactor (CSTR) has been studied using the computational fluid dynamics (CFD) method [12].

The geometric model is first validated with the literature date available from [11] for the same fluids as taken by the authors. Khapre and Munshi [6] used the same

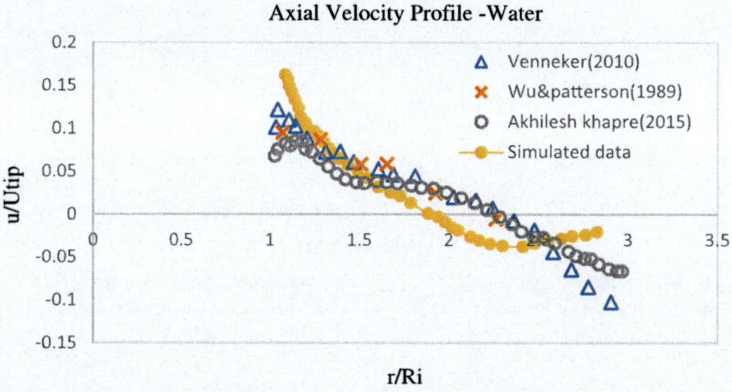

Fig. 3 Axial velocity versus normalized radial distance for water

system for validation but had conducted studies on entropy generation for different impellers and to study the effect of entropy with respect to the blade width. The comparison of the predicted results with the experimental data, Fig. 3, for water as working fluid shows good agreement.

5 Results and Discussion

5.1 Effect on Power Consumed

Every impeller has a unique power curve. From the literature, data are verified that the power curve obtained by simulation is in agreement with the unique power curve of the Rushton turbine impeller. The power consumption is highest in the laminar region and decreases linearly with the Reynolds number, showing the system is in agreement with unique power curve. The hydrodynamics and mixing behavior are largely affected by the impeller design and fluid properties. Power consumption is one of the hugely concerned factors. To minimize the power consumption and to study the effect of blade design, cuts have been introduced into the standard Rushton blade turbine impeller. The power curves for standard Rushton blade turbine, V-cut turbine, and W-cut turbine are shown in Fig. 4a–c at different rotational speeds, for different shear thinning fluids.

5.2 Effect on Mixing Time

Mixing time or blend time is one of the significant parameters to characterize the performance of the mixing tank. It is the time taken to achieve the maximum (99%)

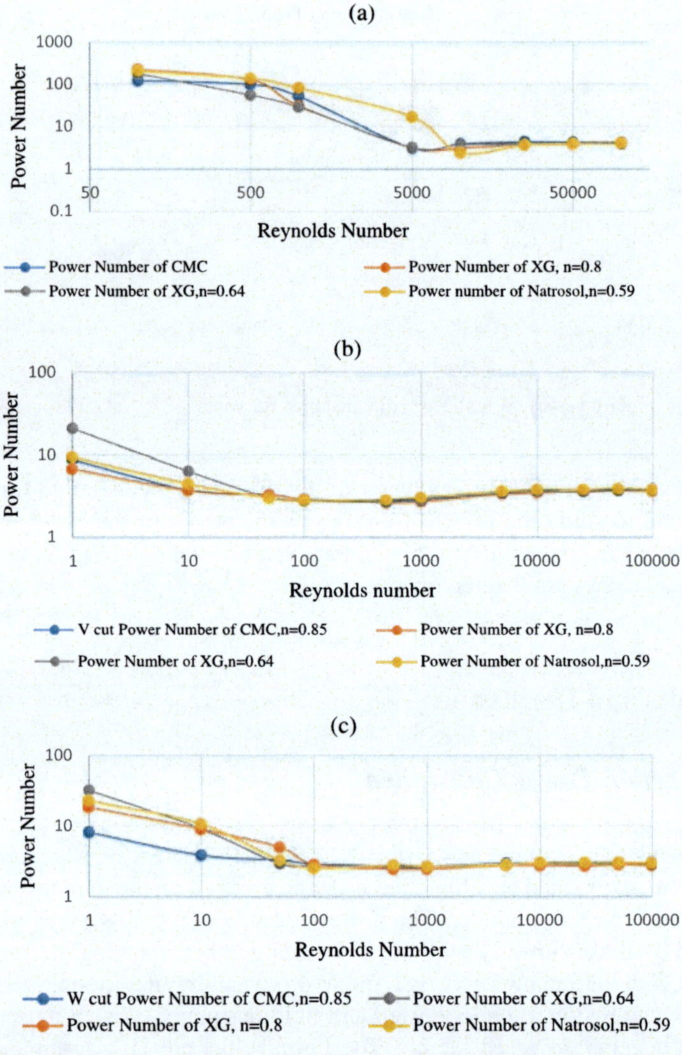

Fig. 4 The power curve for **a** standard Rushton blade turbine, **b** V-cut turbine, **c** W-cut turbine

of the steady-state concentration. The transport of a tracer helps to understand the degree of homogeneity in the agitated tank. Mixing time was predicted using "Transient transport of a neutrally-buoyant tracer (scalar)." The probe locations and tracer injection points are shown in Fig. 5.

The overall mixing time for standard design turbine is observed to be around 12 s as shown in Fig. 6a.

The mixing time for V-cut and W-cut turbines is observed to be around 9 s and 8 s, respectively, as shown in Fig. 6b, c.

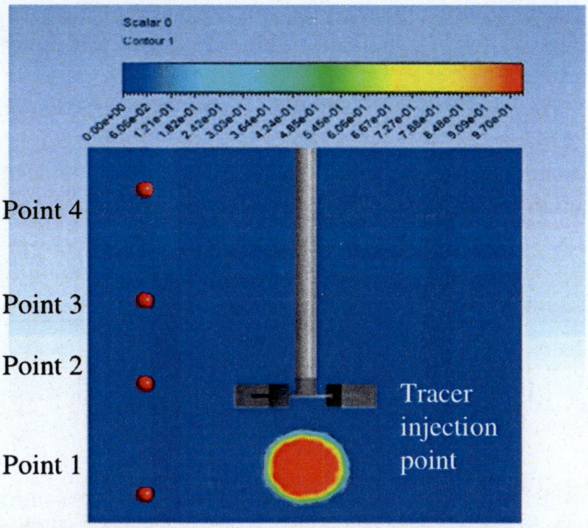

Fig. 5 Tracer injection and probe points inside the tank

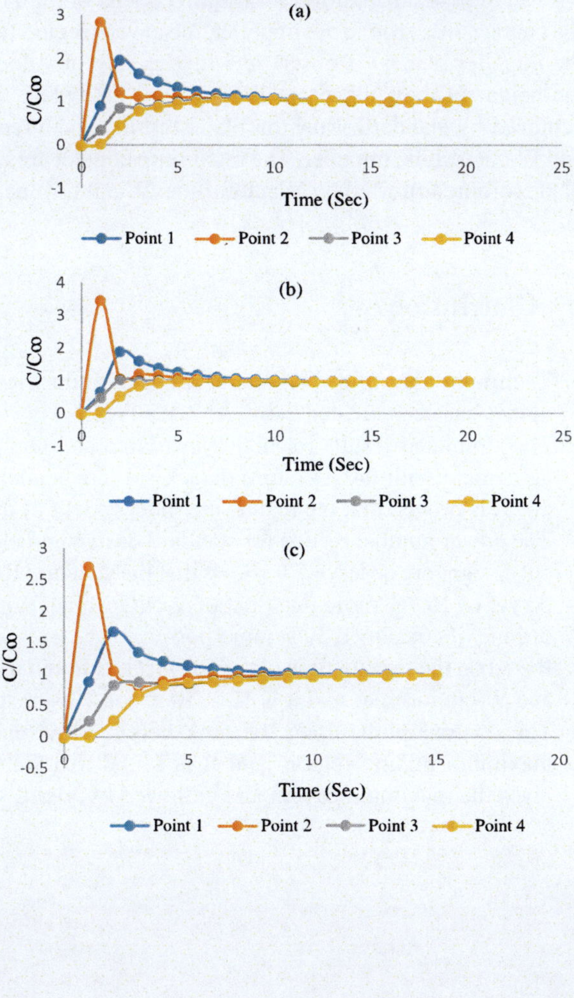

Fig. 6 Response curve of CMC at 180 rpm at point 1, point 2, point 3, and point 4 for **a** Standard Rushton blade turbine, **b** V-cut turbine, **c** W-cut turbine

The highest mixing time is observed for the standard Rushton blade turbine, followed by W-cut blade turbine, and the least time is taken by the V-cut blade turbine.

The reason for this order could be the strong impinging flow of fluid by the V-cut blade turbine allows the flow radially with stronger recirculation loops that are reaching the top surface of the tank. Though standard Rushton blade turbine has better radial flow with maximum cavern size, the time taken to reach the homogeneity is greater. Though the results of mixing time are differed by the value of one to two seconds, to find the best compromise among the mixing time power consumption, the studies of mixing time are considered here.

5.3 Effect on Cavern Size

In the food, polymer and pharmaceutical industries intimate contact between the fluid particles is required to achieve high-quality final products. This can be accomplished with an impeller that allows maximum contact of fluid in the stirred tank. To observe the contact area around the impeller, the cavern region (the well-mixed region around the impeller blades) is observed with the help of velocity contour plots along with the height of the tank. From Fig. 7a–c, it is evident that maximum cavern size is achieved by standard. Rushton blade turbine and V-cut turbine when compared to and W-cut turbine impeller. The maximum contact area is given by standard Rushton blade turbine followed by V-cut turbine, W-cut turbine.

6 Conclusions

- The importance of mixing in the chemical industries is highlighted. A geometric model has been created using ANSYS Fluent.
- The simulated results for both Newtonian and non-Newtonian fluids have good agreement with the literature data. Grid independence test has been conducted, and it is proven that the results are independent of the mesh size.
- The power number results for standard design and modified (V-cut design and W-cut design) are observed. It shows that the standard Rushton turbine has consumed power of 10.7% more than that of V-cut turbine. When compared with the W-cut turbine, it consumed 39% more power.
- Based on the mixing time observations, it is found that for standard design, V-cut, and W-cut turbines and it is 12 s, 10 s, and 13 s, respectively.
- The contour plots along the tank height show the cavern size to predict the maximum mixing region, and it is found that standard Rushton blade turbine gives the maximum cavern size followed by V-cut, W-cut.

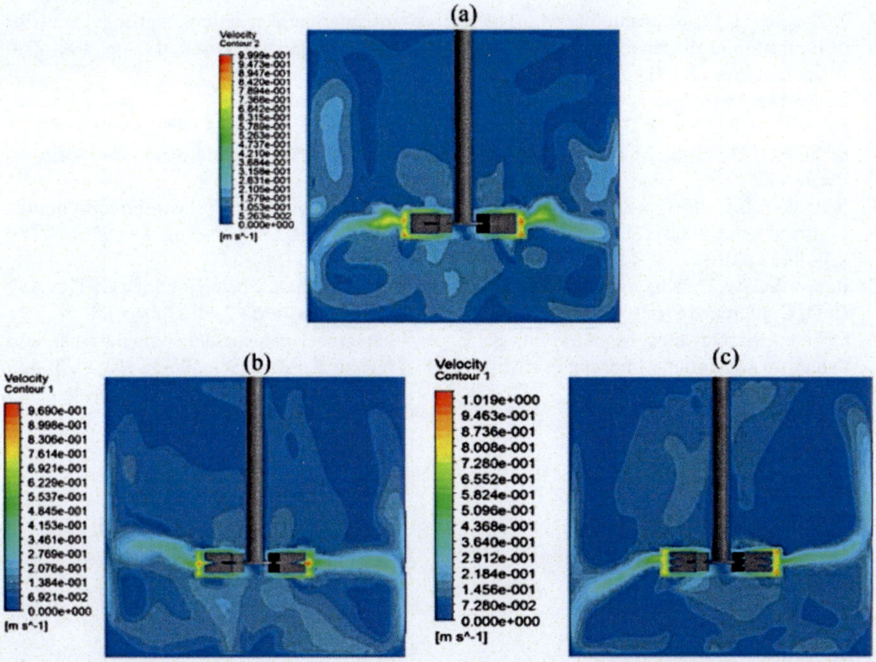

Fig. 7 Velocity contour plots of **a** Standard Rushton blade, **b** V-cut turbine, **c** W-cut blade turbine

- A compromise on both power consumption, cavern size, and mixing time for shear thinning fluid, CMC gives us the following order of preference for efficient mixing operation standard Rushton turbine > V-cut turbine > W-cut turbine.

References

1. Taghavi, M., Zadghaffari, R., Moghaddas, J., Moghaddas, Y.: Experimental and CFD investigation of power consumption in a dual Rushton turbine stirred tank. Chem. Eng. Res. Des. **89**(3), 280–290 (2011)
2. Wang, S., Wu, J., Ohmura, N.: Inclined-shaft agitation for improved viscous mixing. Ind. Eng. Chem. Res. **52**(33), 11741–11751 (2013)
3. Rao, D.A., Sivashanmugam, P.: Experimental and CFD simulation studies on power consumption in mixing using energy saving turbine agitator. J. Ind. Eng. Chem. **16**(1), 157–161 (2010)
4. Cortada-Garcia, M., Dore, V., Mazzei, L., Angeli, P.: Experimental and CFD studies of power consumption in the agitation of highly viscous shear thinning fluids. Chem. Eng. Res. Des. **1**(119), 171–182 (2017)
5. Ameur, H.: Modifications in the Rushton turbine for mixing viscoplastic fluids. J. Food Eng. **1**(233), 117–125 (2018)
6. Khapre, A., Munshi, B.: Numerical investigation of hydrodynamic behavior of shear thinning fluids in stirred tank. J. Taiwan Inst. Chem. Eng. **1**(56), 16–27 (2015)

7. Tavlarides, L.L., Stamatoudis, M.: The analysis of interphase reactions and mass transfer in liquid-liquid dispersions. In: Advances in Chemical Engineering, vol. 11, pp. 199–273. Academic Press (1981)
8. Ansys Fluent 13: User's Guide (2011). U.S.A.: Ansys Inc.
9. Metzner, A.B., Otto, R.E.: Agitation of non-Newtonian fluids. AIChE J. **3**(1), 3–10 (1957)
10. Ferziger, J.H., Perić, M., Street, R.L.: Computational methods for fluid dynamics. Springer, Berlin (2002)
11. Venneker, B.C., Derksen, J.J., Van den Akker, H.E.: Turbulent flow of shear-thinning liquids in stirred tanks—the effects of Reynolds number and flow index. Chem. Eng. Res. Des. **88**(7), 827–843 (2010)
12. Rajavathsavai, D., Khapre, A., Munshi, B.: Study of hydrodynamic behaviour of a CSTR using CFD. Conference Paper · March 2011. https://doi.org/10.13140/RG.2.1.2270.8161/1
13. Ochieng, A., Onyango, M., Kiriamiti, K.: Experimental measurement and computational fluid dynamics simulation of mixing in a stirred tank: a review. S. Afr. J. Sci. **105**(11–12), 421–426 (2009)

Analysis of Room Airflow Characteristics Using CFD Approach

Parthkumar Patel, Ravikumar Karmur, Gautam Choubey, and Sumit Tripathi

Abstract Air conditioners have become an important part of our lives and are widely used in commercial buildings, residential houses, shopping centres and various industries for better thermal comfort and indoor air quality. Room airflow characteristics and locations of ventilation passages play an important role in giving better thermal comfort. In this work, a study has been conducted on the overall effectiveness of air ventilation systems inside a room. A CFD simulation is performed on a 2D horizontal section of a room using Ansys Fluent, which is integrated into Ansys Workbench. For numerical solution of the equations, SIMPLE algorithms are opted for pressure-based solver, while for turbulence modelling, RANS equations are adopted with RNG k-e mathematical model for simulation of the domain. The simulations are performed for two cases: cross-ventilation and corner-ventilation arrangements. The main goal of this study is to determine the temperature and velocity distributions in the room. The effects of humidity are not considered in this work. The results of the study are presented in terms of overall performance of the two ventilation systems through transient simulations. It is observed that a cross-ventilation system gives better thermal comfort compared to the corner-ventilation system in the overall domain. Also, the corner-ventilation arrangement is found to be inefficient up to certain level because of poor distribution of air in the room.

Keywords Room air distribution · Ventilation · RNG model · Ansys Fluent

1 Introduction

Air conditioning systems are widely used to maintain indoor thermal comfort conditions regardless of the outdoor conditions of temperature variations. As the air conditioning systems consume substantial amount of electricity, they significantly

P. Patel · R. Karmur · G. Choubey · S. Tripathi (✉)
Department of Mechanical and Aerospace Engineering, Institute of Infrastructure, Technology, Research and Management, Ahmedabad 380026, India
e-mail: sumittripathi@iitram.ac.in

R. P. Bharti and K. M. Gangawane (eds.), *Recent Trends in Fluid Dynamics Research*, Lecture Notes in Mechanical Engineering,
https://doi.org/10.1007/978-981-16-6928-6_2

contribute to the total power consumption by residential and industry sectors. In fact, heating and cooling contribute to almost half of the total energy consumed by the residential sectors [1]. Along with increasing the efficiency of the machinery systems (such as compressor and heat exchanger), an efficient ventilation system also contributes in enhancing the thermal comfort and decreasing the associated energy consumption. Prediction of airflow patterns inside a ventilated room can help us in designing better systems for thermal comfort. The air flow by mechanical ventilation system mainly depends on location and orientation of inlet and outlet vents, duct geometry, velocities of air at inlet and outlet vents, and room configuration. Badly located positions of vents cannot efficiently distribute air in the room and thus can result in poor air quality with less circulation in the areas away from the vents, such as corners and certain portions of roof and wall. This can lead to reduced thermal comfort and accumulation of toxic and harmful substances which can cause various respiratory diseases and infections [2]. In some cases, simply increasing the ventilation rate inside the room or building can partially solve this issue, but it comes at a price that conflicts with building energy efficiency [3].

Mean age of the air (MAA) and effective ventilation are the main criteria that can be used to determine thermal comfort inside a room [4]. MAA is the average time the air particles spend travelling inside a room, i.e. flow from inlet vent to outlet vent, and better MMA leads to higher air exchange rates. The ventilation efficiency depends on supply air flow and the airflow patterns and is associated with rate of fresh air supply and even airflow distribution inside a room. Chung et al. investigated the ventilation efficiency of different ventilation patterns arranged by two inlets and two outlets diffusers at different locations and observed that ventilation efficiency might be dominantly influenced by location of diffuser rather than air change rate [5]. Khatri et al. used a passive cooling technology to enhance indoor thermal comfort with low energy consumption, and they used three kinds of cooling configurations and analysed uniform temperature distribution and concluded that two cooling mechanisms on the same side of the wall achieved the fastest cooling compared to any other configurations [6]. Wang et al. analysed temperature, velocity and pressure fields for three different heating mechanisms for better thermal performance of indoor air flow and observed that wall-mounted air conditioners with velocity not more than 0.2 m/s can achieve thermal comfort in the shortest time due to forced convection [7]. A healthy human being is very adaptable and gets acclimatized to withstand lower temperatures during winter and higher temperatures during summer, and in the HVAC systems, a DBT of 25 °C, 50% relative humidity and a 0.1–0.2 m/s air velocity are taken as design conditions for maximum comfort [8]. Also, different heating mechanisms have very little effect in indoor pressure that can be neglected. Further, in the numerical simulations of room air flow, it has been widely reported in the literature that renormalization group (RNG) k-ϵ model is quite accurate for the studies near the wall or surfaces and shows good agreement between the experimental observations [3, 9, 10].

In this study, we perform CFD simulation of the air flow inside a room with two inlets and a single outlet for two configurations of the room having different positions of vents. Numerical models for simulation, boundary conditions and position of vents have been determined based on available studies in the literature. We analyse and present the temperature and velocity distributions inside both the room configurations along with vortex formation. The effects of humidity have not been considered in the present study.

2 Model Development

The modelling was done using Ansys integrated tools, the geometry was created in design modeler, and simulations were performed in Fluent. The turbulent behaviour of air inside the room was predicted using Reynolds average Navier–Stokes equations (RANS) method, as used in majority of the indoor airflow applications [11, 12].

2.1 Geometry and Mesh Generation

A two-dimensional horizontal cross-sectional area of a room is considered for airflow study with different inlet and outlet positions of vents as shown in Fig. 1. It is assumed that there are no heat gains or heat loss from walls, there is no heat generation inside the room, and there is no obstruction in the flow, i.e. room is empty. Both the geometries have size of 1m*1m square with two inlet vents and one outlet vent, the size of the inlet vents are 100mm each, and that of outlet vent is 200mm. For cross-ventilation arrangement, the inlet and outlet vents are located on the opposite sides, and that for corner-ventilation arrangement, on the adjacent sides of the wall as shown in Fig. 1a, b, respectively.

The grid convergence analysis was performed using appropriate grids from coarse to finer sizes as shown in Table 1. Grid size of 0.01m was determined as an optimum size for the present study. The rectangular grid element was taken as to better fit the geometry and was kept denser near inlets and outlet region for appropriately predicting the rapid change in velocity and temperature.

2.2 Boundary Conditions

Velocity inlet boundary condition has been used at the two inlets with constant velocity of 0.15 m/s. The inlet flow is considered having 5% turbulent intensity and turbulent viscosity ratio of 10. Temperature of the inlet air flow is given as 290 K. The Reynolds number for velocity in this case is close to 1000. The outlet vent is considered as a pressure-based outlet and the pressure at the outlet is considered as 0

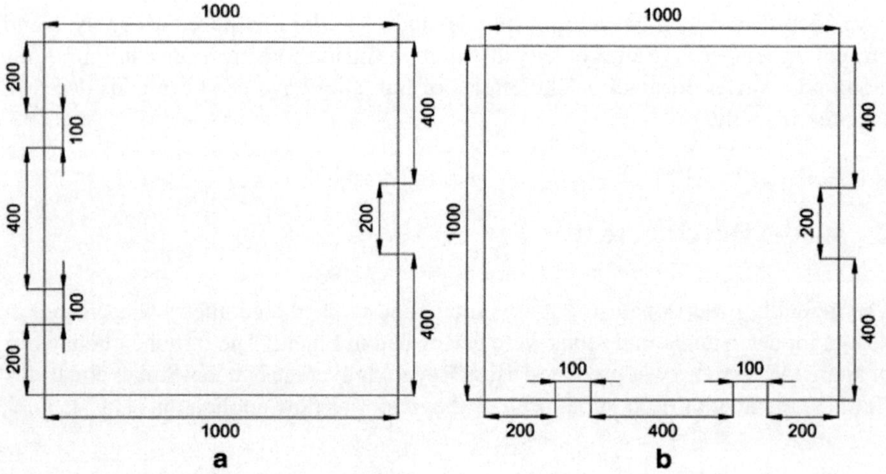

Fig. 1 Model geometry for room configurations: **a** cross-ventilation and **b** corner ventilation (all dimensions are in mm)

Table 1 Mesh size for grid convergence analysis

S. no.	Number of cells	Size (m)
1	679	0.04
2	2569	0.02
3	10090	0.01
4	22121	0.005

pascal (i.e. atmospheric pressure). No-slip boundary condition has been used at the wall as a standard wall function. The initial temperature is taken as 300 k, and heat flux is considered as zero.

2.3 Governing Equations and Solution

It is assumed that there is zero heat flux through the walls of the room and no heat is generated inside the room. The pressure-based solver and RNG k-ϵ model have been used in this study. The main governing equations in pressure-based solver are continuity equation, momentum conservation equation and energy conservation equation. As these equations are widely available in the literature, they are not shown here. The simulation is run for the flow time of 5 min, and different fields like velocity and temperature at different instances for both the cases (cross- as well as corner-ventilation arrangements) are computed and analysed.

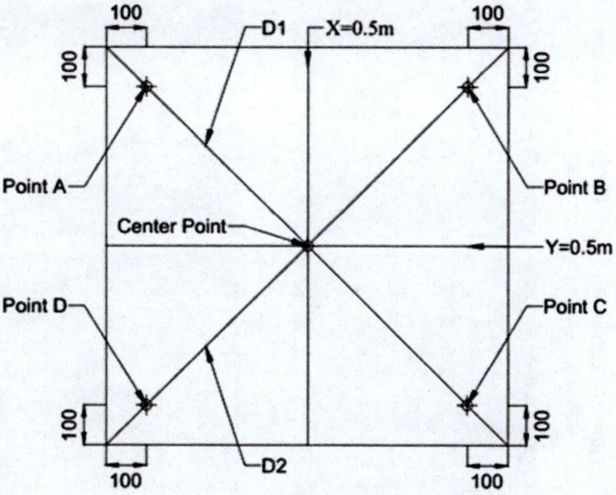

Fig. 2 Reference lines and points in room geometry (all dimensions are in mm)

3 Results and Discussion

In this section, velocity and temperature contours at different time instances are presented for both the cases. The results, in terms of temporal and spatial variations of velocity and temperature, are also presented at different points and lines in the room geometry as shown in Fig. 2.

3.1 *Velocity Contours*

Velocity contours of air flow for at different time instances for both cases are shown in Fig. 3. It can be seen that majority of air is flowing directly from inlet to outlet, without being distributed into the room, which means a substantial portion of cooled air is being exiting the room without giving cooling effect. This results in high energy consumption, making ventilation system inefficient to a certain level. For higher ventilation efficiency, higher should be the airflow distribution throughout the room that will result in better heat exchange effects and thus reducing temperature of the room faster with the same ventilation rate. Here, it can be noted that the air distribution is relatively better in cross-ventilation compared to corner ventilation, primarily because of the positions of inlet and outlet vents. Additionally, the vortex formation in cross-ventilation leads to better interaction of inlet air with the room air, ultimately improving the air quality, while the vortex creation in corner ventilation leads to poor air distribution in a larger portion of the room.

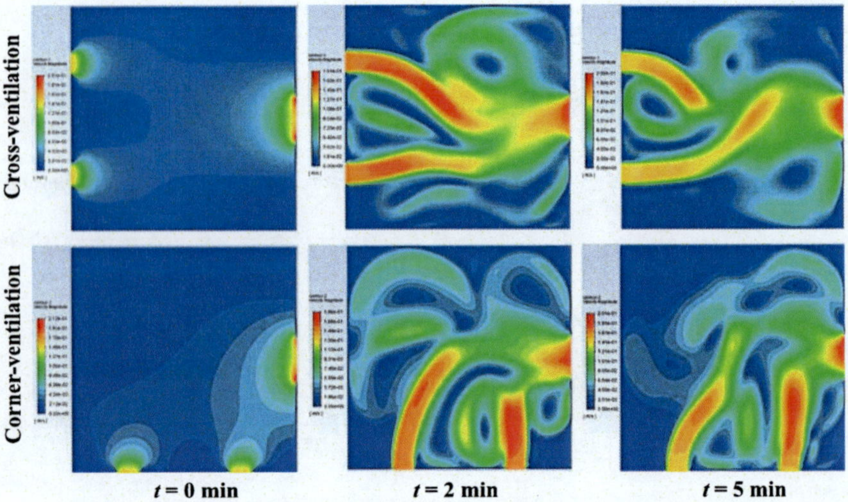

Fig. 3 Velocity contours of air flow at various instances for cross-ventilation and corner-ventilation arrangements

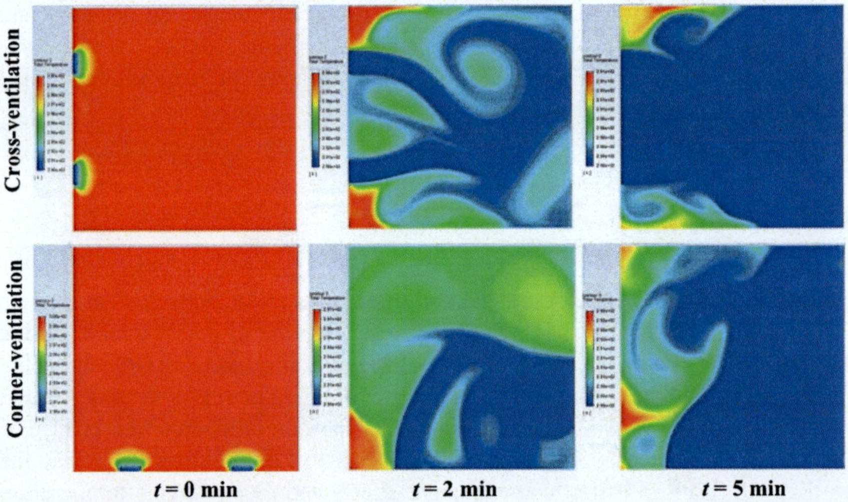

Fig. 4 Temperature contours of air flow at various instances for cross-ventilation and corner-ventilation arrangements

3.2 *Temperature Contours*

The corresponding temperature contours are shown in Fig. 4. It can be noted that, after 5 min, majority of room has attained comfort level temperature in both cases. In cross-ventilation, the temperature is slightly higher than the average room temperature at the inlet side corners of the room where proper air flow is not available. Similarly, in corner ventilation, higher temperature is observed on the left wall due to improper distribution of cool air. Also, this is to note that the rate of cooling is faster in cross-ventilation compared to corner ventilation, which is a result of vortex formation and similarity of flow across horizontal centreline.

3.3 *Comparison of Cross-Ventilation and Corner-Ventilation Arrangements*

The velocity profiles along the centrelines of the room after 5 min for cross-ventilation and corner ventilation are shown in Fig. 5. Some interesting traits are observed in velocity distribution along these lines. It can be seen that along the line y = 0.5, velocity increases with x in both the cases, but the fluctuations are higher in cross-ventilation which is primarily because of the vortex shedding patterns. Also, along the line x = 0.5, velocity at the centre region is higher compared to the wall region in both cases and fluctuations in both cases are also quite similar. It is to be noted that the range of velocities in both cases is very close to each other.

Temporal profiles of temperature at centrepoint of 2D room geometry for both cases are shown in Fig. 6. It can be seen that the rate of temperature reduction is faster in cross-ventilation compared to corner ventilation, and temperature fluctuations are relatively more in corner ventilation. For the given geometry and boundary condi-

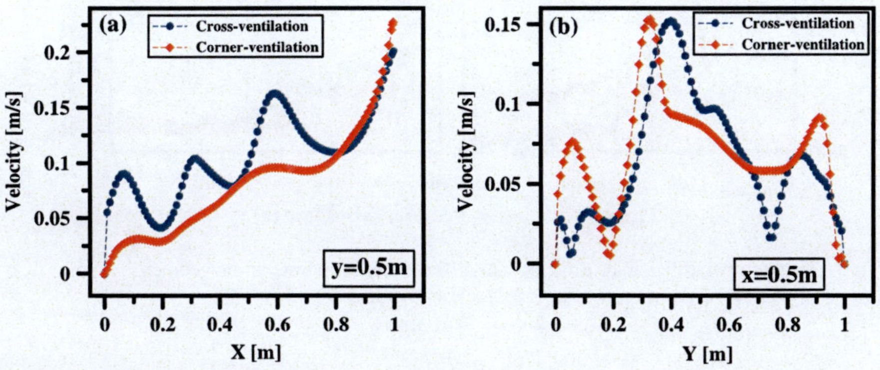

Fig. 5 Velocity profiles along the centrelines of the room: **a** horizontal centreline, **b** vertical centreline

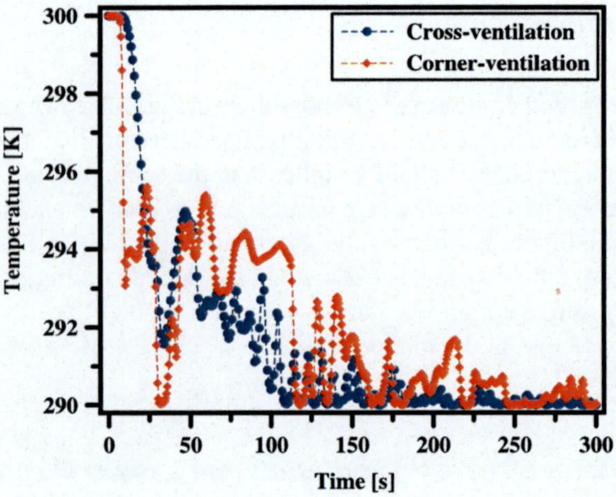

Fig. 6 Temporal variations of temperature at centrepoint of room geometry

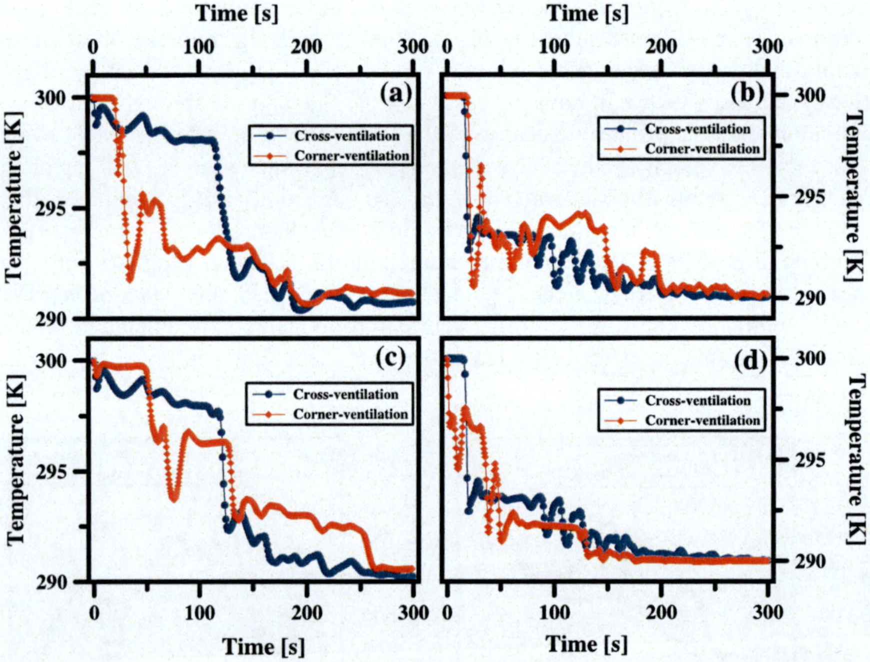

Fig. 7 Total temperature versus time plots at different corner points of the geometry: **a** point A; **b** point B; **c** point D; **d** point C (see Fig. 2 for location details)

tions, significant reduction in temperature is achieved in about 120 sec at centrepoint. Similar plots at different points A, B, C and D are shown in Fig. 7 (locations of these points are shown in Fig. 2). The four points A, B, C and D are near the corners of the room geometry. The temperature profiles at points A and D (and also at B and C) are similar for cross-ventilation due to symmetry. It can be noted that temperature is not reduced at a rate with which it is reducing at centrepoint. It can be clearly observed that around 150 sec, the temperature at these four points is still higher and has potential for further reduction, whereas at centrepoint the corresponding reduction has been achieved relatively faster.

4 Conclusions

A two-dimensional horizontal cross section of a room has been studied with two different configurations having different vent positions. The velocity and temperature fields are compared for faster and uniform temperature distribution for thermal comfort. As observed in temperature vs time graph, the cross-ventilation arrangement attains thermal comfort temperature relatively faster compared to the corner-ventilation arrangement. The vortex formation and shedding were observed in both the arrangements. Although the intensity of vortexes was not enough to create any kind of discomfort, it could affect the quality of air due to the increase of mean age of air. On the other hand, these vortexes help in better interaction of inlet air with room air by covering broader area of the room. It is also observed that in cross-ventilation the temperatures at the left corners were higher than the normal temperature because of lack of air flow at the inlet side of corners. In corner ventilation, there was less air flow on the whole left side of the room, so the temperature at the left wall was higher than normal. Overall, it can be said that if the inlet vents and outlet vent are adjacent to each other than majority of air is passed directly from inlet to outlet leading to lesser air flow throughout the room which eventually decreases ventilation efficiency. Thus, in general, for uniform cooling of major portion of the room, cross-ventilation configuration should be preferred over the corner-ventilation configuration. However, if it is required to cool only specific areas of a ventilated room, the corner vent positions can be preferred.

References

1. Hurak, B.: Computational fluid dynamics analysis of air flow and temperature distribution in buildings. Ph.D. thesis, The Ohio State University (2011)
2. Cao, G., Awbi, H., Yao, R., Fan, Y., Sirén, K., Kosonen, R., Zhang, J.J.: A review of the performance of different ventilation and airflow distribution systems in buildings. Build. Environ. **73**, 171–186 (2014)
3. Posner, J., Buchanan, C., Dunn-Rankin, D.: Measurement and prediction of indoor air flow in a model room. Energy Build. **35**(5), 515–526 (2003)

4. Buratti, C., Palladino, D.: Mean age of air in natural ventilated buildings: experimental evaluation and co2 prediction by artificial neural networks. Appl. Sci. **10**(5), 1730 (2020)
5. Chung, K.C., Hsu, S.P.: Effect of ventilation pattern on room air and contaminant distribution. Build. Environ. **36**(9), 989–998 (2001)
6. Khatri, R., Singh, A.P., Khare, V.R.: Identification of ideal air temperature distribution using different location for air conditioner in a room integrated with eathe-a cfd based approach. Energy Procedia **109**, 11–17 (2017)
7. Wang, Y., Zhang, X., Tian, Z., Li, Y., et al.: Numerical simulation of thermal performance of indoor airflow in heating room. Energy Procedia **158**, 3277–3283 (2019)
8. Sinha, S., Arora, R., Roy, S.: Numerical simulation of two-dimensional room air flow with and without buoyancy. Energy Build. **32**(1), 121–129 (2000)
9. Nielsen, P.V.: The selection of turbulence models for prediction of room airflow (1998)
10. Espi, E., Berne, P., Duverneuil, P.: Using cfd to understand the air circulation in a ventilated room. Comput. Chem. Eng. **22**, S751–S754 (1998)
11. Horikiri, K., Yao, Y., Yao, J., et al.: Numerical simulation of convective airflow in an empty room. Int. J. Energy Environ. **5**(1), 574–581 (2011)
12. Stamou, A., Katsiris, I.: Verification of a cfd model for indoor airflow and heat transfer. Build. Environ. **41**(9), 1171–1181 (2006)

Slow Flow Past a Slip Sphere in Cell Model: Magnetic Effect

Madasu Krishna Prasad and Priya Sarkar

Abstract The problem of the steady, axisymmetric flow past a rigid sphere in a hypothetical spherical cavity under the effect of the magnetic field based on Happel's and Kuwabara's cell model has been investigated analytically. The inner sphere is assumed to be rigid, and the outer one is fictitious. Slip boundary condition applied on the surface of the inner sphere. Happel's and Kuwabara's boundary conditions are applied at the outer sphere surface. The flow field inside the cavity is governed by the Stokes equation. The expression for drag acting on the sphere is obtained. Variation of drag coefficient against various parameters like slip parameter and Hartmann number is studied graphically. Some special cases are also discussed.

Keywords Sphere · MHD · Stokes equation · Cell model · Drag

1 Introduction

The creeping flow of viscous fluid past a sphere is of the interesting area because of its vast applications in various areas of engineering and science like environmental engineering, industrial chemical, biomedical, meteorology, etc. The mathematical analysis to study the behavior of a cluster of spherical particles suspended in a viscous fluid is difficult to solve. But, Happel and Brenner [1] introduced the cell model technique by assuming the shape of the outer surface to be sphere. In the literature, Happel's and Kuwabara's cell models [2, 3] have been widely considered. In Happel's model, on cell surface, there is no friction, i.e., the tangential stresses vanish. In Kuwabara's model, on the outer cell, the vorticity vanishes.

Hydrodynamic interactions of spherical particles in an unbounded and bounded medium are studied in past [4–8]. Faltas and Saad [9] investigated cell models for a slip eccentric sphere. They found that when the slip coefficient is increasing, the drag force also increases. Saad [10] studied micropolar fluid flow past a viscous fluid

M. K. Prasad (✉) · P. Sarkar
Department of Mathematics, National Institute of Technology, 492010 Chhattisgarh, India
e-mail: madaspra.maths@nitrr.ac.in; kpm973@gmail.com

R. P. Bharti and K. M. Gangawane (eds.), *Recent Trends in Fluid Dynamics Research*, Lecture Notes in Mechanical Engineering,
https://doi.org/10.1007/978-981-16-6928-6_3

sphere in a cell model and vice versa. Prasad [11, 12] investigated non-Newtonian fluid flow past a semipermeable sphere and a non-concentric semipermeable sphere in cell models, respectively. Using a cell model technique, Escott and Wilson [13] investigated the rheology of a rigid sphere suspension in a second-order fluid.

Magnetohydrodynamics (MHD) is the study of the behavior of electrically conducting fluids and magnetic properties. The field of MHD was introduced by Hannes Alfven [14]. Many researchers have much attention to MHD because of its vast applications such as geophysics, solar physics, fusion energy research, and industry. MHD influences many man-made and natural flows. Sherief et al. [15] investigated the study of magneto-micropolar pipe flows under transverse magnetic field using slip conditions. Kyrlidis et al. [16] examined the impact of aligned magnetic field on creeping flow past an axisymmetric bodies. Saad [17] studies viscous fluid flow past through porous medium under the transverse magnetic field using cell models. Recently, Prasad and Bucha [18, 19] have studied the effect of magnetic field on semipermeable particle in an unbounded medium and weakly permeable particle in cell models, respectively. The above-mentioned investigations motivated us to study the present problem in the presence of transverse magnetic field.

In this paper, we investigate the magnetic effect on the flow of viscous fluid past a sphere in cell models. We assume that flow is a steady and axisymmetric. We assumed that the inner sphere as rigid and the outer sphere is fictitious. On inner surface, slip condition is applied and Happel's and Kuwabara's conditions are used on the outer surface. Stream function is obtained. The drag exerted on the slip sphere subject to a magnetic field is calculated and special known cases are deduced. The drag coefficient depends on the slip parameter, Hartmann number, and separation parameter. The effect of these parameters is studied graphically.

2 Mathematical Formulation

Consider the viscous fluid past a solid sphere located at the center of fictitious spherical surface, having a uniform velocity U subject to the transverse magnetic field (Fig. 1).
Assumptions of the problem [18]:

- The flow is axisymmetric, slow, and steady.
- The Reynold's number $Re = \frac{Ua}{\nu}$ is consider to be sufficient small.
- The body force is absent.
- The direction of the magnetic field is normal to the flow.
- Assume that Magnetic Reynold's number $Re_m = Ua\mu_h\sigma$ is very small.
  Where electric conductivity and magnetic permeability of the fluid are σ and μ_h, respectively.
- The external electric fluid is absent.
- The induced magnetic field is neglected.

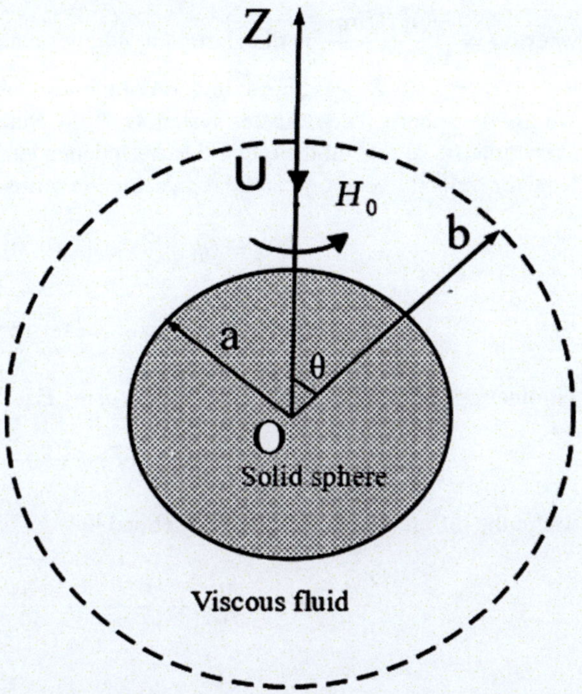

Fig. 1 Viscous fluid past a slip sphere under transverse magnetic field

- Lorentz force **F** is
  $\mathbf{F} = \mathbf{J} \times \mathbf{H}$
  Where the electric current density is **J** and magnetic induction vector is **H**
  $\mathbf{F} = \mu_h^2 \sigma(\mathbf{q} \times \mathbf{H}) \times \mathbf{H}$

The governing equations in a flow region of a incompressible viscous fluid under the magnetic effect are [17, 18]

$$\nabla \cdot \mathbf{q} = 0 \tag{1}$$

$$\nabla p + \mu \nabla \times \nabla \times \mathbf{q} - \mu_h^2 \sigma(\mathbf{q} \times \mathbf{H}) \times \mathbf{H} = 0 \tag{2}$$

where $\mathbf{q}$ is the velocity, p is the pressure, and μ is the viscosity coefficient.
For non-dimensionalize governing equations, we use non-dimensional variables

$$\mathbf{q} = U\tilde{\mathbf{q}}, \quad p = \frac{\mu U}{a}\tilde{p}, \quad \nabla = \frac{\tilde{\nabla}}{a}, \quad \mathbf{H} = H_0 \tilde{\mathbf{H}}$$

Substituting above terms in Eqs. (1) and (2) and then dropping the tildes, we get

$$\nabla p + \nabla \times \nabla \times \mathbf{q} - \alpha^2(\mathbf{q} \times \mathbf{H}) \times \mathbf{H} = 0 \tag{3}$$

where $\alpha = \sqrt{\dfrac{\mu_h^2 H_0^2 \sigma a^2}{\mu}}$ is the Hartmann number and α is non-negative. If $\alpha = 0$, Eq. (3) represents Stokes approximation of the viscous fluid.
We use the spherical coordinates system (r, θ, ϕ). Since the fluid flow is steady and axisymmetric so that all the quantities are independent of ϕ.
Now the velocity vector $\mathbf{q}$ can be represented in terms of stream function $\psi(r, \theta)$ as

$$\mathbf{q} = (q_r(r, \theta), q_\theta(r, \theta), 0) \tag{4}$$

$$q_r = -\frac{1}{r^2 \sin\theta}\frac{\partial \psi}{\partial \theta}, \quad q_\theta = \frac{1}{r \sin\theta}\frac{\partial \psi}{\partial r} \tag{5}$$

Eliminating the pressure term from Eq. (3) using Eq. (5), we get

$$E^2\left(E^2 - \alpha^2\right)\psi = 0 \tag{6}$$

where the differential operator E^2 is defined by

$$E^2 = \frac{\partial^2}{\partial r^2} + \frac{1}{r^2}\frac{\partial^2}{\partial \theta^2} - \frac{\cot\theta}{r^2}\frac{\partial}{\partial \theta} \tag{7}$$

3 Solution of the Problem

By separation of variables, the solution of Eq. (6) is given by [1, 17, 19]

$$\psi = \frac{1}{2}\left(Ar^2 + Br^{-1} + C\sqrt{r}K_{3/2}(\alpha r) + D\sqrt{r}I_{3/2}(\alpha r)\right)\sin^2\theta \tag{8}$$

where A, B, C, and D are arbitrary constants. $I_{3/2}(\alpha r)$ and $K_{3/2}(\alpha r)$ are modified Bessel functions of the first and second kind, respectively.

4 Boundary Conditions

On the inner surface of sphere $r = a$,

$$\mathbf{q}\cdot\mathbf{n} = 0, \tag{9}$$

$$(\mathbf{n}\cdot t)\cdot\mathbf{s} = \beta_1\mathbf{q}\cdot\mathbf{s}, \tag{10}$$

where **s**, **n** are tangential and unit normal vectors at the inner sphere surface. β_1 is the slip coefficient of the sphere. The coefficient β_1 depends on the nature of fluid and geometry. The standard no-slip condition is obtained as $\beta_1 \longrightarrow \infty$.
On the cell surface $r = b$,

$$q_r + U\cos\theta = 0, \tag{11}$$

Happel's model :

$$t_{r\theta} = 0, \tag{12}$$

Kuwabara's model:

$$E^2\psi = 0. \tag{13}$$

Solving Eqs. (9)–(12) for Happel's model and Eqs. (9)–(11) and Eq. (13) for Kuwabara's model, we get arbitrary constants $A, B, C,$ and D
Happel's model:

$$\begin{aligned}
A &= -\frac{\left(-6\eta^4\alpha G_5(\beta+2) + \sqrt{\eta}G_{17}\right)}{2\eta\Delta_1},\\
B &= \frac{6\eta^2 G_8 + 2\eta\alpha G_7 + \alpha^2 G_8}{2\sqrt{\eta}\Delta_1},\\
C &= -\frac{6\eta^4 G_{11} + 3\sqrt{\eta}G_{12}(\beta+2)}{2\eta\Delta_1},\\
D &= \frac{-6\eta^4 G_{13} + 3\sqrt{\eta}G_{14}(\beta+2)}{2\eta\Delta_1}.
\end{aligned} \tag{14}$$

Kuwabara's model:

$$\begin{aligned}
A &= \frac{-\beta\alpha G_3 + \alpha^2 G_4 - 2\alpha G_3}{\Delta_2},\\
B &= \frac{G_{18}}{\Delta_2},\\
C &= \frac{-3J_4(\beta+2)}{\Delta_2},\\
D &= \frac{3J_2(\beta+2)}{\Delta_2}.
\end{aligned} \tag{15}$$

The expressions S_{i_1}, J_{i_1}'s, i_1 =1–4, G_{i_2}'s, i_2 =1–21, and Δ_{i_3}'s, i_3 =1–5 are defined in the Appendix.

5 Drag Force

The drag exerted by the viscous fluid on the rigid sphere under the magnetic field with fictitious spherical surface is evaluated by this formula

$$F = 2\pi a^2 \int_0^{\pi} r^2 \left(t_{rr} \cos\theta - t_{r\theta} \sin\theta\right)\big|_{r=1} \sin\theta d\theta \tag{16}$$

$$F = \frac{2}{3}\pi\mu a U \alpha^2 \left(-2A + B - 2C K_{3/2}(\alpha) - 2D I_{3/2}(\alpha)\right) \tag{17}$$

Happel's model:

$$F = \frac{6\eta^2\alpha^2 G_8 + 2\eta\alpha^3 G_7 + \alpha^4 G_8}{\sqrt{\eta}\Delta_1} \tag{18}$$

Kuwabara's model:

$$F = \frac{2\alpha^2 G_{18}}{\Delta_2} \tag{19}$$

6 Special Cases

6.1 *Case 1*

Absence of magnetic effect ($\alpha = 0$) in Eqs. (18) and (19), we get the drag exerted on the slip sphere located at the center of the spherical cavity
Happel's model:

$$F = 4\pi\mu a U \frac{2(\beta - 3)\eta^5 + 3(\beta + 2)}{\Delta_3} \tag{20}$$

Kuwabara's model:

$$F = \frac{30\pi\mu a U(\beta + 2)}{\Delta_4} \tag{21}$$

This result is same as Faltas and Saad [9].

6.2 *Case 2*

If $\beta \longrightarrow \infty$ in Eqs. (18) and (19), under the magnetic effect, we get the drag exerted on the no-slip solid sphere located at the center of spherical cavity

Happel's model:

$$F = -2\pi\mu a U \left(\frac{\alpha(2\alpha\eta G_{19} + G_{20}L)}{\Delta_5}\right) \tag{22}$$

Kuwabara's model:

$$F = 2\pi\mu a U \left(\frac{\alpha^2 G_{20}}{\eta^3 G_{20} - \alpha G_3}\right) \tag{23}$$

which is identical to the results obtained by Saad [17].

6.3 Case 3

As the separation parameter $\eta \longrightarrow 0$ in Eqs. (18) and (19), we get the drag exerted on the slip sphere in an infinite medium

$$F_\infty = -\pi\mu a U \left(\frac{2(\alpha^2 + 3\alpha + 3)\beta + 2\alpha^3 + 6(\alpha^2 + 2\alpha + 2)}{\beta + \alpha + 3}\right) \tag{24}$$

6.4 Case 4

Absence of magnetic effect ($\alpha = 0, \quad$ and $\quad \eta \longrightarrow 0$) in Eq. (24), the drag on the slip sphere in an unbounded medium ,

$$F_\infty = -6\pi\mu a U \frac{(\beta + 2)}{\beta + 3} \tag{25}$$

6.5 Case 5

If $\eta \longrightarrow 0$ and $\beta \longrightarrow \infty$ in Eq. (24), the drag exerted on no-slip sphere in a viscous fluid in presence of the magnetic field,

$$F_\infty = -2\pi\mu a U(\alpha^2 + 3\alpha + 3) \tag{26}$$

This result is similar to Prasad and Bucha [19].

6.6 Case 6

If $\eta \longrightarrow 0$ and $\beta \longrightarrow \infty$ in Eq. (24),

$$F_{\infty} = -6\pi\mu a U \tag{27}$$

This result is Stokes drag on sphere [1].

7 Graphical Representation

The results of the wall correction factor W for Happel's and Kuwabara's model which is the ratio of drag force F experienced by the viscous fluid on the closed boundary under the effect of a transverse magnetic field to the drag force F_{∞} exerted by a sphere in an unbounded medium are depicted graphically in Figs. 2, 3, 4, and 5 for various parameters as follows

- Separation parameter: $\eta = a/b \quad (0 < \eta < 1)$,
- Slip parameter: β
- Hartmann number (magnetic parameter): α.

Figure 2 shows the variation of wall correction factor W against separation parameter η for different values of magnetic parameter α with $\beta = 3$. It is observed that when the value of the Hartmann number increases, the wall correction factor decreases, which shows that whenever the value of the magnetic effect increases it reduces the ease of flow. Here, we also noticed that the wall correction factor for Kuwabara's model is slightly higher on a small scale compared to Happel's model and difference

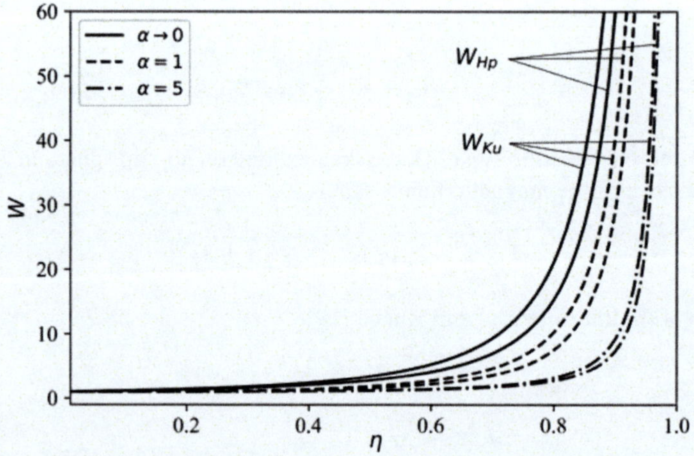

Fig. 2 Magnetic effect α on W versus η, $\beta = 3$

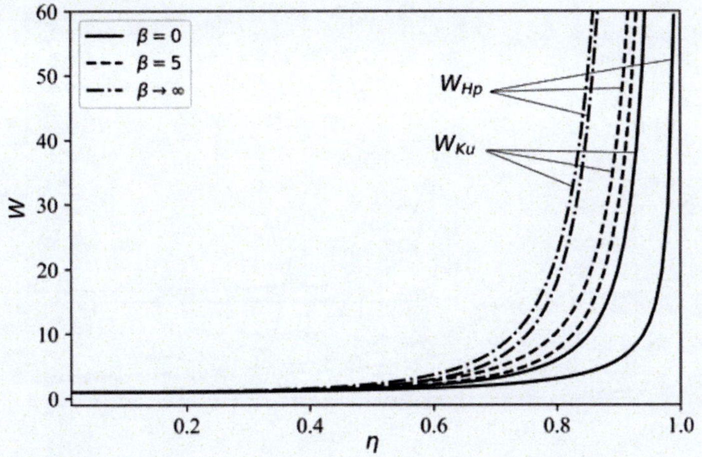

Fig. 3 Effect of slip β on W versus η, $\alpha = 5$

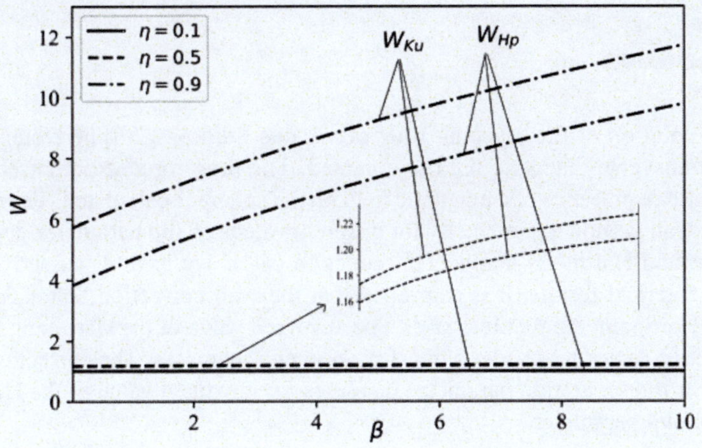

Fig. 4 Effect of separation parameter η on W versus β, $\alpha = 5$

is decreased as the magnetic effect increases. This result is similar to Prasad and Bucha [19].

The effect of β on W is discussed in Fig. 3 with $\alpha = 5$. It is clearly shown that increasing values of slip parameter β, W increases, for any fixed value of α. Figure 4 presents the effect of surrounding fluid bounded by a spherical boundary η on W against β with $\alpha = 5$. W increases as η increases, for any fixed values of α.

Figure 5 illustrates the boundary effect on W against α with $\beta = 3$. W increases as η increases. In Figs. 2, 3, 4, and 5, we have noticed that Kuwabara's model anticipates a stronger effect on the wall correction factor than Happel's model does. This is happening because in the cell model, Kuwabara's model produces a large energy dissipation.

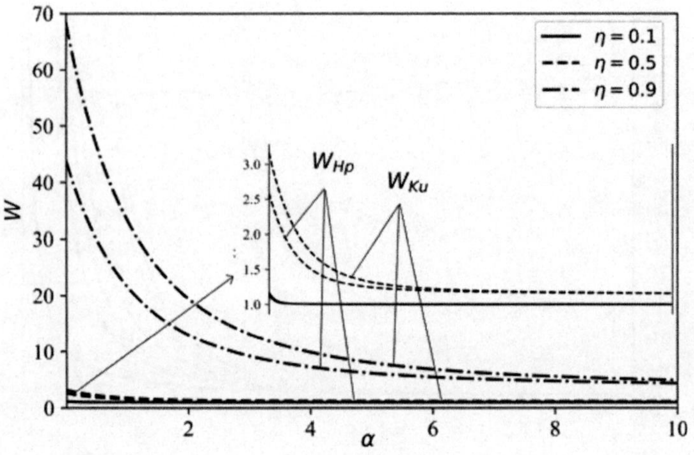

Fig. 5 Effect of separation parameter η on W versus α, $\beta = 3$

8 Conclusion

An exact solution of the creeping flow of viscous fluid past a sphere subject to a uniform transverse magnetic field is obtained. The drag force experienced by the viscous fluid in presence of a magnetic field on the sphere is calculated. The effect of boundary wall is shown graphically for different values of slip parameter, separation parameter, and Hartmann number. In reduction cases, we have discussed all well-known theoretical results. It is examined that the wall correction factor decreases when the Hartmann number increases. For any fixed value of the Hartmann number, the correction factor increases as the slip parameter increases. The correction factor increases as the separation parameter increases for any fixed value of the Hartmann number or slip parameter.

Appendix

The expressions S_{i_1}, J_{i_1}'s, i_1 =1–4, G_{i_2}'s, i_2 =1–21, and Δ_{i_3}'s, i_3 =1–5 in Eqs. (14) and (15) are defined as

$$
\begin{aligned}
&S_1 = K_{1/2}(\alpha), \quad S_2 = K_{3/2}(\alpha), \quad S_3 = I_{1/2}(\alpha), \quad S_4 = I_{3/2}(\alpha),\\
&J_1 = K_{1/2}(\alpha/\eta), \quad J_2 = K_{3/2}(\alpha/\eta), \quad J_3 = I_{1/2}(\alpha/\eta), \quad J_4 = I_{3/2}(\alpha/\eta),\\
&G_1 = J_1S_3 - J_3S_1, \quad G_2 = J_1S_4 + J_3S_2,\\
&G_3 = J_2S_3 + J_4S_1, \quad G_4 = J_2S_4 - J_4S_2,\\
&G_5 = S_1S_4 + S_2S_3, \quad G_6 = J_1J_4 + J_2J_3,\\
&G_7 = \alpha G_1(\beta+2) - 3G_2(\beta+2) - \alpha^2 G_2,\\
&G_8 = \alpha G_3(\beta+2) - 3G_4(\beta+2) - \alpha^2 G_4,\\
&G_9 = G_3(\beta+2) - \alpha G_4, \quad G_{10} = G_1(\beta+2) - \alpha G_2,\\
&G_{11} = (\alpha S_3 - 3S_4)(\beta+2) - \alpha^2 S_4,\\
&G_{12} = 6\eta^2 J_4 - 3\alpha\eta J_3 + \alpha^2 J_4,\\
&G_{13} = (\alpha S_1 + 3S_2)(\beta+2) + \alpha^2 S_2,\\
&G_{14} = 6\eta^2 J_2 + 3\alpha\eta J_1 + \alpha^2 J_2,\\
&G_{15} = 2\eta^4 G_7 + \alpha\eta^3 G_8 - 6\eta^2 G_9 - 2\alpha\eta G_{10} - \alpha^2 G_9,\\
&G_{16} = (\eta G_5 + G_6)(\beta+2),\\
&G_{17} = 6\alpha\eta^2 G_9 + 2\alpha^2\eta G_{10} + \alpha^3 G_9,\\
&G_{18} = \beta(\alpha G_3 - 3G_4) - \alpha^2 G_4 + 2\alpha G_3 - 6G_4,\\
&G_{19} = \alpha G_1 - 3G_2, \quad G_{20} = \alpha G_3 - 3G_4,\\
&G_{21} = \eta G_5 + G_6, \quad L = 6\eta^2 + \alpha^2,\\
&\Delta_1 = \frac{\alpha\sqrt{\eta}G_{15} + 6\alpha\eta^3 G_{16}}{2\eta},\\
&\Delta_2 = \eta^3 G_{18} - \alpha\beta G_3 + \alpha^2 G_4 - 2\alpha G_3,\\
&\Delta_3 = (\eta-1)^2\left(2(\beta-3)\eta^4 + (\beta-6)\eta^3 - 6\eta^2 - (\beta+6)\eta - 2(\beta+3)\right),\\
&\Delta_4 = (\beta-3)\eta^6 - 5\beta\eta^3 + 9(\beta+2)\eta - 5(\beta+3),\\
&\Delta_5 = G_3 Z - 2\eta^4 G19 - \eta(\alpha\eta^2 G_{20} + 6\eta^{3/2} G_{21} - 2\alpha G_1).
\end{aligned}
$$

References

1. Happel, J., Brenner, H.: Low reynolds number hydrodynamics prentice-hall. Inc., Englewood Cliffs, NJ **331** (1965)
2. Happel, J.: Viscous flow in multiparticle systems: slow motion of fluids relative to beds of spherical particles. AIChE J. **4**(2), 197–201 (1958)

3. Kuwabara, S.: The forces experienced by randomly distributed parallel circular cylinders or spheres in a viscous flow at small reynolds numbers. J. Phys. Soc. Japan **14**(4), 527–532 (1959)
4. Chen, S.H., Keh, H.J.: Axisymmetric motion of two spherical particles with slip surfaces. J. Colloid Interface Sci. **171**, 63–72 (1995)
5. Keh, H.J., Chen, S.H.: Low-reynolds-number hydrodynamic interactions in a suspension of spherical particles with slip surfaces. Chem. Eng. Sci. **52**(11), 1789–1805 (1997)
6. Keh, H., Chang, J.: Boundary effects on the creeping flow and thermophoretic motions of an aerosol particle in a spherical cavity. Chem. Eng. Sci. **53**, 2365–2377 (1998)
7. Ramkissoon, H., Rahaman, K.: Wall effects with slip. Z. Angew. Math. Mech. **83**(11), 773–778 (2003)
8. Keh, J.H., Lee, T.C.: Axisymmetric creeping motion of a slip spherical particle in a nonconcentric spherical cavity. Theor. Comput. Fluid Dyn. **24**, 497–510 (2010)
9. Faltas, M., Saad, E.: Stokes flow past an assemblage of slip eccentric spherical particle-in-cell models. Math. Methods Appl. Sci. **34**(13), 1594–1605 (2011)
10. Saad, E.: Cell models for micropolar flow past a viscous fluid sphere. Meccanica **47**(8), 2055–2068 (2012)
11. Prasad, M.K.: Cell models for non-newtonian fluid past a semipermeable sphere. Int. J. Math. Eng. Manag. Sci. **4**(6), 1352–1361 (2019)
12. Prasad, M.K.: Boundary effects of a nonconcentric semipermeable sphere using happel and kuwabara cell models. Appl. Comput. Mech. **15**(1) (2021)
13. Escott, L.J., Wilson, H.J.: Investigation into the rheology of a solid sphere suspension in second-order fluid using a cell model. Phys. Rev. Fluids **5**(8), 083,301 (2020)
14. Alfvén, H.: Existence of electromagnetic-hydrodynamic waves. Nature **150**(3805), 405–406 (1942)
15. Sherief, H., Faltas, M., El-Sapa, S.: Pipe flow of magneto-micropolar fluids with slip. Can. J. Phys. **95**(10), 885–893 (2017)
16. Kyrlidis, A., Brown, R.A., Walker, J.S.: Creeping flow of a conducting fluid past axisymmetric bodies in the presence of an aligned magnetic field. Phys. Fluids A **2**(12), 2230–2239 (1990)
17. Saad, E.I.: Effect of magnetic fields on the motion of porous particles for happel and kuwabara models. J. Porous Media **21**(7) (2018)
18. Prasad, M.K., Bucha, T.: Effect of magnetic field on the steady viscous fluid flow around a semipermeable spherical particle. Int. J. Appl. Comput. Math. **5**(3), 1–10 (2019)
19. Prasad, M.K., Bucha, T.: Magnetohydrodynamic creeping flow around a weakly permeable spherical particle in cell models. Pramana **94**(1), 1–10 (2020)

Inertial Migration of Cylindrical Particle in Stepped Channel—A Numerical Study

Manjappatta Pazhiyottumana Neeraj, Ranjith Maniyeri, and Sangmo Kang

Abstract Inertial migration of solid rigid particle in fluid flow occurs by the virtue of pure mechanical forces and it can play a pivotal role in separation techniques. The present computational study tries to capture the inertial migration dynamics of single rigid neutrally buoyant cylindrical particle in fluid flow which is residing in a stepped (sudden contraction) channel by determining the equilibrium position and migration time. The immersed boundary method based on feedback forcing scheme is used to develop the numerical model. The particle performs both translation and rotation motion according to the fluid flow condition and is modelled as rigid immersed boundary and the governing fluid momentum, and continuity equations are discretized using finite volume method in a staggered grid system and solved using semi-implicit fractional step algorithm. The study is mainly performed for centre and off-centre initial positions and its influence on the equilibrium position and migration time. It is observed that the equilibrium position is dependent on the initial position of release of particle. As initial position shifts from centre of channel, the particle equilibrium position also shifts accordingly. Further, the effect of height and length of step (contraction portion) on lateral migration is explored. The equilibrium position is found to be shifting towards the upper wall with decrease in height of step. However, the change in height of step does not have any significant effect on migration time of particle. It is identified that the increase in length of step reduces the migration time of particle although the equilibrium position remains same.

Keywords Equilibrium position · Feedback forcing · Immersed boundary method · Inertial migration · Migration time

M. P. Neeraj · R. Maniyeri (✉)
Biophysics Laboratory, Department of Mechanical Engineering, National Institute of Technology Karnataka (NITK), Surathkal 575025, India
e-mail: mranji1@nitk.edu.in

S. Kang
Department of Mechanical Engineering, Dong-A University, Saha-gu 604-714, Republic of Korea

R. P. Bharti and K. M. Gangawane (eds.), *Recent Trends in Fluid Dynamics Research*, Lecture Notes in Mechanical Engineering,
https://doi.org/10.1007/978-981-16-6928-6_4

1 Introduction

The particles suspended in a flowing fluid migrate in lateral direction by the virtue of two lift forces. The curvature of fluid velocity profile provides the shear induced lift whereas the wall induced lift acts in opposite direction due to the bulk shear flow near the wall. The balance of these two forces creates an equilibrium position for the particle after which it stops moving in lateral direction. The time taken to attain the equilibrium position is called migration time. This phenomena of lateral movement is known as inertial migration. Segre and Silberberg observed the inertial migration for the first time in 1961 [1] in an experiment where polymethyl methacrylate spherical particles which are suspended in fluid medium having mixtures of glycerol, butanediol and water were found to travel to 0.6 times the radius of pipe, and this phenomena was later termed as Segre-Silberberg effect. The numerical methods like arbitrary Lagrangian-Eulerian (ALE) [2, 3] and fictitious domain method [4] were used to simulate migration of neutrally buoyant particles in fluid flow, whereas the lattice Boltzmann method (LBM) was utilized by Inamuro et al. [5] for capturing the motion of cylindrical particle in Newtonian fluid. Since the simulation of fluid structure interaction is the most important part in study of inertial migration, the immersed boundary method (IBM) can be proposed. IBM was put forward by Peskin [6] which was incorporated for the simulation of heart interaction with blood flow. Shin et al. [7] presented feedback forcing-based IBM by combining the virtual boundary method which was put forward by Goldstein et al. [8] with Peskin's regularized delta function. Our previous work captured the single cylindrical shaped rigid particle migration in Poiseuille flow in a straight channel [9]. The feedback forcing-based immersed boundary method was used in this study for the simulation of inertial migration of rigid cylindrical particle, and the effect of Reynolds number and initial position of release on lateral migration was addressed in the same work. Through literature review, it is inferred that the study of inertial migration of particles in non-straight channels is limited. The specialty of these type of channels is the presence of secondary flows inside it. This secondary flows apply additional force on the particles flowing inside the channel [10] and it will help the particle to attain the equilibrium at a faster rate compared to that in straight channel. So the non-straight channels, like serpentine [11], bifurcated [12] and stepped channel, can play an important role in inertial migration and increase separation rate of bio-particles like bacteria cells. Hence, in the present work, the migration of single neutrally buoyant rigid cylindrical particle in stepped channel is simulated using immersed boundary method based on feedback forcing scheme. Further, the effect of initial position of release of particle, height and length of contraction section (step) on lateral migration is analyzed.

2 Methodology

The migration of single cylindrical particle in a stepped channel is illustrated in Fig. 1.

Here, H is the height of channel, L is the length of channel, $\mathbf{H_S}$ **and** $\mathbf{L_S}$ are height and length of contraction portion (step) in the channel and d is the diameter of particle. $\mathbf{F_S}$ **and** $\mathbf{F_W}$ denote the shear and wall induced lifts, respectively. The fluid enters from left boundary with constant velocity. The Navier–Stokes and continuity equations are non-dimensionalised and are shown in (1) and (2).

$$\nabla.\boldsymbol{v} = 0 \quad (1)$$

$$\frac{\partial \boldsymbol{v}}{\partial t} + \boldsymbol{v}.\nabla\boldsymbol{v} = -\nabla p + \frac{1}{Re}\nabla^2\boldsymbol{v} + \boldsymbol{f}(\boldsymbol{x}) \quad (2)$$

where t, **v**, p, **f**, **x** and Re are the time, velocity of flow, pressure, Eulerian force, Eulerian coordinate and Reynolds number, respectively. The discretized Navier–Stokes equation is shown in Eq. (3). Equations (1) and (2) are discretized using the semi-implicit fractional step method and finite volume scheme.

$$\frac{\boldsymbol{v}_{i,j}^{n+1} - \boldsymbol{v}_{i,j}^{n}}{dt} + N\left(\boldsymbol{v}_{i,j}^{n}\right) = -\left(\frac{\partial p}{\partial x_i}\right)^n + \frac{1}{2Re}\left(\nabla^2\boldsymbol{v}^{n+1} + \nabla^2\boldsymbol{v}^{n}\right) \quad (3)$$

where n represents the time step, and i and j denote the fluid points. The convection term is indicated by N. The velocity of immersed boundary $\mathbf{V}_{\text{ib}}$ is obtained by Eq. (4).

$$\boldsymbol{V}_{ib}(\boldsymbol{X}) = \int \boldsymbol{v}(\boldsymbol{x})\delta(\boldsymbol{x} - \boldsymbol{X}(s))d\boldsymbol{x} \quad (4)$$

The fluid grid size is denoted by **dx**. The feedback forcing law is utilized to calculate the Lagrangian force **F** [7]. **X** is used to represent the coordinate of particle.

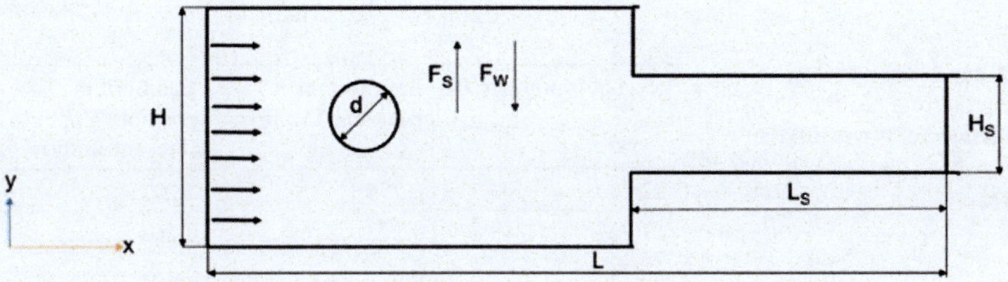

Fig. 1 Physical model

$$F(s) = \alpha \int (V_{ib} - V)dt + \beta(V_{ib} - V) \tag{5}$$

The large negative constants α and β are used, and **V** is the velocity of particle and computed by [13].

$$\mathbf{V} = \mathbf{VP} + \boldsymbol{\omega}\mathbf{P}(\mathbf{X} - \mathrm{Xc}) \tag{6}$$

where the centre of particle is denoted by Xc. In (6), the translational velocity is represented by $\mathbf{V_P}$ and the rotational speed is denoted by $\boldsymbol{\omega}_\mathbf{P}$.

3 Results and Discussion

The migration of a neutrally buoyant cylindrical particle in stepped channel of dimension 10 × 1 is captured with the aid of IBM based on feedback forcing scheme. The particle and the step are modelled as immersed boundary with Lagrangian grid size of 0.001. The length and height of contraction portion are 5 and 0.5, respectively, and the fluid grid size is taken as 0.05. The diameter of particle is 0.25, and the values of α and β are -4000.0 and -600.0. A FORTRAN code is developed. All the quantities are non-dimensionalised.

3.1 *Validation*

The developed code is validated by simulating the migration of particle in straight channel of 50 × 1 dimensions and comparing that with the results of Inamuro et al. [5]. The particle is released from same initial position (1.0, 0.4) in all three cases. The equilibrium position observed from the simulations are tabulated and compared with the results of [5] in Table 1. The results are in good agreement. The equilibrium position is denoted by $\mathbf{Y_{eq}}$.

Table 1 Comparison of equilibrium positions $\mathbf{Y_{eq}}$ of particle for three different Reynolds number with that of [5]

Reynolds number, Re	Equilibrium position, $\mathbf{Y_{eq}}$ [5]	Equilibrium position, $\mathbf{Y_{eq}}$ (Present study)
12.786	0.275	**0.297**
27.727	0.273	**0.268**
96.74	0.271	**0.239**

3.2 *Migration of Particle in Stepped Channel*

The particle migration in stepped channel is simulated further. The particle is released from an initial position of (2.5, 0.5). The initial and equilibrium position of particle are shown in Fig. 2a and b, respectively. The diameter of particle is 0.2.

The particle attains equilibrium position at (8.401, 0.5171). The velocity contour and vector plot are shown in Figs. 3a and b and 4.

The particle attains equilibrium position almost close to the centre of channel and it can be observed from both Figs. 2, 3 and 4. The lift forces are balanced at equilibrium position and it can be seen from the variation of coefficient of lift force with respect to lateral position. The lift coefficient becomes zero at equilibrium position (Fig. 5). The y coordinate of the centre of particle is represented by Y_{CC}.

Further, the effect of initial position of release of particle on the equilibrium position is analyzed and it is illustrated in Fig. 6 (Re = 12.786). The diameter of particle is kept constant for all cases as 0.2. The x coordinate of centre of particle is denoted by X_{CC}.

The equilibrium position shifts off-centre with respect to the initial position. In the next stage, the effect of height of contraction portion on equilibrium position is analyzed and it is shown in Fig. 7.

The particle equilibrium position shifts slightly towards the upper wall when the height of contraction portion is lowered. The velocity of flow increases when the height of contraction portion decreases. Due to this fact, the shear force on particle increases and the shear induced lift becomes more dominant than wall induced lift. This may be the reason for the particle to migrate towards the top wall when height of contraction portion reduces. Finally, the effect of length of contraction portion on the lateral migration is addressed (Fig. 8).

As we can depict from Fig. 8 that the equilibrium position of particle does not change significantly with respect to variation in length of contraction portion. This may be due to the fact that the velocity of flow remains unchanged with length of contraction portion. However, the migration time decreases with increase in length of contraction portion (shown in Table 2). The length of entrance region decreases with increase in length of contraction portion because the total length of channel remains unchanged and it may be the cause for reduction in migration time. The migration time is represented by $\mathbf{t_{eq}}$.

4 Conclusions

Two-dimensional computational model to study the inertial migration of single rigid neutrally buoyant cylindrical particle in stepped channel is developed using immersed boundary method based on feedback forcing scheme. The developed model is validated with the previous works of inertial migration in straight channel. Further, the influence of initial position of release of particle, height and length of contraction

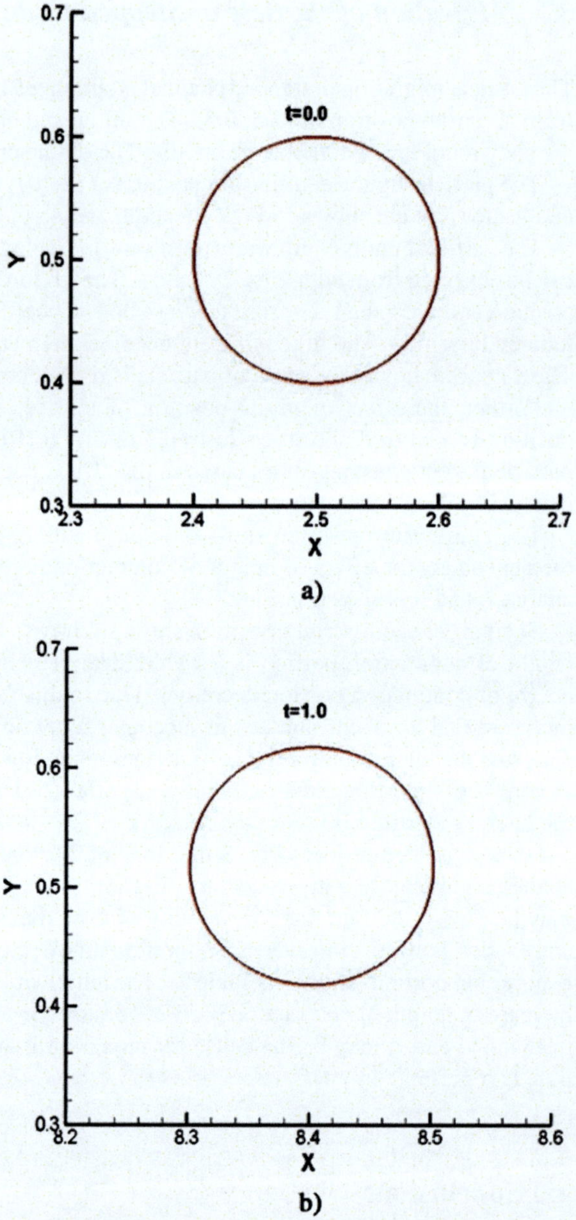

Fig. 2 **a** The initial position of the cylindrical particle in stepped channel (2.5, 0.5). **b** Final position = (8.401, 0.5171) (Re = 12.786, d = 0.2)

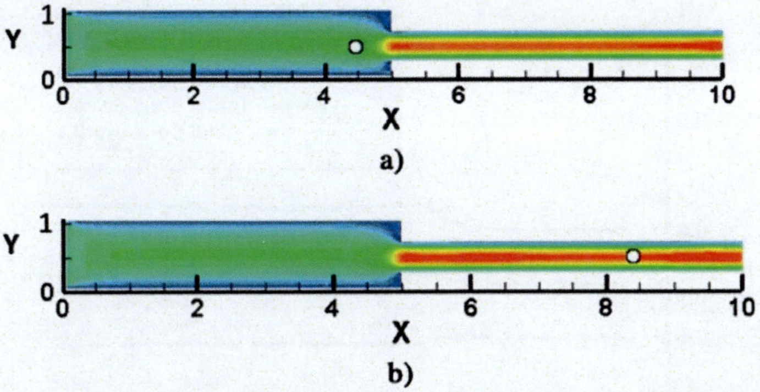

Fig. 3 The positions of particle (**a**) at time t = 0.5 and (**b**) t = 1.0 in the fluid flow (u velocity contour for an initial position of (2.5, 0.5) and Re = 12.786)

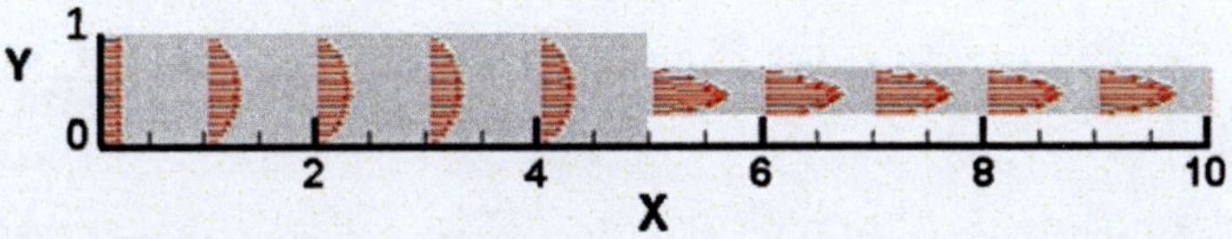

Fig. 4 The vector plot in the stepped channel for Re = 12.786

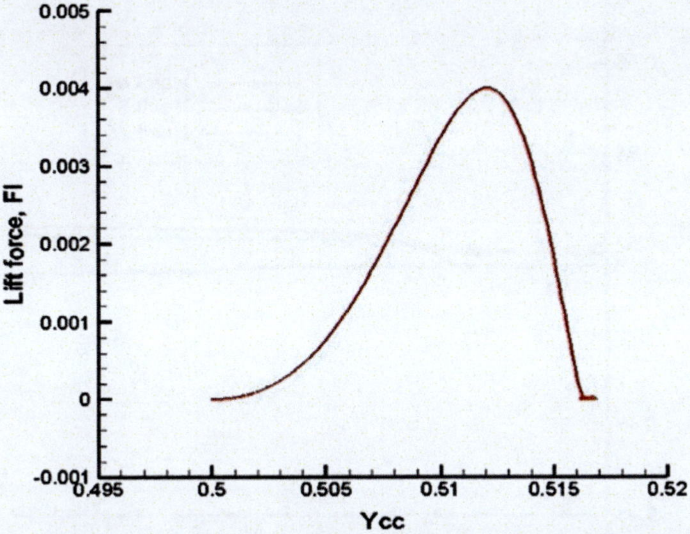

Fig. 5 The variation of lift coefficient with lateral position of centre of cylinder for Re = 12.786

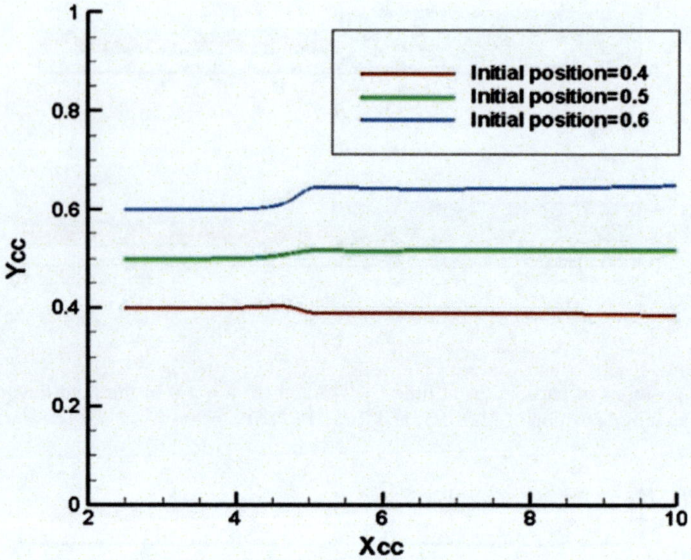

Fig. 6 Trajectory of centre of particle for three initial positions of release (Re = 12.786)

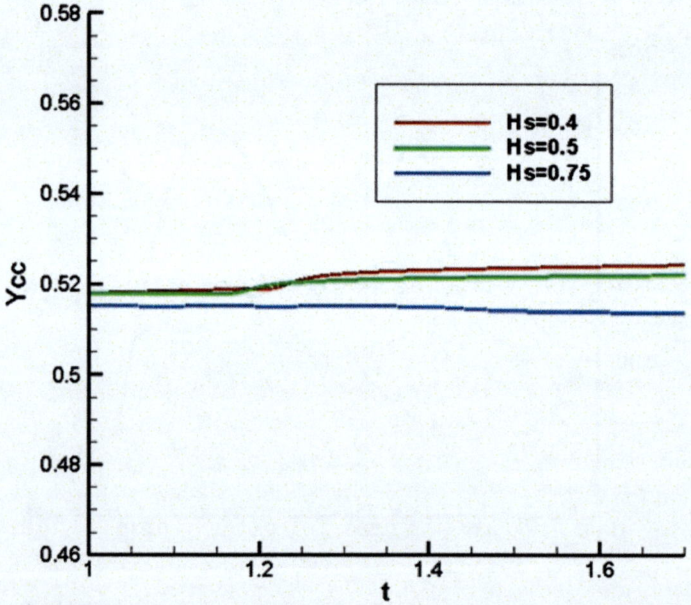

Fig. 7 The trajectory of particle with time for different height of contraction region (Re = 12.786, initial position = 0.5)

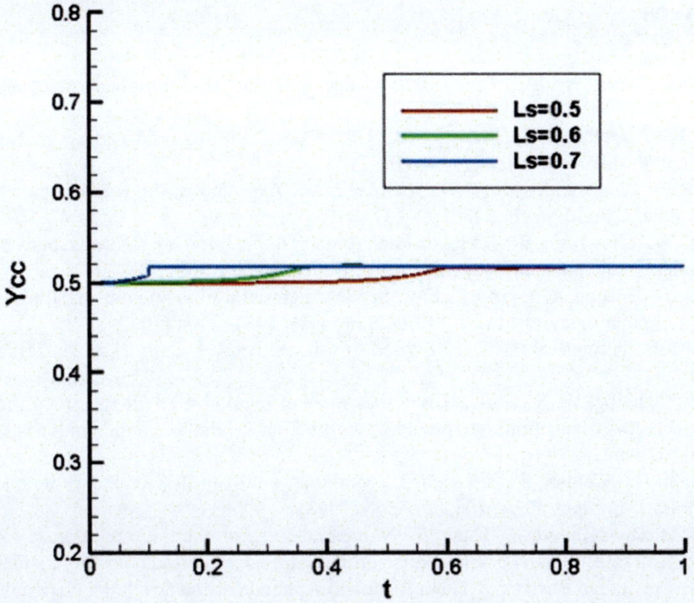

Fig. 8 The trajectory of particle with time for different length of contraction region (Re = 12.786, initial position = 0.5)

Table 2 Variation of migration time with length of contraction region

Length of contraction region L_S	Migration time $\mathbf{t_{eq}}$
7.0	0.119
6.0	0.423
5.0	0.534

portion on inertial migration is studied. The particle equilibrium position shifts off-centre with respect to its initial release position and it moves towards the upper wall with respect to the reduction in height of contraction portion. This may be due to the increase in velocity of flow when height of contraction portion is lowered. The shear induced lift becomes higher and becomes more dominant over wall induced lift when velocity of flow increases and hence the equilibrium position shifts slightly towards the upper wall. However, the migration time reduces with increase in length of contraction portion. The developed computational model can be extended to study the migration of single and multiple particles in different shaped non-straight channels and also for the three-dimensional simulation of inertial migration of particles.

References

1. Segre, G., Silberberg, A.: Radial particle displacements in Poiseuille flow of suspensions. Nature **189**, 209–210 (1961)
2. Hu, H.H., Joseph, D.D., Crochet, M.J.: Direct simulation of fluid particle motions. Theor. Computat. Fluid Dyn. **3**(5), 285–306 (1992)
3. Yang, B.H., Wang, J., Joseph, D.D., Hu, H.H., Pan, T.W., Glowinski, R.: Migration of a sphere in tube flow. J. Fluid Mech. **540**, 109–131 (2005)
4. Pan, T.-W., Glowinski, R.: Direct simulation of the motion of neutrally buoyant circular cylinders in plane poiseuille flow. J. Comput. Phys. **181**(1), 260–279 (2002)
5. Inamuro, T., Maeba, K., Ogino, F.: Flow between parallel walls containing the lines of neutrally buoyant circular cylinders. Int. J. Multiph. Flow **26**, 1981–2004 (2000)
6. Peskin, C.S.: Numerical analysis of blood flow in the heart. J. Comput. Phys. **25**(3), 220–252 (1977)
7. Shin, S.J., Huang, W.-X., Sung, H.J.: Assessment of regularized delta functions and feedback forcing schemes for an immersed boundary method. Int. J. Numer. Meth. Fluids **58**(3), 263–286 (2008)
8. Goldstein, D., Handler, R., Sirovich, L.: Modeling a no-slip flow boundary with an external force field. J. Comput. Phys. **105**, 354–366 (1993)
9. Neeraj, M. P., Maniyeri, R., Kang, S.: Numerical study on inertial migration of a cylindrical particle in straight channel using feedback forcing based immersed boundary method. In: 65th Congress of Indian Society of Theoretical and Applied Mechanics. https://istam.iitkgp.ac.in/resources/2020/proceedings/Full_paper/FM/91fullpaper.pdf (2020).
10. Tang, W., Zhu, S., Jiang, D., Zhu, L., Yang, J., Xiang, N.: Channel innovations for inertial microfluidics. Lab on chip **20**(19), 3485–3502 (2020)
11. Liu, Y., Li, Q., Nie, D.: Two-dimensional numerical study on the migration of particle in a serpentine channel. J. Nanotechnol. **2615404**, 1–10 (2018)
12. Wang, X., Zandi, M., Ho, C.-C., Kavalc, N., Papautsky, I.: Single stream inertial focusing in a straight microchannel. Lab Chip **15**, 1812–1821 (2015)
13. Yuan, C., Pan, Z., Wu, H.: Inertial migration of single particle in square microchannel over wide ranges of re and particle sizes. Microfluid. Nanofluid. **22**(102), 1–13 (2018)

Effect of Turbulence Model on the Hydrodynamics of Gas–solid Fluidized Bed

Mona Mary Varghese and Teja Reddy Vakamalla

Abstract Fluidized bed reactors have been extensively employed in processing industries as it provides perfect mixing, efficient operation, and large heat and mass transfer rates. Understanding the particle–fluid interaction inside the bed is a significant parameter for the effective operation of the fluidized bed. This work aims to study the effect of the turbulence model on the mean solids volume fraction and mean flow field at different operating parameters (static bed height, inlet velocity). In the current numerical study, the unsteady multiphase simulations are performed in a three-dimensional fluidized bed (Gao et al. 2012) using the two-fluid model (TFM) with the kinetic theory of granular flow (KTGF) option. k-Ɛ is selected to model the turbulence. Gidaspow, Syamlal and O'Brien and energy minimization multi-scale (EMMS) drag models are considered for modeling the interphase momentum exchange coefficient. The three-dimensional models could capture the flow behavior inside the turbulent fluidized bed. The numerically predicted time-averaged solid volume fraction fits well with the experimental data at the center compared to the wall using the incorporation of EMMS drag with the k-Ɛ turbulence model. Similar to the experiments, a dense region is observed with descending particles near the wall and the dilute region near the center portion of the bed. It can be noted that close numerical predictions can be obtained using the selection of an appropriate drag model and turbulence model.

M. M. Varghese · T. R. Vakamalla (✉)
Department of Chemical Engineering, NIT Calicut, Calicut-Mukkam road, Kattangal, Kozhikode 673601, India
e-mail: teja@nitc.ac.in

R. P. Bharti and K. M. Gangawane (eds.), *Recent Trends in Fluid Dynamics Research*, Lecture Notes in Mechanical Engineering,
https://doi.org/10.1007/978-981-16-6928-6_5

Keywords CFD · Drag · Gas–solid fluidized bed · Hydrodynamics · Turbulence model

Nomenclature

$C_{1\varepsilon}$, $C_{2\varepsilon}$, $C_{3\varepsilon}$ [-]	Model constant
C_d[-]	Drag coefficient
d_p[m]	Particle diameter
e[-]	Restitution coefficient
g[m s^{-2}]	Acceleration due to gravity
g_0[-]	Radial distribution coefficient
H[m]	Axial position
H_0[m]	Static bed height
I[-]	Stress tensor
J_{vis}[W]	Dissipation rate due to viscous damping
J_{slip}[W]	Generation rate due to viscous damping
k[J kg^{-1}]	Turbulent kinetic energy
k_s[kg $m^{-1}s^{-1}$]	Diffusion coefficient for granular energy
P[Pa]	Pressure
Re_p[-]	Reynolds number of particle
U_g[m s^{-1}]	Superficial gas velocity
u[m s^{-1}]	Velocity

Greek symbols

α [-]	Volume fraction
β [kg $m^{-3}s^{-1}$]	Interphase momentum transfer coefficient
γ[kg $m^{-1}s^{-3}$]	Collisional energy dissipation
λ[kg $m^{-1}s^{-1}$]	Bulk viscosity
μ[kg $m^{-1}s^{-1}$]	Shear viscosity
ρ[kg m^{-3}]	Density
τ [Pa]	Stress tensor
Θ [m^2s^{-2}]	Granular temperature

1 Introduction

Multiphase reactors are considered as the heart of many industrial processes, namely chemical, petrochemical, mineral processing, petroleum refining, and pharmaceutical industries [1]. Turbulent fluidized bed is the most prominent multiphase reactor having advantages like uniform temperature distribution, adequate particle mixing, and large surface area of contact. The benefits also include low maintenance, minimum construction cost, and the capability to fluidize a wide range of particles [2]. The continuous increase in the gas velocity causes the bubbles to disappear and the turbulent motion of the solid cluster is observed in the turbulent fluidized bed [3]. Turbulent fluidized bed reactors are one of the important units for large applications in physical and chemical processes, examples of which include the drying process, Fischer–Tropsch synthesis, production of acrylonitrile, etc.

Even though many studies are conducted in these areas, the fluidized bed reactor remains an active research area as many of the essential hydrodynamic parameters are yet to be explored. This knowledge of parameters will guide to proper design, scale-up, and will be useful for validating the numerical model predictions. The current studies on the hydrodynamic properties of turbulent fluidized beds emphasize mainly on gas–solid mixing, solid concentration, particle velocity, etc., [4]. The understanding of the hydrodynamic parameter, namely the solid concentration is critical for the scale-up, design, optimization, and modeling of industrial units. It also influences the gas–solid mixing, performance of the reactor, heat and mass transfer rate, and mainly helps to understand the interaction between two phases [5].

With the advancement of parallel computing, numerical simulation of fluidized beds has become a powerful method for analyzing the internal flow characteristics. This model helps researchers to understand the phenomena inside different types of the fluidized bed in detail as obtaining the experimental is difficult at different operating conditions [6]. The simulated results are validated with the experimental measurements of Gao et al. [7].

2 Literature Review

In the last 2 decades, CFD has become an important research tool for better understanding the fluidization process while obtaining internal fluid dynamics of the gas–solid system [8]. Mainly, TFM and Eulerian–Lagrangian approaches are used to model gas–solid flow. The Eulerian–Lagrangian approach considers the gas phase as continuous, while the particles are tracked using the Lagrangian approach and are not employed in industrial applications due to large computational costs. The TFM considers both phases as continuous and interpenetrating and is more appropriate for a fluidized bed system [9]. The coupling between phases is ensured using the momentum transfer term where the drag force is dominant. Different drag correlations are used to compute the interphase momentum exchange coefficients.

In the literature, several computational studies figured out the importance of drag force between the particle and fluid. Many researchers carried out simulation studies in bubbling fluidized beds, but only limited numerical studies are available [10, 11] in turbulent fluidized beds. Taghipour et al. [12] focused on the hydrodynamic study of a 2D gas–solid turbulent fluidized bed with spherical glass beads of diameter 250 – 300 μm. The adopted drag models of Syamlal and O'Brien, Gidaspow, and Wen & Yu showed a good agreement for the values of mean pressure drop, bed expansion, and qualitative gas–solid flow pattern. They suggested that further investigation of the experimental and computational study is needed for better validation. Further, Lundberg and Halvorsen [13] studied the effect of homogeneous drag models using two-dimensional grids and the experimental results were best predicted by RUC, Hill—Koch-Ladd, and Gidaspow models. Modifications based on the RUC model were performed to analyze the influence of multiple particle phases and wall functions on bubble frequencies.

As the homogeneous drag models fail to consider the effect of mesoscale structures, Li and Kwank [14] developed an EMMS model for heterogeneous flow that mainly predicts the heterogeneity inside the fluidization system. The results obtained using the EMMS model are more realistic and accurate, whereas the conventional Gidaspow and Syamlal and O'Brien drag consider a completely homogeneous condition within the computational grid. Wang et al. [15] selected the EMMS drag model to predict the hydrodynamics of a turbulent fluidized bed with FCC particles. The model suggested by Ullah et al. [16] showed close validation with experimental results using the EMMS drag model for time-averaged solid volume fraction. Shah et al. [17] compared the results obtained using EMMS and Gidaspow drag models with experimental data. Time-averaged radial and axial solid volume fraction profiles were examined using different models and found that EMMS predicted well compared to other drag models. The simulation study conducted by Shi et al. [18] and Varghese et al. [19] in a 2D and 3D turbulent fluidized showed that a combination of different numerical parameters for 2D simulation resulted in accurate predictions. Whereas, it is noted that the simulations conducted using the 3D model were less sensitive to the variation in numerical parameters. Further, it is observed that the 3D simulations are rather very few in the literature as a result of high computational costs. Thus, the hydrodynamics in the turbulent gas–solid fluidized bed was investigated computationally in this work using the 3D model.

3 Scope of Work

From the previous works, it is noted that a large number of the studies consider the system as homogeneous, neglecting the influence of heterogeneous flow structures mainly in the turbulent fluidized bed. Firstly, this study is based on the experimental work of Gao et al. [7] and mainly focuses on understanding the solid flow properties inside the gas–solid turbulent fluidized bed for different operating conditions. It also aims to understand the importance of drag models on the turbulent fluidization regime by assessing the flow predictions, i.e., solids volume fraction. The drag models considered for this study include the Gidaspow, Syamlal and O'Brien, and EMMS models. The above drag models are tested to find their capability in predicting accurate flow dynamics in turbulent fluidization conditions. This work mainly focuses on understanding the importance of selecting an appropriate drag model and turbulence model on the numerical prediction of hydrodynamic behavior in a turbulent fluidized bed. Further, the CFD predicted flow properties (solids volume fraction and particle velocity) are validated against the experimental measurements of Gao et al. [7].

4 Model Description

4.1 Governing Equation

Here, both the phases are modeled using the Eulerian-Eulerian two-fluid approach. Both the phases are assumed to be an interpenetrating continuum. The phase changes and chemical reactions are not considered in this study. The hydrodynamic model used for gas–solid fluidization primarily uses the principles of conservation of mass and momentum. Different ways are used to formulate the two-fluid flow model depending on the averaging procedure and closure law. Here, the integral balance of mass and momentum for a fixed control volume is adopted for both continuous and discrete phases. Anderson & Jackson [20] and Ishii [21] derived governing equations for two-fluid flow.

The mass conservation equation for gas and solid phase:

$$\frac{\partial(\alpha_g \rho_g)}{\partial t} + \nabla.(\alpha_g \rho_g u_g) = 0 \tag{1}$$

$$\frac{\partial(\alpha_s \rho_s)}{\partial t} + \nabla.(\alpha_s \rho_s u_s) = 0 \tag{2}$$

Here, $\alpha_g + \alpha_s = 1$.

where subscript g-gas phase and s-solid phase.

This equation is used for calculating the volume fraction for each phase. The momentum equation:

For gas phase,

$$\frac{\partial(\alpha_g \rho_g u_g)}{\partial t} + \nabla.(\alpha_g \rho_g u_g u_g) = -\alpha_g \nabla P + \nabla.(\alpha_g \tau_g) + \beta(u_s - u_g) + \alpha_g \rho_g g \tag{3}$$

For solid phase,

$$\frac{\partial(\alpha_s \rho_s u_s)}{\partial t} + \nabla.(\alpha_s \rho_s u_s u_s) = -\alpha_s \nabla P - \nabla P_s + \nabla.(\alpha_s \tau_s) + \beta(u_g - u_s) + \alpha_s \rho_s g \tag{4}$$

Further, the constitutive relations for the solid phase stress are derived by Lun et al. [22]. The equation used is as follows:

$$P_s = \alpha_s \rho_s \Theta + 2\alpha_s^2 \rho_s g_0 \Theta (1 + e) \tag{5}$$

The radial distribution function is given by the equation

$$g_0 = \left[1 - \left(\frac{\alpha_s}{\alpha_{s,\max}}\right)^{1/3}\right]^{-1} \tag{6}$$

$\alpha_{s,\max}$- maximum volume fraction of particles.

The solid bulk viscosity that considers the resistance of granular particles is calculated by Lun et al. [23] from the equation

$$\lambda_s = \frac{4}{3}\alpha_s^2 \rho_s d_p g_0 (1 + e)\sqrt{\frac{\Theta}{\Pi}} \tag{7}$$

The transport equation of granular temperature is given by the equation

$$\frac{3}{2}\left[\frac{\partial(\alpha_s \rho_s \Theta)}{\partial t} + \nabla.(\alpha_s \rho_s u_s \Theta)\right] = (-P_s I + \tau_s) : \nabla u_s$$
$$+ \nabla.(k_s \nabla \Theta) - \gamma_s + J_{vis} + J_{slip} \tag{8}$$

where the first term on the right-hand side corresponds to the generation of granular energy; the diffusion of granular temperature is denoted by the second term; the dissipation of granular energy is represented by the third term; the last term corresponds to the energy exchange between the phases.

Turbulence model:

The standard k-Ɛ model is given by the equation

$$\frac{\partial(\rho k)}{\partial t} + \frac{\partial(\rho k u_i)}{\partial x_i} = \frac{\partial}{\partial x_j}\left[\left(\mu + \frac{\mu_t}{\sigma_k}\right)\frac{\partial k}{\partial x_j}\right] + G_k + G_b - \rho_\varepsilon - Y_M + S_k$$

$$\frac{\partial(\rho \epsilon)}{\partial t} + \frac{\partial(\rho \varepsilon u_i)}{\partial x_i} = \frac{\partial}{\partial x_j}\left[\left(\mu + \frac{\mu_t}{\sigma_k}\right)\frac{\partial \varepsilon}{\partial x_j}\right] + C_{1\varepsilon}\frac{\varepsilon}{k}(C_{1k} + C_{3\varepsilon}G_b) - C_{2\varepsilon}\rho\frac{\varepsilon^2}{k} + S_\varepsilon$$

4.2 Interphase Momentum Exchange

The drag force correlations used in the simulations are as follows:

1. Gidaspow Drag Model [24]:

It is a homogeneous drag model obtained by combining the Wen and Yu [25] and Ergun [26] equations. The equation used to evaluate the interphase momentum transfer coefficient is

$$\beta = \frac{150\,\alpha_s^2\mu_g}{\alpha_g d_p^2} + \frac{1.75\,\alpha_s\,\rho_g\,|u_g - u_s|}{d_p}\ \text{for}\ \alpha_g \leq 0.8$$

$$\beta = \frac{3}{4}\frac{\alpha_s\,\alpha_g}{d_p}\rho_g|u_g - u_s|\,C_{d,isol}\,\alpha_g^{-2.65}\ \text{for}\ \alpha_g > 0.8$$

Where

$$C_{d,isol} = \begin{cases} 0.44 \\ {}^{24}\!/_{\mathrm{Re}_p}\left(1 + 0.15\mathrm{Re}_p^{0.687}\right) \end{cases} \text{for}\ \begin{matrix} \mathrm{Re}_p \geq 1000 \\ \mathrm{Re}_p < 1000 \end{matrix} \tag{9}$$

2. Syamlal and O'Brien Drag Model [27]:

This equation is based on the terminal velocity of a single particle in a fluid.

$$\beta = \frac{3}{4}\frac{\alpha_s\,\alpha_g}{u_{rs}^2 d_p}\rho_g C_{d,isol}|u_g - u_s|$$

Where

$$C_{d,isol} = \left(0.63 + \frac{4.8}{\sqrt{{}^{\mathrm{Re}_p}\!/_{u_{rs}}}}\right)^2$$

$$u_{rs} = 0.5\left(A - 0.06\mathrm{Re}_p + \sqrt{(0.06\mathrm{Re}_p)^2 + 0.12\mathrm{Re}_p(2B - A) + A^2}\right)$$

$$A = \alpha_g^{4.14}$$

$$B = \begin{cases} 0.8\alpha_g^{1.28} \\ \alpha_g^{2.65} \end{cases} \text{for}\ \begin{matrix} \alpha_g \leq 0.85 \\ \alpha_g > 0.85 \end{matrix} \tag{10}$$

3. Energy Minimization Multiscale (EMMS) Drag Model [14, 28]:

It is based on energy minimization for the transport and suspension of particles.

$$\beta = \frac{150\,(\alpha_s)^2\mu_g}{(1 - \alpha_s)d_p^2} + \frac{1.75\,\alpha_s\,\rho_g\,|u_g - u_s|}{d_p}\ \text{for}\ \alpha_s \geq 0.26$$

$$\beta = \frac{3}{4}\frac{\alpha_s\,(1-\alpha_s)}{d_p}\rho_g\left|u_g - u_s\right| C_{d,isol}\ \omega(\alpha_s)\ \text{for}\ \alpha_s < 0.26$$

Where

$$\omega(\alpha_s) = -0.5760 + \frac{0.0214}{4((1-\alpha_s) - 0.7463)^2 + 0.0044}\ \text{for}\ 0.18 \le \alpha_s < 0.26$$

$$\omega(\alpha_s) = -0.0101 + \frac{0.0038}{4((1-\alpha_s) - 0.7789)^2 + 0.0040}\ \text{for}\ 0.03 \le \alpha_s < 0.18$$

$$\omega(\alpha_s) = -31.8295 + 32.8295(1-\alpha_s)\ \text{for}\ \alpha_s < 0.03 \tag{11}$$

4.3 Numerics

The experimental measurements of Gao et al. [7] are used to validate the present 3D simulation study conducted in a cylindrical turbulent fluidized bed with dimensions as shown in Fig. 1a. The glass beads with a mean diameter of 130 μm and a density of 2400 kg/m^{3}are used in this study. The experiments are conducted using the multifunctional optical fiber probe in the Gao et al. [7] study.

The granular option enabled TFM model coupled with the laminar, and the k-Ɛ turbulence model is used for the simulation study in turbulent fluidized beds. The interphase drag force models of Gidaspow [24], EMMS [14], and Syamlal and O'Brien [27] are considered to model the interphase momentum transfer coefficient. The new drag models are introduced using user defined functions (UDFs) in ANSYS Fluent 20.1. The particle properties are described using the KTGF.

The fine mesh is created near the wall to accurately predict the velocity distribution near-wall region. The grid independence test is performed with three different meshes of size varying from 42,680 to 1,07,520 nodes. The corresponding grid independence test data using the EMMS drag model are shown in Fig. 1b.

The boundary conditions of the velocity inlet and the pressure outlet are specified at the column inlet and the outlet. For the gas phase, a no-slip condition is used at the walls. The partial slip for the solid phase with a specularity coefficient of 0.6 is used based on Passalacqua and Marmo [29] for the turbulent fluidized bed. The particle restitution coefficient that represents the granular phase collisions is given a value of 0.95 [30]. The convective terms of the transport equations are solved using the quadratic upstream interpolation for convective kinematics (QUICK) scheme. A second-order implicit scheme is selected for the time discretization. The semi-implicit method for pressure linked equations (SIMPLE) algorithm is considered for the pressure and velocity coupling. For the turbulence calculation, the laminar and k-Ɛ models are used. A maximum of 20 iterations per time step is applied, and

Table 1 Summary of variables used in simulation studies

Variables	Values
Inlet gas velocity, ms^{-1}	0.94, 1.10, 1.25
Particle diameter, μm	139
Static bed height, m	0.096, 0.151, 0.204
The angle of internal friction, °	28
Time step, s	10^{-4}
Simulation time, s	30
Iterations per time step	20
Restitution coefficient	0.95
Specularity coefficient	0.6
Coefficient of lift	0.25
Turbulence model	Laminar and K-Ɛ
Drag model	Gidaspow, Syamlal and O' brien, EMMS

the residual is specified to be 0.0001 for the convergence criteria. The summary of variables used for the numerical simulations is provided in Table 1.

5 Results and Discussion

5.1 Grid Independence Test

Figure 1b represents the time-averaged void fraction for different grid sizes. It is calculated for a superficial velocity of 1.25 m/s and a static bed height of 0.151 m. The results are plotted at an axial position of 0.338 m. It is observed that both the 70 k and 100 k mesh are giving close predictions with the experimental results. While the coarser-sized mesh (~ 42 k) could not capture the correct behavior of particles inside the bed. So it is concluded that the medium size mesh (~70 k) can be selected for further study. Also, due to the large computational time, the use of finer mesh is not recommended.

5.2 Effect of Drag Models

Initially, a 3D cylindrical column with a uniform inlet superficial velocity of 1.25 m/s is considered for running the numerical simulation. The gas is injected uniformly through the bottom of the bed. A static bed height of 0.151 m is maintained before initializing the simulation. As time passes, it is noted that the bed gets expanded, which leads to the complete mixing of solid particles.

The time-averaged experimental void fraction data show a symmetric behavior in the radial direction. When Gidaspow, Syamlal and O'Brien, and EMMS models are compared, it is observed that the EMMS drag model gives close predictions. The conventional drag model is largely deviated and fails to include the effect of clusters. The presence of particle clusters is mainly observed near the wall because of the low particle velocity at the wall. Also, it is noted that the solid distribution is dilute in the middle compared to the dense portion near the wall. A large deviation is noticed with the Gidaspow drag model as it causes higher bed expansion and neglects the presence of mesoscale structures. At a higher inlet velocity, more phase interactions and larger turbulence (Fig. 2) are present. This can be clearly analyzed by using the structure-dependent EMMS drag model.

5.3 *Effect of Turbulence Models*

Fluctuating and chaotic behavior are the main characteristics of turbulent flow. The effect of turbulence increases with an increase in velocity. It is clear from Fig. 3 that both the particle clusters and bubbles are present inside the turbulent fluidized bed. Simulations are performed with the selected EMMS drag model using the laminar and k-ε turbulence models. In the laminar flow, no turbulence parameter is present in the set of equations while the k-ε model is an empirical model based on the model transport equations for the turbulence kinetic energy and the dissipation rate. The standard k-ε model used to model the turbulence in the gas phase shows superior results. It is also noted that the k-ε model reveals better predictions of solid volume fraction both in the core and near the wall region. Moreover, the k-ε model is used for a wide range of applications in turbulent flows due to its easy convergence and low memory requirement. It is also noted that the laminar model is giving better predictions at the top dilute region of the bed.

5.4 *Comparison of Simulated and Experimental Results*

Radial time-averaged solid volume fraction. The results obtained for time-averaged solid volume fraction using the EMMS drag model and k-ε turbulence model in the radial direction are compared against the experimental data and shown in Fig. 4. It is noted that better predictions are observed near the center of the column. The increase in velocity of gas causes the bubbles to rise and coalesce, forming larger bubbles. It finally breaks at the free surface. The solid distribution with the dilute center and a dense region near the wall can be observed from Fig. 3. It is noted that the simulation results agree well at the bottom dense region, while the upper dilute zone results are underestimated. The EMMS model shows close predictions with

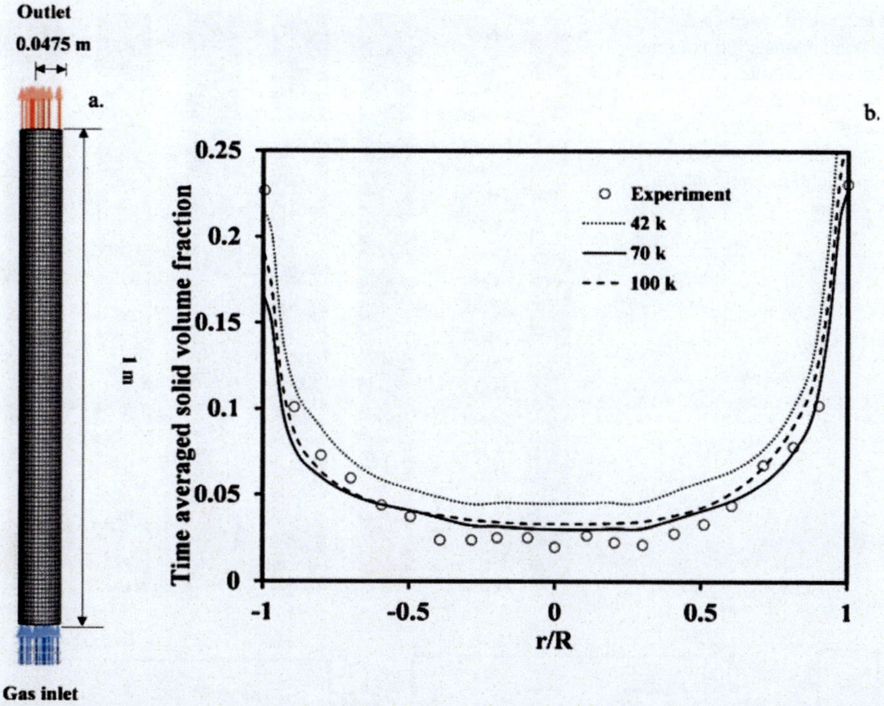

Fig. 1 **a** 3D mesh used for turbulent fluidized bed simulations, **b** mesh independence study in the turbulent fluidized bed for a velocity of 1.25 m/s and a static bed height of 0.151 m [7]

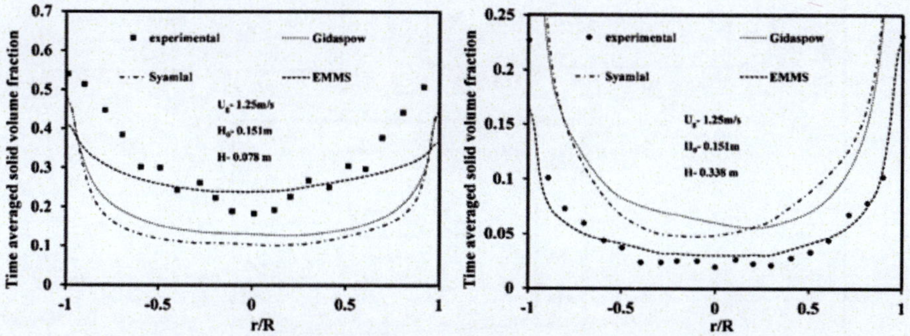

Fig. 2 CFD predicted radial variation of time-averaged solid volume fraction compared against experimental data [7] using different drag models at two different axial heights with 1.25 m/s of inlet gas velocity

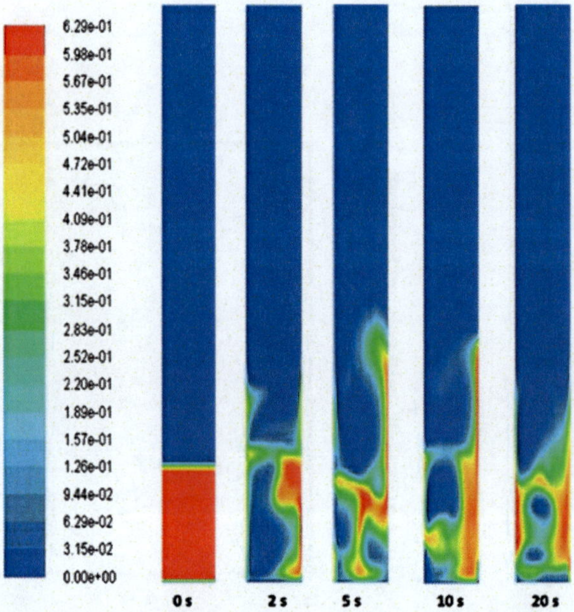

Fig. 3 CFD predicted instantaneous solid volume fraction contours of the 3D fluidized bed using the EMMS drag model with k-ε turbulence model for the superficial gas velocity of 1.25 m/s

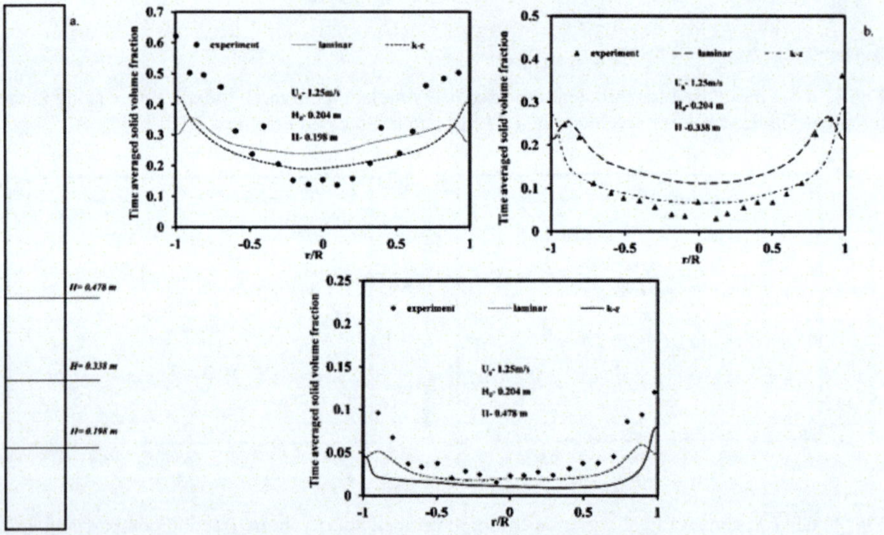

Fig. 4 **a** Outline of fluidized bed indicating different axial positions, **b** CFD predicted time-averaged solid volume fraction compared against experimental data [7] along the radial direction using two turbulence models at different axial heights for the inlet gas velocity of 1.25 m/s

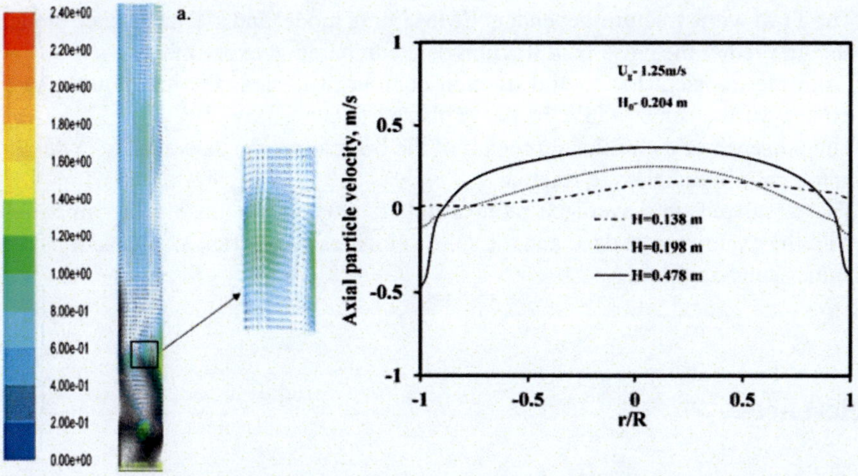

Fig. 5 **a** Void fraction contour and solid velocity vector in a 3D turbulent fluidized bed for the superficial velocity of 1.25 m/s, **b** Radial distribution of the axial solid velocities at different axial positions

the experimental results as it considers the effect of the mesoscale structure formed inside the bed.

It is found that the predicted solid volume fraction data are lower near the wall, whereas it is higher near the center compared to the experimental measurements. This may be due to the assumptions of the mean particle diameter instead of particle size distribution [7].

Particle velocity profile. The properties displayed by a fluidized bed mainly depend upon the velocity of particles, i.e., the movement of particles inside the bed. The solid particles always tend to move away from the center of the bed. This is because the gas in the form of bubbles tends to move upward and coalesces to form a larger bubble. Figure 5a shows the void fraction contour and solid velocity vector for a superficial velocity of 1.25 m/s using the EMMS drag model. In all the cases, high voidage is observed at the center and the particle-rich phase is found adjacent to the wall. In general, the velocity is positive near the center and approaches negative toward the wall. The negative velocities near the wall are expected due to the particle's back mixing. It is noticed from Fig. 5b that the radial profile of axial velocity increases with the axial position. The rise in particle velocity also improves the mixing of particles inside the bed.

6 Conclusions

- The characteristics of the gas–solid flow behavior in the turbulent fluidized bed are investigated using CFD simulations.

- The TFM with structure-dependent EMMS drag model and k-Ɛ turbulence model able to predict the close void fraction data compared to experiments.
- From the transient local solid fraction contour, it is clear that the dilute phase exists near the center while the dense phase exists near the wall.
- The presence of the dilute top region of the bed causes the discrepancy in height in the case of radial solid fraction.
- The simulated time-averaged radial solid fraction shows close approximations with the experimental data, and the solid velocities at different axial positions are numerically analyzed.

References

1. Rüdisüli, M., Schildhauer, T.J., Biollaz, S.M.,van Ommen, J.R.J.P.T.: Scale-up of bubbling fluidized bed reactors—A review. Powder Technology 217, 21–38 (2012)
2. Daizo, K., Levenspiel, O.: Fluidization engineering, 2nd edn. Butterworth Publishers, United States (1991)
3. Drake, J.: Hydrodynamic characterization of 3D fluidized beds using noninvasive techniques. Graduate Theses and Dissertations. Iowa State University (2011)
4. Ellis, N., Bi, H.T., Lim, J., Grace, J.: Hydrodynamics of Turbulent Fluidized Beds of Different Diameters. Powder Technol. **141**, 124–136 (2004)
5. Mostoufi, N., Chaouki, J.: Local Solid Mixing in Gas-Solid Fluidized Beds. Powder Technol. **114**, 23–31 (2001)
6. Zhou, L., Zhang, L., Bai, L., Shi, W., Li, W., Wang, C., Agarwal, R.: Experimental study and transient CFD/DEM simulation in a fluidized bed based on different drag models. RSC Adv. **7**(21), 12764–12774 (2017)
7. Gao, X., Wu, C., Cheng, Y.-W., Wang, L.-J., Li, X.: Experimental and numerical investigation of solid behavior in a gas–solid turbulent fluidized bed. Powder Technol. **228**, 1–13 (2012)
8. Sau, D.C., Biswal, K.C.: Computational fluid dynamics and experimental study of the hydrodynamics of a gas–solid tapered fluidized bed. Appl. Math. Model. **35**(5), 2265–2278 (2011)
9. Hamzehei, M.: CFD modeling and simulation of hydrodynamics in a fluidized bed dryer with experimental validation. International Scholarly Research Notices 2011, (2011)
10. Chang, J., Wu, Z., Wang, X., Liu, W.: Two- and three-dimensional hydrodynamic modeling of a pseudo-2D turbulent fluidized bed with Geldart B particle. Powder Technol. **351**, 159–168 (2019)
11. Wu, Y., Shi, X., Gao, J.,Lan, X.: A four-zone drag model based on cluster for simulating gas-solids flow in turbulent fluidized beds. Chemical Engineering and Processing—Process Intensification 155, 108056 (2020)
12. Taghipour, F., Ellis, N., Wong, C.: Experimental and computational study of gas–solid fluidized bed hydrodynamics. Chem. Eng. Sci. **60**(24), 6857–6867 (2005)
13. Lundberg, J.,Halvorsen, B.M.: A review of some exsisting drag models describing the interaction between phases in a bubbling fluidized bed. Proc 49th Scandinavian Conference on Simulation and Modeling (SIMS 2008), pp. 7–8. Oslo, Norway (2008)
14. Li, J., Kwauk, M.: Particle-Fluid Two-Phase Flow: the Energy-Minimization Multi-Scale Method. Metallurgical Industrial Press, Beijing (1994)
15. Wang, B., Li, T., Sun, Q.-W., Ying, W.-Y., Fang, D.-Y.: Experimental Investigation on Solid Concentration in Gas-Solid Circulating Fluidized Bed for Methanol-to-Olefins Process. Int. J. Chem. Eng. **4**(8), 494–500 (2010)

16. Ullah, A., Jamil, I., Hamid, A., Hong, K.: EMMS mixture model with size distribution for two-fluid simulation of riser flows. Particuology **38**, 165–173 (2018)
17. Shah, M.T., Utikar, R.P., Tade, M.O.,Pareek, V.K.J.C.E.J.: Hydrodynamics of an FCC riser using energy minimization multiscale drag model. Chem. Eng. J. **168** (2), 812–821 (2011)
18. Shi, H., Komrakova, A., Nikrityuk, P.J.P.T.: Fluidized beds modeling: Validation of 2D and 3D simulations against experiments. Powder Technol. **343**, 479–494 (2019)
19. Varghese, M.M., Vakamalla, T.R., Mantravadi, B., Mangadoddy, N.: Effect of Drag Models on the Numerical Simulations of Bubbling and Turbulent Fluidized Beds. Chem. Eng. Technol. **44**(5), 865–874 (2021)
20. Anderson, T.B.,Jackson, R.: Fluid Mechanical Description of Fluidized Beds. Equations of Motion. Industrial & Engineering Chemistry Fundamentals **6** (4), 527–539 (1967).
21. Ishii, M.: Thermo-fluid dynamic theory of two-phase flow. Eyrolles, [Paris] (1975).
22. Van der Hoef, M.A., Ye, M., Van Sint Annaland, M., Andrews, A.T., Sundaresan, S.,Kuipers, J.A.M.: Multiscale Modeling of Gas-Fluidized Beds. Academic Press (2006)
23. Lun, C.K.K., Savage, S.B., Jeffrey, D.J., Chepurniy, N.: Kinetic theories for granular flow: inelastic particles in Couette flow and slightly inelastic particles in a general flowfield. J. Fluid Mech. **140**, 223–256 (1984)
24. Gidaspow, D.: Multiphase flow and fluidization: continuum and kinetic theory descriptions. Academic press (1994)
25. Wen, C.Y.: Mechanics of fluidization. Chemical engineering progress symposium series, pp. 100–111. (1966)
26. Ergun, S.: Fluid flow through packed columns. Chem. Eng. Prog. **48**, 89–94 (1952)
27. Syamlal, M.,O'Brien, T.: The derivation of a drag coefficient formula from velocity-voidage correlations. Technical Note, US Department of energy, Office of Fossil Energy, West Virginia, (1987)
28. Yang, N., Wang, W., Ge, W., Wang, L.,Li, J.: Simulation of heterogeneous structure in a circulating fluidized-bed riser by combining the two-fluid model with the EMMS approach. Industrial Engineering Chemistry Research 43 (18), 5548–5561 (2004)
29. Passalacqua, A.,Marmo, L.J.C.E.S.: A critical comparison of frictional stress models applied to the simulation of bubbling fluidized beds. Chemical Engineering Science **64** (12), 2795–2806 (2009)
30. Loha, C., Chattopadhyay, H., Chatterjee, P.K.: Effect of coefficient of restitution in Euler-Euler CFD simulation of fluidized-bed hydrodynamics. Particuology **15**, 170–177 (2014)

Steady Flow of Power-Law Fluids Past an Inclined Elliptic Cylinder

Prateek Gupta, Deepak Kumar, and Akhilesh K. Sahu

Abstract The laminar flow of incompressible power-law fluids over an elliptic cylinder in cross-flow arrangement is studied numerically. The elliptic cylinder (aspect ratio, $e = 0.2$), is placed at different angle of attack (α) to the flow direction. The two-dimensional governing equations are solved using finite volume method-based solver *FLUENT*. Numerical results are presented for moderate Reynolds number ($5 \leq Re \leq 40$). The power-law index of the fluid is varied from $0.4 \leq n \leq 1.6$ for $\alpha = 0°, 30°, 45°, 60°$ and $90°$. The effect of the Reynolds number (Re), fluid behavior (n), and angle of attack (α) on the flow field, and engineering parameters are studied in detail. The flow patterns in the form of streamlines, and aero-dynamics coefficient, i.e., drag (C_D) and lift (C_L) coefficient are shown for the aforementioned parameters. Variations in drag ratio (C_{DP}/C_{DF}) depict how individual components of C_D get affected by Re, α and n.

Keywords Elliptic cylinder · Power-law fluids · Angle of attack

Nomenclature

a	semi-major axis of the elliptic cylinder (in meters)
b	semi-minor axis of the elliptic cylinder (in meters)
C_D	Total Coefficient of Drag
C_{DP}	Pressure component of C_D
C_{DF}	Friction component of C_D
C'	C_{DP}/C_{DF}
C_L	Lift coefficient
e	Aspect ratio of the cylinder (b/a)
L_u	Upstream length (in meters)

P. Gupta · D. Kumar · A. K. Sahu (✉)
Department of Chemical Engineering, National Institute of Technology Rourkela, Odisha 769008 Rourkela, India
e-mail: sahuak@nitrkl.ac.in

R. P. Bharti and K. M. Gangawane (eds.), *Recent Trends in Fluid Dynamics Research*, Lecture Notes in Mechanical Engineering,
https://doi.org/10.1007/978-981-16-6928-6_6

L_d	Downstream length (in meters)
H	Total height of the domain (in meters)
D	Diameter of the elliptic cylinder (=2a)
F_D	Drag force per unit length of cylinder (N/m)
F_L	Lift force per unit length of cylinder (N/m)
n	Power-law index
Re	Reynolds number

Greek symbols

α	Angle of inclination (in degrees)
ρ	Density (kg/m^3)
μ	Dynamic viscosity ($N.s/m^2$)

1 Introduction

Various researchers have studied the phenomena of flow around bluff body of various shapes [1]. It is an old and interesting problem in the field of fluid dynamics. Flow around a circular cylinder serves as an exemplary model for understanding the flow dynamics and explains many fundamental theories in fluid mechanics such as vortex formation, vortex shedding, boundary-layer separation, Potential flow theory, etc. Bluff body replicates many applications in engineering processes. Among various bluff bodies, elliptic cylinder is considered as a unique bluff body because of its physical appearances and importance. Elliptic cylinder is characterized by its aspect ratio, $e(= b/a)$, which is defined as ratio of minor axis (b) to the major axis (a) of the cylinder. It can behave as a flat plate at infinitely low aspect ratio and a circular cylinder at unity aspect ratio. Because of geometrical flexibility of the elliptic cylinder, it has huge application in aero-dynamics, mechanical, chemical and food processing industries, where solid fluids interaction take place [2]. Also, it has been demonstrated conclusively from time to time that elliptic cylinders have better aerodynamic capabilities than circular cylinders. A group of elliptic cylinders kept in close proximity has higher space economy due to their compact design [3]. Modeling of flow in heat exchangers, namely tubular and pin-type heat exchanger is a prime example of its industrial use. To investigate the structural integrity of chimneys with elliptic cross section, cross flow of air around elliptical cylinder can be considered as a model problem. Flow around elliptic airfoils can also be modeled by elliptic cylinders [4]. These diverse and interesting applications of elliptic cylinders are reason enough to keep it the focal point of our study.

We consider a Power-law fluid as the working fluid to make this study more desirable from an industrial perspective. Most of the substances which we see in chemical and food processing industry exhibit non-Newtonian flow characteristics. Typical examples of the substance exhibiting non-Newtonian fluid behavior are Food stuffs, greases, lubricating oils, paints, printing colors, inks, pharmaceutical products, etc. Thus, in current work, a numerical investigation has been performed to see the flow behavior of non-Newtonian power-law fluids past an elliptic cylinder inclined to the freestream direction. However, before presenting and assessing the new results, a concise summary of some relevant works is discussed in order to accentuate the importance and novelty of the current work.

2 Previous Work

Various researchers have investigated and presented enormous literature on flow past an elliptic cylinder. In initial studies, researchers investigated the flow around elliptic cylinder using analytical methods. Imai [5] came up with an analytical technique of solving Oseen's linearized equations. He varied the cylinder aspect ratio from $0 \leq e \leq 1$ and kept the Reynolds number very low, $Re = 0.1$ and 1. Hasimoto [6] also came up with an analytical solution of the linearized Oseen's equations and analyzed the flow past an elliptic cylinder. Both investigations were limited to very low values of Reynolds number. With the development of CFD in early 70's researchers shifted from analytical solution to Numerical solution. Epstein and Masliyah [7] numerically solved the Navier–Stokes (N-S) equations and analyzed flow past elliptical cylinder by varying the aspect ratio, $0.2 \leq e \leq 1$ and Re up to 90. They kept the major axis of the cylinder at right angle to the freestream direction. Subsequently, Lugt and Haussling [8] presented numerical results for flow past an elliptic cylinder inclined at 45° to the freestream. They observed steady flow for $Re =$ 15 and 30, and vortex shedding at $Re = 200$. Vorticity-stream function formulation for the incompressible N-S equations was used by D'Alessio and Dennis [9] to describe flow around an elliptic cylinder. They presented lift and drag coefficient results for $Re = 5$ and 20 with angle of inclination varying between 0° and 90°. Badr et al. [10] numerically investigated fluid flow and convective heat transfer from an elliptic cylinder inclined to the fluctuating freestream. They varied Re from 20 to 500 and found that freestream fluctuations facilitate the vortex shedding. Dennis and Young [11] investigated the flow around an inclined elliptic cylinder for two aspect ratios - 1:5 and 1:10, with $Re \leq 40$. A concise review of all the studies for power-law fluid flow over a circular cylinder is presented by Sivakumar et al. [12]. They studied the global and local flow characteristics of the power-law fluids across an elliptic cylinder. They presented results in the form of C_{DP}, C_{DF}, C_D, streamlines and vorticity contours for the following parameter space $0.2 \leq e \leq 5$;$0.2 \leq n \leq 1.8$; $0.01 \leq Re \leq 40$. Paul et al. [13] numerically investigated the unsteady flow field past an inclined elliptic cylinder. They analyzed the effect of different Re, α and e on various engineering parameters ($\overline{C_D}$, $\overline{C_L}$ and St) by plotting contours of pressure and streamlines.

In summary, voluminous literature is available on the interaction of a Newtonian fluid with an elliptic cylinder placed at different inclinations to the freestream. Detailed research has also been conducted on the flow characteristics of non-Newtonian fluids past an elliptic cylinder kept in a direction parallel and normal to the freestream. There are no results available on the flow phenomena of a non-Newtonian fluid around an inclined elliptic cylinder. We have tried to fill that gap herein. The parameter space consists of $5 \leq Re \leq 40$; $0.4 \leq n \leq 1.6$ and $0° \leq \alpha \leq 90°$.

3 Mathematical Modeling

In this work, we numerically investigate the two-dimensional, steady and incompressible laminar flow of power-law fluids over an elliptic cylinder inclined to the freestream direction. The aspect ratio (or "thickness") of the cylinder is considered $e = 0.2$ throughout the investigation. Lower the aspect ratio higher the ellipticity is the reason behind considering this particular aspect ratio. Fluid flows toward the cylinder from left to right with uniform velocity V_∞ (Fig. 1). To solve the unconfined nature of the flow, we introduced here a computational domain in the shape of a rectangular box with appropriate boundary conditions (Fig. 1). The elliptic cylinder is positioned at the middle of the box. The size of the rectangular box is considered in such a way that, it does not influence the flow around the cylinder and it requires less computational resources.

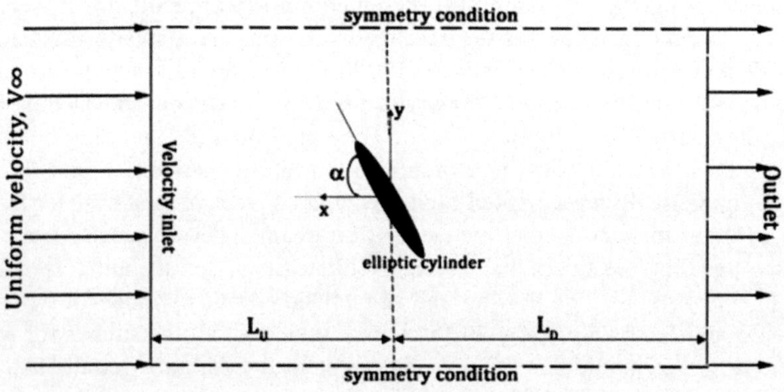

Fig. 1 Schematic representation of the Computational domain

3.1 Governing Equations

Two-dimensional governing equations describing the flow can be written as:
Continuity equation:

$$\frac{\partial u}{\partial x} + \frac{\partial v}{\partial y} = 0, \tag{1}$$

Momentum equation:

$$\rho\left(u\frac{\partial u}{\partial x} + v\frac{\partial u}{\partial y}\right) = -\frac{\partial p}{\partial x} + \frac{\partial \tau_{xx}}{\partial x} + \frac{\partial \tau_{xy}}{\partial y} \tag{2}$$

$$\rho\left(u\frac{\partial v}{\partial x} + v\frac{\partial v}{\partial y}\right) = -\frac{\partial p}{\partial y} + \frac{\partial \tau_{yx}}{\partial x} + \frac{\partial \tau_{yy}}{\partial y} \tag{3}$$

For power-law fluids the relation between stress tensor (τ_{ij}) and rate of deformation tensor (S_{ij}) can be written as:

$\tau_{ij} = 2\eta S_{ij}$, where η is the apparent viscosity and the deformation tensor (S_{ij}) is defined as:$S_{ij} = \frac{1}{2}\left(\frac{\partial u_i}{\partial x_j} + \frac{\partial u_j}{\partial x_i}\right)$, and the apparent viscosity (η)for power-law fluids can be calculated by:$\eta = m\left\{2\left[\left(\frac{\partial u}{\partial x}\right)^2 + \left(\frac{\partial v}{\partial y}\right)^2\right] + \left(\frac{\partial u}{\partial y} + \frac{\partial v}{\partial x}\right)^2\right\}^{\frac{(n-1)}{2}}$, where m is the consistency index of the flow and n is the power-law index of the fluids (shear-thinning fluids, $n < 1$; Newtonian fluids, $n = 1$; and shear-thickening fluid, $n > 1$).

3.2 Boundary Conditions

The boundary conditions for this flow configuration can be written as follows:

(1) At the inlet plane: $u = V_\infty/; and/; v = 0$.
(2) On the surface of the cylinder: The standard no-slip boundary condition is applied.
(3) At the top and bottom walls of the computational domain: shear stress and normal velocity is zero which can be mathematically described as: $\frac{\partial u}{\partial y} = 0 and v = 0$.
(4) At the outlet plane: The default outflow boundary condition option in $FLUENT$ is used, which can be written as; $\frac{\partial u}{\partial x} = 0/; and/; \frac{\partial v}{\partial x} = 0$.

Other non-dimensional numbers are defined as follows:

- Reynolds number,

$$Re = \frac{\rho (2a)^n V_\infty^{2-n}}{m} \tag{4}$$

- Total drag coefficient (C_D) is the sum of pressure (or form) drag coefficient and friction drag coefficient: $C_{DP} = \frac{2F_{DP}}{\rho V_\infty^2 (2a)}$ and $C_{DF} = \frac{2F_{DF}}{\rho V_\infty^2 (2a)}$

$$C_D = C_{DP} + C_{DF} = \frac{2F_D}{\rho V_\infty^2 (2a)} \tag{5}$$

- Similarly, Lift coefficient is defined as:

$$C_L = \frac{2F_L}{\rho V_\infty^2 (2a)} \tag{6}$$

4 Numerical Method

The governing equations have been solved using FVM-based commercial CFD software *FLUENT (ANSYS 15)*. The computational domain and mesh were generated using ANSYS 15. All-Triangles Method (ATM) scheme was employed over the entire domain to generate an efficient grid with economical distribution of cells. Two-dimensional pressure-based solver was used to solve the incompressible flow equations. The pressure term was discretized by a second order method, and a second order upwind scheme was used to discretize the convective terms. SIMPLE scheme was used to resolve the pressure–velocity coupling. The non-Newtonian power-law model was employed for handling viscosity. *FLUENT* uses a point-by-point Gauss–Seidel smoother in synchrony with the AMG cycle to reduce the CPU time. A convergence criterion of 10^{-7} was prescribed for the residuals of continuity and x- and y-components of the velocity. Computational domain size and grid distribution within the computational domain plays an important role in accuracy and stability of a numerical solution. A poor choice of computational parameters can lead to imprecise and unstable solution. In present solution, we characterized the domain size by L_u, L_d, and H as shown in Fig. 1. After doing rigorous domain independence test, we finalized the size of the domain by taking $L_u = 30a$, $L_d = 40a$, and $H = 80a$. This domain size can be considered as nice conciliation between accuracy and computational time. It is followed by a grid independence test which ensures that the solution does not depend on the grid cell size.

5 Result and Discussion

5.1 Validation Process

To establish the present numerical methodology, a validation is discussed in this section. The present results have been validated with existing results published by

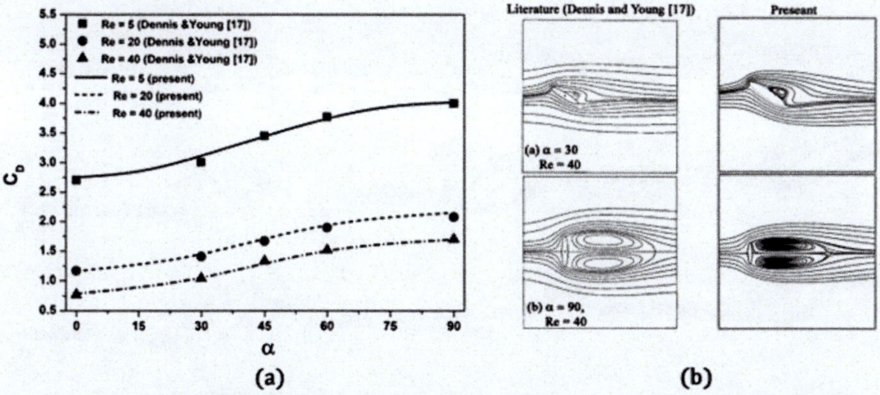

Fig. 2 Comparison of present results with literature [11] for $e = 0.2$, $Re = 5, 20$ and 40

Dennis and Young [11]. Figure 2(a) has summarized the comparison of C_D values for $n = 1$. It can be observed from the figure that present results have a very good agreement with previous results. Additionally, we validate the reliability of our grid and domain by comparing streamline patterns with Dennis & Young [11]. This comparison also shows a nice match with the existing results. The compatibility between the present and existing results provides us assurance in the credibility of the numerical methodology adopted here.

5.2 Flow Phenomenon

The flow behavior near the cylinder is presented in the form of streamline patterns around the cylinder. The streamline patterns presented in this section are for two values of $Re(= 5\,and\,40)$ and $n(= 0.4, 1.0\,\&\,1.6)$, respectively. The investigation here is confined for moderate values of Reynolds number varying between $5 \le Re \le 40$, and for this range of Re, the flow is considered to be laminar and steady [6]. The streamline patterns are shown in Figs. 3 and 4 for $Re = 5$ and 40, $n = 0.4$, 1 and 1.6, and $\alpha = 0^{\circ}, 30^{\circ}, 45^{\circ}, 60^{\circ}, and\, 90^{\circ}$.

Effect of α: For a constant value of n and Re, the flow pattern depends strongly on the inclination. The first observation that we make is that the stagnation point changes position with α. It moves from the tip of elliptic cylinder at 0° to the center-point of the windward side at 90°. At 0° inclination, there is no wake formation behind the cylinder for all cases considered here. However, at 90°, the downstream contains two symmetric wakes. To explain this development, we consider the case with $n = 1$ and $Re = 40$. As α starts to increase, boundary-layer separation commences. Initially, for small inclination, it gives rise to a single attached vortex. As the inclination increases, this vortex grows in size along with the appearance of a second eddy below it which is detached from the cylinder surface. This detached eddy is weaker in strength and

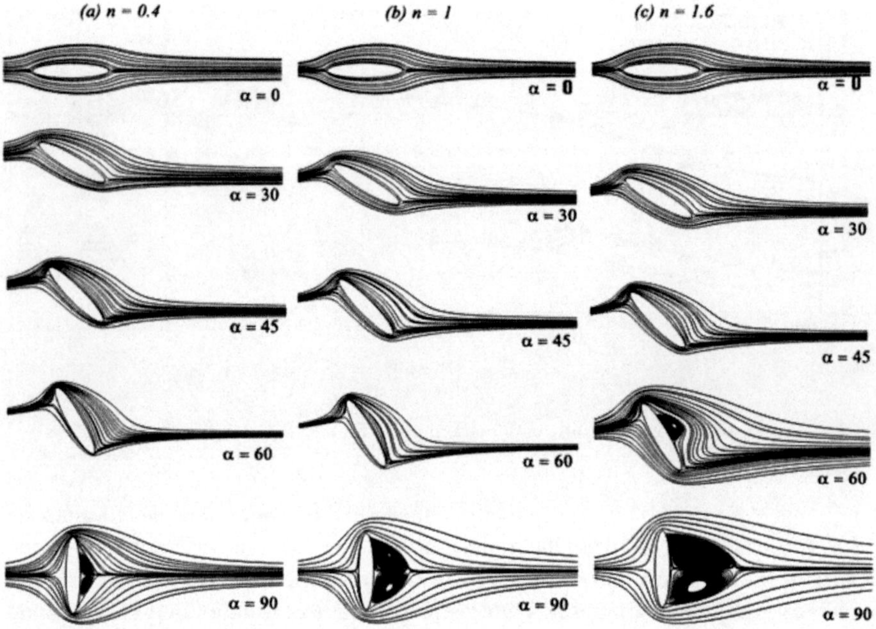

Fig. 3 Streamline patterns in the vicinity of the cylinder for $Re = 5$

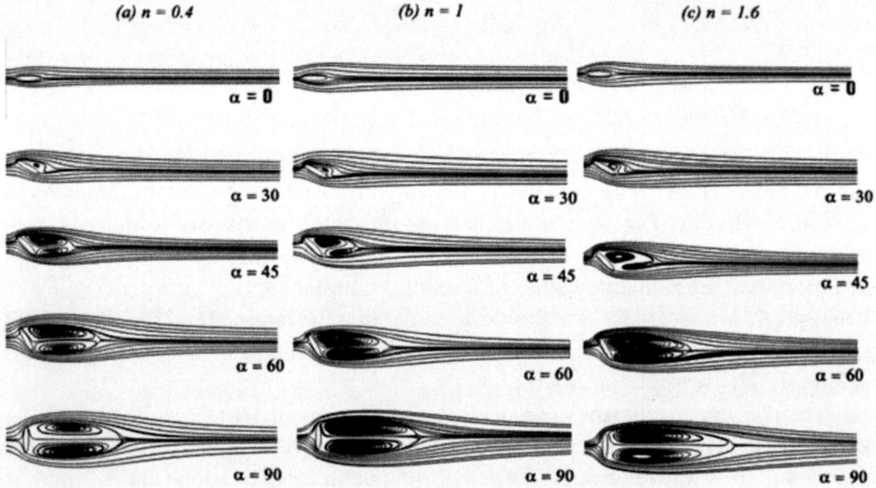

Fig. 4 Streamline patterns in the vicinity of the cylinder for $\mathrm{Re} = 40$

appears to feed off the attached vortex. As the bluffness continues to increase, the detached vortex grows in size and continues to move toward the cylinder surface. At some α just before 90°, it attaches itself to the cylinder surface. From this point onward, the bottom recirculatory zone continues to grow until the flow becomes symmetric. The vortices become exactly equal in strength and are referred to as *twin-vortex*. A similar development process happens for other values of n and Re.

Effect of Re and n: For a given value of n and α, the flow pattern varies immensely across Re. Higher the value of Re, more is the dominance of inertial forces. With increase in this fluid inertia, its tendency to separate from the cylinder surface increases. For instance, consider the case of $n = 0.4$ and $\alpha = 90°$. At $Re =$ 5, the separation is somewhat delayed, evident from two small recirculating zones behind the cylinder. As Re increases to 10, these recirculating zones become more prominent in both stream wise and lateral direction. The size of the recirculating zone further amplifies at $Re = 40$. It can be observed that the length of the separation bubble increases with Re, which is again attributed to an increase in inertial forces. Finally, fixing α and Re gives us insight as to how the flow pattern varies across n. Firstly, the tendency for flow-separation increases in shear-thickening fluid ($n > 1$). A typical example of this observation is in Fig. 3. For $Re = 5$ and $\alpha = 60°$, one can notice that a small vortex is visible for $n = 1.6$, whereas no vortex is visible for $n = 0.4$ and $n = 1$. Secondly, the length of recirculation zone increases with increase in n. This observation is clearly visible at $\alpha = 90°$ for all values of Re. A possible explanation as to why the flow-separation is delayed for $n < 1$ is as follows: For $n < 1$; $\mu \propto [1/\left(\frac{du}{dy}\right)]$ where $\left(\frac{du}{dy}\right)$ is the shear rate.

We know that shear rate is maximum near the cylinder surface which implies that μ is minimum. Now, for a given shear rate, μ around the cylinder is less in case of $n < 1$ as compared to $n > 1$. This suggests that viscous resistance near the cylinder is low for $n < 1$. It is well known that this viscous resistance (viscosity) is what causes the flow to separate. As it is low in case of $n < 1$, flow-separation is delayed.

Variation of C_D with Re, α, and n: Fig. 5(i) represents the variation of C_D with Re, α and n. It can be clearly observed from the graphs that for all values of Re and n considered.here drag coefficient (C_D) linearly increases with α. This can be simply justified from the fact that as α increases from 0° to 90°, the elliptic cylinder becomes bluffer, which result into enhancement of flow-separation even at very low Reynolds number which leads to increase the pressure drag (C_{DP}). As a result, the value of C_D increases with a strong contribution from C_{DP}.This can be validated from the C_D curves. After comparing Figs. 5(i) (a)–(c) one can clearly say that at a fixed value of n and α, C_D decreases with increment in Re. This follows from the fact that at low Reynolds number, viscous forces dominate, which result into more resistance to the flow near the cylinder. An interesting thing to note is the nature of the graphs at different Re. At lower Reynolds number,$Re = 5$, C_D becomes higher with a decrement in the value of n. However, at higher Reynolds number,$Re = 40$, opposite behavior can be seen, i.e.,C_D increases with increment in the value of n. At somehow moderate Reynolds number, $Re = 10$ a complex behavior can be seen (see Fig. 5(i) (b)). The proportionality between C_D and n is

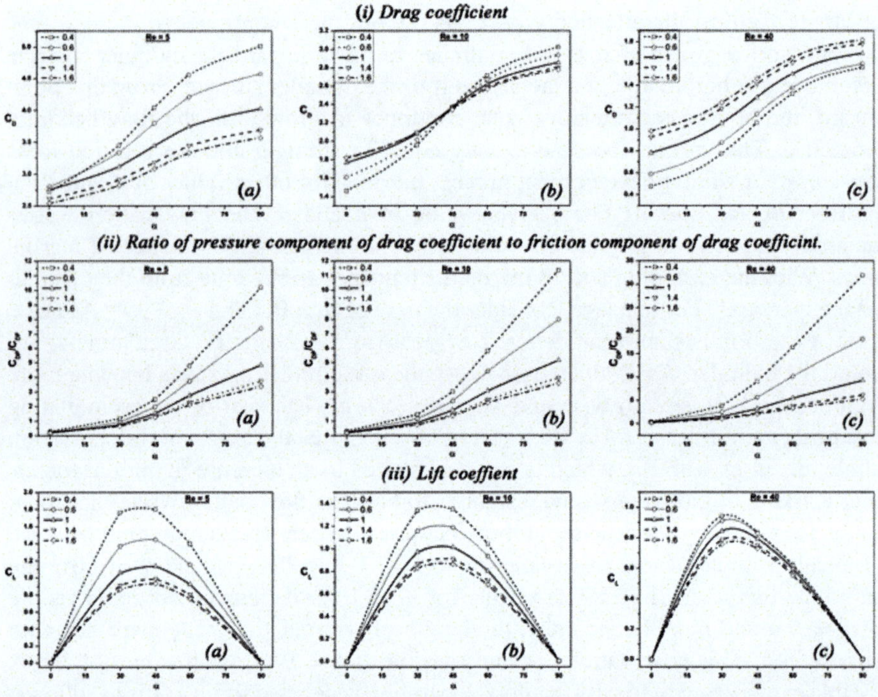

Fig. 5 Representation of variation of (**i**) C_D (**ii**) C_{DP}/C_{DF} and (**iii**) C_L with angle of attack, α

direct, for the α varying between $0^\circ \leq \alpha \leq 45^\circ$, beyond which opposite behavior can be seen, i.e.,C_D decreases with increment in n.

Lift coefficient: The variation of Lift coefficient (C_L) with Re, α and n is shown in Fig. 5(iii). It can be observed from the graphs that for all values of Re and n, C_L curve shows non-monotonic behavior with α, with a maximum lying in between $\alpha = 30^\circ - 45^\circ$. The inclination corresponding to the highest value of C_L is called critical angle of attack (α_c). This value of α_c varies slightly with different values of n and Re. We notice that C_L increases up to $\alpha = \alpha_c$, beyond which it starts decreasing and the elliptic cylinder is said to be "in stall," during which the boundary-layer detaches itself at both the edges of the cylinder and gives rise to a recirculating zone behind the cylinder. In addition, the value of lift coefficient is zero at $\alpha = 0^\circ$ and 90°. Also, we notice that irrespective of Re, the value of C_L has an inverse relation with power-law index at any given inclination. A possible reason as to why C_L follows an inverse relation with n is as follows: for shear-thinning fluids ($n < 1$), the tendency of the flow to detach from the cylinder surface is low. As we discussed earlier that boundary-layer separation at $\alpha > \alpha_c$ is what causes the C_L value to decrease. Now, since shear-thinning fluids ($n < 1$) have a lower tendency for flow detachment, it in turn increases the lift. Furthermore, n exerts more influence on the C_L value at lower value of Re. This is simply due to the fact that at higher Re, inertial forces start dominating and the value of n has little effect on the flow field around the cylinder.

Ratio of Form Drag and Friction Drag: Fig. 5 (ii) illustrates the variation of $C_{DP}/C_{DF}(= C^{'})$ with Re, n and α. We observe that across different values of Re, the nature of the graph remains same. At $\alpha = 0°$, $C^{'} < 1$; which tells us that friction drag dominates over its pressure counterpart. This is because at zero inclination the cylinder acts like a streamlined body, with minimal pressure drop across its edges. Whereas, the contact area between fluid and the cylinder surface is maximum, which result into more friction drag. But with increase in α, the pressure drop across the cylinder increases dramatically, that is why $C^{'}$ gives a sudden rise in its value. Furthermore, at higher values of α, the influence of power-law index on the value of $C^{'}$ is more prominent in the shear-thinning ($n < 1$) regime.

6 Conclusion

Flow field around an elliptic cylinder inclined to the freestream direction was investigated numerically at different values of Reynolds number (Re), Power-law index (n), and inclination (α). Streamlines have been plotted which show that the flow is steady for the studied range of Re. The flow pattern shows strong dependence on Re, α and n. It is observed that the downstream vortices grow in size with increase in Re, α and n. The propensity of the flow to disengage from the cylinder surface increases in case of shear-thickening fluids ($n > 1$). Graphs have been plotted which display the trends in C_D, C_{DP}/C_{DF}, and C_L with changes in Re, α and n. The drag coefficient increases monotonically with α, although its dependence on n changes with Re. The C_{DP}/C_{DF} curve shows that its ratio increases drastically with α, irrespective of n and Re. C_L vs α curves show a maximum at $\alpha = \alpha_c$; where α_c is the critical angle of attack.

References

1. Chhabra, R.P. "Fluid Flow and Heat Transfer from Circular and Noncircular Cylinders Submerged in Non-Newtonian Liquids," in Advances in Heat Transfer, Elsevier Inc., **43**, pp. 289–417 (2011)
2. Sen, S.: Surface pressure and viscous forces on inclined elliptic cylinders in steady flow. Sādhanā **45**(1), 172 (2020)
3. Park, J.M., Kim, O.J., Kim, S.J., Shin, Y.C.: Heat transfer characteristics of circular and elliptic cylinders in cross flow. Adv. Mech. Eng. **7**(11), 1–8 (2015)
4. Sun, W., Gao, Z., Du, Y., Xu, F.: Mechanism of unconventional aerodynamic characteristics of an elliptic airfoil. Chinese J. Aeronaut. **28**(3), 687–694 (2015)
5. Isao, I.: A new method of solving Oseen's equations and its application to the flow past an inclined elliptic cylinder. Proc. R. Soc. London. Ser. A. Math. Phys. Sci. **224**(1157), 141–160 (1954)
6. Hasimoto, H.: On the Flow of a Viscous Fluid past an Inclined Elliptic Cylinder at Small Reynolds Numbers. J. Phys. Soc. Japan **8**(5), 653–661 (1953)
7. Epstein, N., Masliyah, J.H.: "Creeping flow through clusters of spheroids and elliptical cylinders," Chem. Eng. J., **3**(C), 169–175, (1972)

8. Lugt, H.J., Haussling, H.J.: Laminar flow past an abruptly accelerated elliptic cylinder at 45° incidence. J. Fluid Mech. **65**(4), 711–734 (1974)
9. D'Alessio, S.J.D., Dennis, S.C.R.: A vorticity model for viscous flow past a cylinder. Comput. Fluids **23**(2), 279–293 (1994)
10. Ahmad, E.H., Badr, H.M.: Mixed convection from an elliptic tube placed in a fluctuating free stream. Int. J. Eng. Sci. **39**(6), 669–693 (2001)
11. Dennis, S.C.R., Young, P.J.S.: Steady flow past an elliptic cylinder inclined to the stream. J. Eng. Math. **47**(2), 101–120 (2003)
12. Sivakumar, P., Bharti, R.P., Chhabra, R.P.: Steady flow of power-law fluids across an unconfined elliptical cylinder. Chem. Eng. Sci. **62**(6), 1682–1702 (2007)
13. Paul, I., Arul Prakash, K., Vengadesan, S.: "Numerical analysis of laminar fluid flow characteristics past an elliptic cylinder," Int. J. Numer. Methods Heat Fluid Flow, **24**(7), 1570–1594, (2014)

Thermal Analysis of Flow Across Two Tandem Triangular Bluff Bodies in Unsteady Regime

Richa Agarwal and Ravikant R. Gupta

Abstract Thermal behavior of flow across sharp edge bluff bodies (such as rectangular, triangular bluff bodies, etc.) is repeatedly used for the designing of novel heat exchange systems, support structures, impellers, etc. However, the triangular bluff bodies are primarily studied for their use in the designing, construction, and working of a vortex flow meter. Thus, the present paper focuses on the two-dimensional transfer of heat by forced convection by the flow of Newtonian fluid around two isothermal tandem isosceles triangular bluff bodies placed in a horizontal channel with adiabatic walls. The effect of gap space (i.e., gap between two triangular bluff bodies) ranging from 1 to 4 for Prandtl number of 0.71 (air) and Reynolds number of 100 is investigated, by keeping the blockage ratio fixed as 25%. Simulation of the present problem is carried out by solving governing equations, i.e., equation for conservation of mass, conservation of momentum, and conservation of energy, along with suitable boundary conditions with the SIMPLE method by using a finite volume-based solver. Contours of streamline and isotherms help in understanding the flow and temperature fields across the two triangular cylinders, respectively. The average Nusselt number, mean drag coefficient, etc., are calculated. It is found that the values of the average Nusselt number and the mean drag coefficient, both, decline with the declining gap space between the two tandem bluff bodies. The changes in the average Nusselt number and mean drag coefficient values are more significant for the second triangular bluff body than the first one.

Keywords Triangular bluff body · CFD · Thermal analysis

Nomenclature

B	Triangle width, m
C_D	Drag coefficient, (= 2FD / ρumax2 B)
C_{D1}	Drag coefficient of the first triangular cylinder (from inlet)

R. Agarwal (✉) · R. R. Gupta
Department of Chemical Engineering, Banasthali University, Banasthali 304022, India

R. P. Bharti and K. M. Gangawane (eds.), *Recent Trends in Fluid Dynamics Research*, Lecture Notes in Mechanical Engineering,
https://doi.org/10.1007/978-981-16-6928-6_7

C_{D2}	Drag coefficient of the second triangular cylinder (from inlet)
c_P	Specific heat of the fluid, J kg^{-1} K^{-1}
F_D	Drag force per unit length of cylinder, N m^{-1}
h	average heat transfer coefficient, W m^{-2} K^{-1}
H	Channel width, m
k	Thermal conductivity of the fluid, W m^{-1} K^{-1}
L	Length of the computational domain, m
N_u	Nusselt number, (= hB / k)d
N_{u1}	Nusselt number for the first cylinder (from inlet)d
N_{u2}	Nusselt number for the second cylinder (from inlet)d
P_r	Prandtl number, (= μcp / k)d
p^*	Pressure, N m–2d
p	Dimensionless pressure, $(= p^* / \rho_{umax}{}^2)$d
Re	Reynolds number, $(= \rho u_{max} B / \mu)$d
S	Gap space between the two triangles, md
S/B	Nondimensional gap spaced
t	Time $_s$d
T	Temperature, K
T_w	Temperature at the surface of the cylinder, K
T_∞	Fluid temperature at the inlet, K
u	Nondimensional x-directional component of velocity, $(= u^*/u_{max})$
u_{max}	Maximum velocity at the inlet of the channel, ms^{-1}
v	Nondimensional y-directional component of velocity, $(= v^*/u_{max})$
u^*, v^*	x and y directional components of velocity, respectively, m s–1
x^*, y^*	Cartesian coordinates, m
x,y	Nondimensional coordinates
X_u	Upstream distance of the cylinder, md
X_d	Upstream distance of the cylinder, md
θ	Nondimensional temperature, $(= (T - T_\infty) / (T_w - T_\infty))$
β	Blockage ratio (= B/H)
τ	Nondimensional time, $(= t / (B/u_{max}))$
ρ	Density, kg m^{-3}
μ	Viscosity, m^2 s^{-1}

1 Introduction

Bluff bodies can be defined as obstacles which when placed in a fluid stream separates the flow over a significant part of their surface. Flow past bluff bodies have many engineering applications such as cooling of electronic components, heat exchangers, and flow metering devices. In particular, some of the examples for the application of triangular bluff bodies are vortex meter, piezometer, solar cells, extended surfaces, etc. Triangular cylinders (which are the sharp edged bluff bodies) are one of the primary configuration of the bluff bodies. Numerous numerical and experimental studies have been conducted for the flow past bluff bodies [1–3, etc.]. While to our best knowledge, flow and transfer of heat across two or more triangular bluff bodies are not studied much by the researchers. Hence, the present work emphasis on the flow and transfer of heat by forced convection around two isosceles triangular bluff bodies in tandem arrangement in a confined laminar flow regime.

Abbassi et al. [4] when studied the heat transfer of air (Pr = 0.71) by forced convection, found that the triangular prism present in a horizontal confined domain effected the heat transfer and flow pattern weakly for symmetric flow ($Re < 45$) while the average Nusselt number ascended with the Reynolds number for periodic flow ($Re > 45$). Zielinska and Wesfreid [5] found the critical Reynolds number to be equal to 38.3 when they performed a numerical analysis of the flow beyond an equilateral triangular bluff body for the Reynolds number varying from 34 to 50 and a fixed blockage ratio of 0.067, which was further validated by Wesfreid et al. [6] experiments. Jackson [7] founds the critical Reynolds number to be equal to 34.318 when the author studied two-dimensional laminar flow around various shape bluff bodies to identify the emergence of periodic behavior. The corresponding Strouhal number was found to be equal to 0.1355. De and Dalal [8] used finite volume method to investigate the hydrodynamic and thermal characteristics for confined flow around an equilateral triangular bluff body for the range of blockage ratio $1/12 \leq \beta \leq 1/3$ and Reynolds number $80 \leq Re \leq 200$. The Strouhal number and the RMS value of lift coefficient increased significantly with Reynolds number and blockage ratio, whereas overall Nusselt number was found to be almost constant for different blockage ratios. The flow was found to be more inclined toward the wake instability and similar to unconfined flow for lower blockage ratios. Srikanth et al. [9] studied the effect Reynolds number (1–80) on the flow and heat transfer of air ($Pr = 0.71$) around an equilateral triangular bluff body placed in a confined horizontal plane for a blockage ratio of 1/4. It was outlined that the critical Reynolds number lies between 58 and 59. Also, while the value of mean drag coefficient was found to ascend with the rise in Reynolds number, whereas the value of average Nusselt number and the wake length was found to descend with the rise in the Reynolds number. Zeitoun et al. [10] studied the heat transfer by forced convection across an equilateral triangular cylinder for two contrasting arrangements in unconfined cross-flow of air ($Pr = 0.71$) for Reynolds number up to 200 and the blockage ratio ($\beta = 1/80$). Critical Reynolds number is reported to be 34.7 for base facing arrangement and 38.03 for vertex facing arrangement. Dhiman and Shyam [11] investigated the unsteady heat transfer for flow of air ($Pr = 0.71$) around an equilateral triangular bluff body for the Reynolds number range (50–150). Results showed that the time-averaged Nusselt number rises monotonically with rising Reynolds number.

Chattopadhyay [12] numerically investigated the confined transfer of heat in the presence of a triangular bluff body in the turbulent wake up to Reynolds number of 40,000. A rise in averaged Nusselt number of about 15% was found in the channel. Many other authors have worked on the flow across a triangular bluff body in the presence of the turbulent wake, i.e., Sharif and Gu [13]; Tatsuno et al. [14]; Srigrarom and Koh [15]; El-Sherbiny [16]; Camarri et al. [17]; Nakagawa [18]; El-Wahed et al.

[19]. But flow past two triangular cylinders have been the focus of a few authors (whether laminar [20–22] or turbulent flow regime). Mohsenzedh et al. [22] studied the effect of gap space at three different Reynolds number (100, 250, and 350) for the forced convective incompressible flow around two isosceles triangular bluff bodied in tandem arrangement in a confined domain. From the above discussion, it can be summarized that enough literature is available for the flow and heat transfer across single triangular cylinder at $Re = 100$ and $\beta = 1/4$ but a few is available for two triangular cylinders at the same Re and β. Hence, the main aim of this study is to explore the hydrodynamic and thermal behavior of the flow of Newtonian fluid across two tandem isosceles triangular obstacles at $Re = 100$, $Pr = 0.71$, and $\beta = 1/4$ for different gap space between them.

2 Model Description

A horizontal domain is considered in which two-dimensional flow of a Newtonian fluid around two tandem isosceles triangular bluff bodies with 90° apex angle is investigated, as shown in Fig. 1. A fully developed velocity field is presumed at the inlet of the channel such that u_{max} is the maximal velocity, and T_∞ is the temperature which is persistent at the inlet. While T_w is the temperature of the wall of the two tandem triangular bluff bodies such that $T_w > T_\infty$. $H/4$ was selected as the rear width (B) of triangular obstacles, and $\beta = 1/4$ is used here ([4, 9, 22]) also studied the same blockage ratio. From the inlet, the dimensionless upstream distance to place the first triangular obstacle was taken as X_u/B and X_d/B and was the dimensionless downstream distance taken for the second cylinder from the outlet. The dimensionless gap (S/B) between the two triangular cylinders was varied from 1 to 4. The dependence

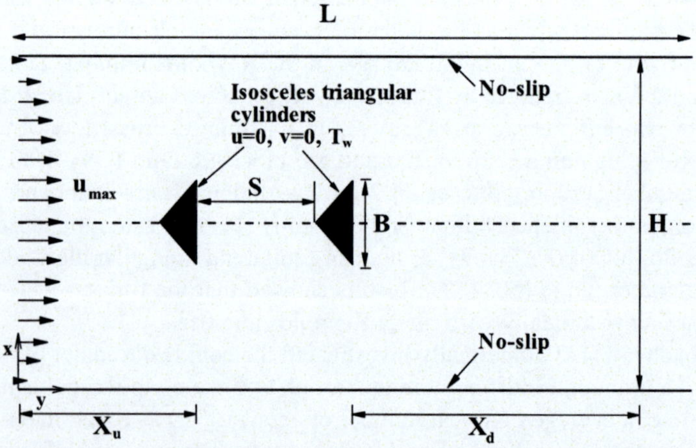

Fig. 1 Diagrammatic of flow across two isosceles triangular bluff bodies in tandem arrangement

of the physical properties of the fluid such as thermal conductivity, viscosity, and density on temperature is negligible as the difference in the temperature of the triangular obstacles, and the fluid is assumed to be very small (i.e., $(T_w - T_\infty) = 2$ K). In the present study, Eqs. (1)–(4) represents the dimensionless form of the continuity, the x- and y- components of momentum and energy equations for the 2D incompressible and laminar flow where the Prandtl number and Reynolds number are given by $Pr = \mu c_p/k$ and $Re = Bu_{max}\rho/\mu$, respectively.

Equation of continuity:

$$\frac{\partial u}{\partial x} + \frac{\partial v}{\partial y} = 0 \tag{1}$$

Equation for conservation of momentum in the x-direction:

$$\frac{\partial u}{\partial \tau} + \frac{\partial uu}{\partial x} + \frac{\partial vu}{\partial y} = -\frac{\partial p}{\partial x} + \frac{1}{\mathrm{Re}}\left(\frac{\partial^2 u}{\partial x^2} + \frac{\partial^2 u}{\partial y^2}\right) \tag{2}$$

Equation for conservation of momentum in the y-direction:

$$\frac{\partial v}{\partial \tau} + \frac{\partial uv}{\partial x} + \frac{\partial vv}{\partial y} = -\frac{\partial p}{\partial y} + \frac{1}{\mathrm{Re}}\left(\frac{\partial^2 v}{\partial x^2} + \frac{\partial^2 v}{\partial y^2}\right) \tag{3}$$

Equation for conservation of energy:

$$\frac{\partial \theta}{\partial \tau} + \frac{\partial u\theta}{\partial x} + \frac{\partial v\theta}{\partial y} = \frac{1}{\mathrm{Re\,Pr}}\left(\frac{\partial^2 \theta}{\partial x^2} + \frac{\partial^2 \theta}{\partial y^2}\right) \tag{4}$$

The dimensionless form of variables is

$$x = \frac{x^*}{B}, y = \frac{y^*}{B}, \tau = \frac{tu_{\max}}{B}, u = \frac{u^*}{u_{\max}}, v = \frac{v^*}{u_{\max}},$$

$$p = \frac{p^*}{\rho u_{\max}^2}, \theta = \frac{T - T_\infty}{T_w - T_\infty} \tag{5}$$

The following dimensionless boundary conditions are used:

- At the inlet boundary, $u = 1 - \left[\left|1 - \frac{2\beta y}{B}\right|\right]^2$ (fully developed flow is assumed with an inlet velocity u_{max}), $(0 \leq y \leq \beta)$ and $v = 0$; $\theta = 0$.
- On the two walls of the channel, $u = 0$; $v = 0$; $\partial\theta/\partial y = 0$.
- On the triangular bluff bodies surfaces, $u = 0$; $v = 0$; $\theta = 1$.
- At the outlet boundary, the default outflow boundary condition in ANSYS Fluent is used. This condition presumes a zero-diffusion flux for all variables of the flow, $\partial u/\partial x = 0$; $\partial v/\partial x = 0$; $\partial\theta/\partial x = 0$.

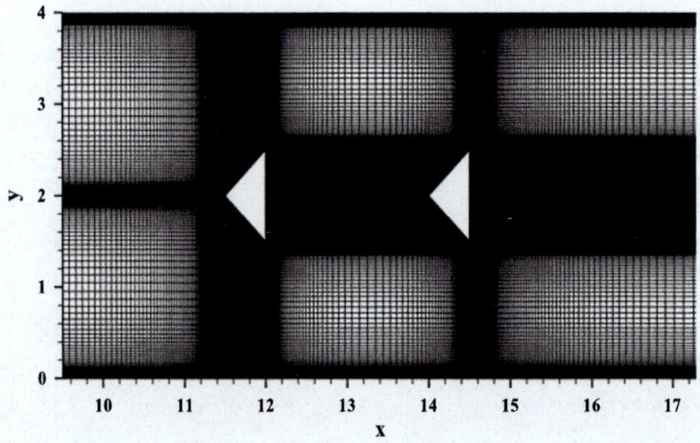

Fig. 2 Magnified view of structure of grid for the blockage ratio of 1/4

3 Structure of Grid and Analytical Procedure

In the present work, ANSYS Fluent is used to generate a computational grid as shown Fig. 2 which consists of 95,530 cells. There is ultrafine distribution of the grid next to the triangular bluff bodies, while the distribution of grid is coarse in the region away. The upstream and downstream distances used are $X_u = 12B$ and $X_d = 20B$, respectively.

Governing equations are solved by finite volume method-based CFD solver ANSYS Fluent, along with boundary conditions. The mathematical problem is simplified by SIMPLE method where the second order upwind scheme allotted for the discretization of convective term, while the discretization of diffusive terms is done by central difference scheme. Gauss-Siedel iterative scheme is used to solve the resulting algebraic equations.

4 Productive Outcomes

In current work, the flow of fluid and heat transfer beyond two equal sized isothermal tandem isosceles triangular bluff bodies laid in a horizontal channel with adiabatic walls are studied for $Re = 100$, $Pr = 0.71$ (air), and fixed blockage ratio, $\beta = 0.25$. The effect of gap space range 1–4 (i.e., space between two cylinders, which is also in accordance of literature [22, 23]) on the physical parameters such as individual and overall drag coefficient and Nusselt number) is investigated.

Table 1 Validation of the present drag results time-averaged Nusselt number and mean drag coefficient at Re = 100

Source	Re	mean drag coefficient	Time-averaged Nusselt number
Present	100	1.7	5.56
S. Srikant et. al. [9]	100	1.67	5.56
De & dalal [8]	100	1.68	5.3

4.1 Validation of Results

The present work is validated with that of Srikanth et al. [9] and De & Dalal [8] for the flow beyond a triangular bluff body in a confined domain (Table 1). An excellent agreement between the two studies exists for the values of time-averaged Nusselt number and mean drag coefficient at Re = 100. At Re = 100, the percentage relative deviation in time-averaged Nusselt number is around 0.3% and 5.26% and that of the mean drag coefficient is around 1.4% and 1.94% with respect to [9] and [8], respectively. The reason behind the greater deviation in Nusselt number value when the present results are compared with that of [8] is the use of coarser grid by [8], i.e., a grid consisting of 93,640 cells such that each side of the triangular obstacle consists of 100 control volumes is used in present study while a grid consisting of 28, 000 cells such that each side of the triangular obstacle consists of 70 control volume is used by [8]. Thus, on the grounds of the accessible literature, the results of the present work are believed to be reliable less than 2–3%. This validates the procedure for the solution methodology for the present study.

4.2 Flow Patterns

The 2D streamline contours beyond the two isosceles triangular obstacles in tandem arrangement are presented to learn the dependence of characteristics of flow, e.g., drag coefficient and flow field on the gap space (*S/B*). The streamline plots for gap ratios (*S/B* = 1–4) at *Re* = 100 are shown in Figs. 3a–d. For all *S/B*, separation of flow is observed to occur from the rear circumference of the triangular bluff bodied. For *S/B* = 1, the two vortices formed behind the first triangular cylinder are symmetric whereas an unsteady vortex formation occurs at the back of the rear triangular bluff body as shown in Fig. 3a. Thus, overall system behaves unsteady at S/B = 1. As we increase the gap space, i.e., for *S/B* = 2 (Fig. 3b), two vortices are formed in the wake of both the triangular bluff bodies which are symmetric and above and below the mid plane which corresponds to the steady state wake. For *S/B* = 3, a pair of steady symmetric vortices is found to form behind the first triangular cylinders but unsteady wake appears at the back of the rear obstacle (Fig. 3c). On further increasing the gap space, i.e., *S/B* = 4, flow becomes erratic with vanishing of symmetric wake at the back of both the triangular bluff bodies (Fig. 3d) which is identical to Karman

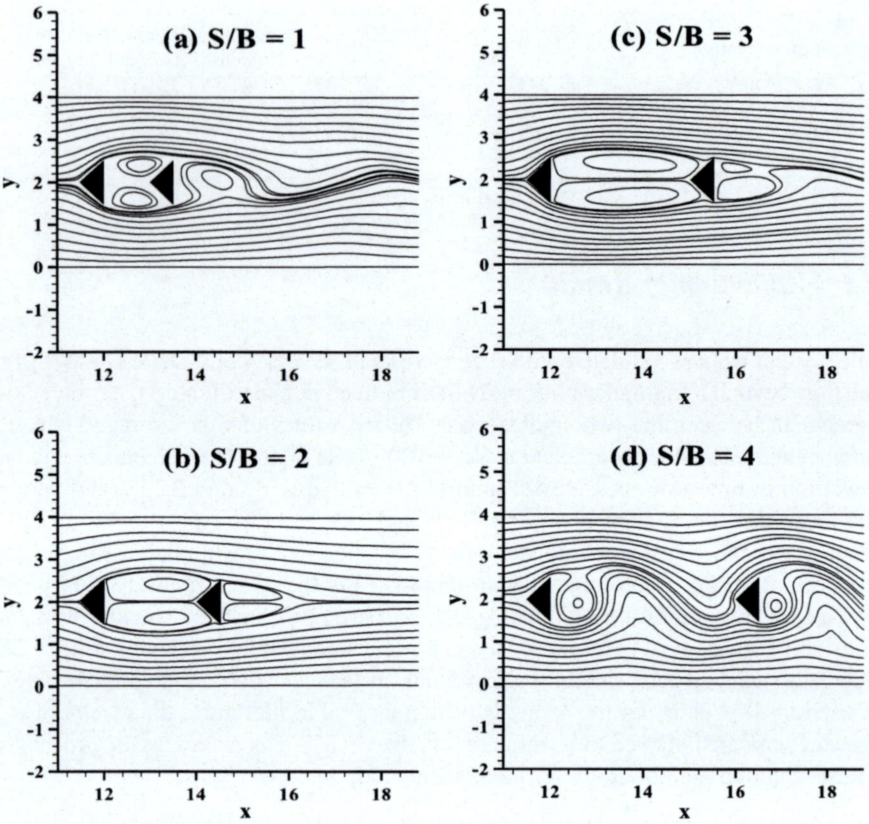

Fig. 3 Streamlines at different *S/B* for $Re = 100$

vortex street. Thus, the findings suggest that the flow is changing from symmetric to (periodic) unsteady between the gap spacing (S/B) 3 and 4 for the first triangular bluff body and between 2 and 3 for the second triangular bluff body. However, the objective of the present study is not to investigate the onset of transition from a steady to time-periodic regime.

4.3 Isotherm Patterns

The effect of gap spacing on the convective transfer of heat over the two triangular bluff bodies is studied by presenting the temperature contours, shown in Figs. 4a–d. As we can see, for $S/B = 1$ and 3, a steady temperature profile is maintained at the back of the upstream cylinder (Fig. 4a and c) whereas an unsteady temperature profile is seen at the back of the rear obstacle. A steady isotherm profile around

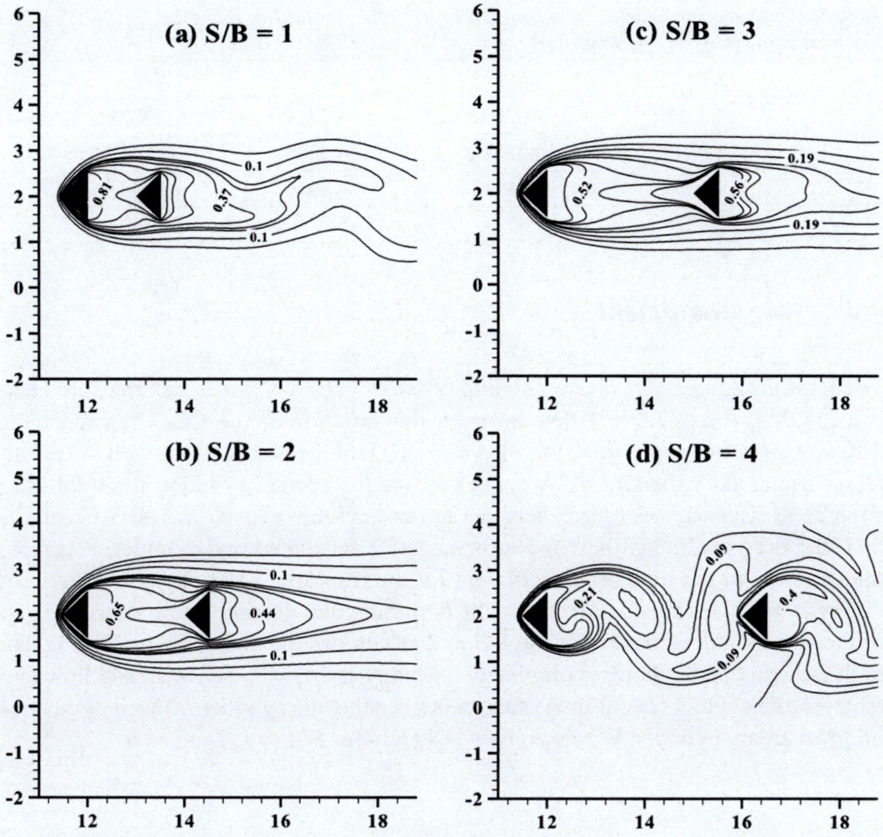

Fig. 4 Temperature profile at different *S/B* for *Re* = 100

both the tandem bluff bodies is observed for *S/B* = 2 (Fig. 4b). While temperature profile becomes unsteady for *S/B* = 4 as shown in Fig. 4d. A temperature street can be observed between the gap of the two obstacles and also at the back of the rear triangular obstacle. Reason being, at a Reynolds number of 100 and for S/B = 1–3, the flow surpasses the two triangular bodies easily without much hindrance by the two bluff bodies, i.e., separation takes place at the apex of first triangular cylinder and it recombines at the rear of the second triangular body. Thus, without any disturbance forming the locked recirculation between the two triangular bluff bodies and the effect of the unsteady flow can be seen at the rear of the second triangular body, i.e., temperature contours become time-periodic for flow field. While on increasing the S/B, the effect of second bluff body on the flow of the fluid can also be seen as the flow separates at the apex of the first bluff body and recombines in the locked region between the two bodies. Thus, making the temperature contours time-periodic for flow field. A temperature street is formed behind the two triangular cylinders, which is very similar to Karman vortex street.

Table 2 Total drag coefficient and average Nusselt number for the first triangle (C_{D1}, N_{u1}) and the second triangle (C_{D2}, N_{u2}), respectively

No.	C_{D1}	C_{D2}	N_{u1}	N_{u2}
1	1.7638	−0.2818	4.5313	2.3537
2	1.6843	0.0207	4.7268	2.7265
3	1.6653	0.3454	4.9016	2.943
4	2.2276	1.8597	5.92	4.7903

4.4 Drag Coefficient

The total drag force (C_D) is summation of two forces, i.e., pressure drag force and viscous drag force. Table 1 demonstrates the variation of the total drag force for different values of gap ($S/B = 1$ to 4) at $Re = 100$ for the first (C_{D1}) as well as second (C_{D2}) triangular cylinders. As we can see, for the second cylinder, the total drag coefficient rises with rising gap between the two cylinders for $Re = 100$. Also at $S/B = 1$, the drag coefficient is found to be negative for the second cylinder. Whereas, the variation of the total drag coefficient for the front triangular obstacle is entirely different, i.e., the drag coefficient for the first triangular descends with the rising gap of the two cylinders except for the $S/B = 4$ which has the highest drag. This is due to the reason that the flow is completely unsteady at S/B = 4, i.e., unsteady flow and temperature contours occur in the space between the two cylinders as well as the rear of downstream cylinder. Whereas, flow is steady for $S/B = 1$, 2, and 3.

4.5 Average Nusselt Number

The dependence of the Nusselt number on the gap space (S/B = 1–4) at Reynolds numbers of 100 for $\beta = 0.25$ is given in Table 2. The Nusselt number for both the cylinders increases with increasing the gap between the two triangular cylinders. The vortex formation in the gap distance has a dominant effect on the heat transfer over the two triangles. Hence, by increasing the gap spacing (S/B = 1–4) between the two triangles, vortices shed in this region which further causes the enhancement of the Nusselt number for both first and second triangular cylinders.

5 Conclusion

Present study comprises of two-dimensional incompressible flow and heat transfer across two isosceles triangular bluff bodies in tandem arrangement in a confined flow regime has been examined at Reynolds number of $Re = 100$ and $Pr = 0.71$ and a fixed β of 0.25 for different gaps (S/B = 1–4) between the two cylinders. The temperature and flow fields across the tandem triangular bluff bodies are presented by the contours of isotherms and streamline, respectively. A rise in total drag coefficient with rising gap space was observed. While the variation of the total drag coefficient for the front triangular obstacle is entirely different, i.e., the drag coefficient for the first obstacle declines with the rising gap between the two cylinders except for the $S/B = 4$, which has the highest drag. As far as Nusselt number is concerned, it increases for both the cylinders with increasing the gap between the two cylinders. The changes in the values of average Nusselt number and mean drag coefficient appear to be more significant for the rear cylinder than the first one. Furthermore, convective heat transfer around multiple bluff bodies (more than one) with different structures (circular, square, elliptical, etc.) further need to be explored as there are many real applications to it such as tube bundles in heat exchangers, solar cells, cooling elements, etc.

References

1. Chakrabarty, D., Brahma, R.: Experimental study of fluid flow and heat transfer from a square prism approaching the wall of a wind tunnel. J Eng Phys Thermophy **82**, 697 (2009). https://doi.org/10.1007/s10891-009-0258-x
2. Lange, C.F., Durst, F., Breuer, M.: Momentum and heat transfer from cylinders in laminar cross flow at $10^{-4} \leq \mathrm{Re} \leq 200$. Int. J. Heat Mass Transfer. **41**, 3409–3430 (1998)
3. Okajima, A.: Strouhal numbers of rectangular cylinders. J. Fluid. Mech. **123**, 379–398 (1982)
4. Abbassi, H., Turki, S., Nasrallah, S.B.: Numerical investigation of forced convection in a plane channel with a built-in triangular prism. Int. J. Thermal Science. **40**, 649–658 (2001). https://doi.org/10.1016/S1290-0729(01)01254-6
5. Zielinska, B.J., Wesfreid, J.E.: On the spatial structure of global modes in wake flow. Physics Fluids. **7**, 1418–1424 (1995)
6. Wesfreid, J.E., GoujonDurand, S., Zielinska, B.J.: Global mode behavior of the stream-wise velocity in wakes. Phys. Fluids. **6**, 1343–1357 (1996). https://doi.org/10.1051/jp2:1996135
7. Jackson, C.P.: A finite-element study of the onset of vortex shedding in flow past variously shaped bodies. J. Fluid Mechanics **182**, 23–45 (1987). https://doi.org/10.1017/S0022112087002234
8. De, A.K., Dalal, A.: Numerical study of laminar forced convection fluid flow and heat transfer from a triangular cylinder placed in a channel. J. Heat transfer. **129**, 646–656 (2007). https://doi.org/10.1115/1.2712848
9. Srikanth, S., Dhiman, A.K., Bijjam, S.: Confined flow and heat transfer across a triangular cylinder in a channel. Int. J. Thermal Sci. **49**, 2191–2200 (2010). https://doi.org/10.1016/j.ijthermalsci.2010.06.010
10. Zeitoun, O., Ali, M., Nuhait, A.: Convective heat transfer around a triangular cylinder in an air cross flow. Int. J. Thermal Sci. **50**, 1685–1697 (2011)
11. Dhiman, A., Shyam, R.: Unsteady heat transfer from an equilateral triangular cylinder in the unconfined flow regime. ISRN Mechanical Engineering **2011**, 1–13 (2011). https://doi.org/10.5402/2011/932738
12. Chattopadhyay, H.: Augmentation of heat transfer. Int. J. Thermal Science. **46**, 501–505 (2007). https://doi.org/10.1016/j.ijthermalsci.2006.07.003
13. Sharif, M.A.R., Gu, Z.: Turbulent flow simulation behind a v-shaped flame stabilizer using a nonlinear ke3 model and a smoothing algorithm. Chem. Eng. Commun. **189**, 471–488 (2002). https://doi.org/10.1080/00986440212092

14. Tatsuno, M., Takayama, T., Tatsuno, M., Tatsuno, M.: On the stable posture of a triangular or a square cylinder about its central axis in a uniform flow. Fluid Dyn. Res. **6**, 201–207 (1990)
15. Srigrarom, S., Koh, A.K.G.: Flow field of self-excited rotationally oscillating equilateral triangular cylinder. J. Fluids struct. **24**, 750–755 (2008). https://doi.org/10.1016/j.jfluidstructs.2007.10.015
16. ElSherbiny, S.: Flow separation and reattachment over the sides of a 90^0 triangular prism. J. Wind Eng. Ind. Aerodynamics. **11**, 393–403 (1983)
17. Camarri, S., Salvetti, M.V., Buresti, G.: Large-eddy simulation of the flow around a triangular prism with moderate aspect ratio. J. Wind eng. Ind. Aerodynamics. **94**, 309–322 (2006). https://doi.org/10.1016/j.jweia.2006.01.003
18. Nakagawa, T.: Vortex shedding mechanism from a triangular prism in a subsonic flow. Fluid Dynamics Res. **5**, 69–81 (1989). https://doi.org/10.1007/BF01376926
19. ElWahed, A.K., Johnson, M.W., Sproston, J.L.: Numerical study of vortex shedding from different shaped bluff bodies. J. Flow meas. Instrum. **4**, 233–240 (1993). https://doi.org/10.1016/0955-5986(93)90030-M
20. Agarwal, R., Dhiman, A.: Flow and heat transfer phenomena across two confined tandem heated triangular bluff bodies. Numer Heat Transf Part A Appl. **66**, 1020–1047 (2014)
21. Agarwal, R., Dhiman, A.: Confined flow and heat transfer phenomena of non-Newtonian shear-thinning fluids across a pair of tandem triangular bluff bodies. Numer Heat Transf Part A Appl. **68**, 174–204 (2015)
22. Mohsenzedh, A., Farhadi, M., Sedighi, K.: Convective cooling of tandem heated triangular cylinders placed in a channel. Therm. Sci. **14**, 183–197 (2010). https://doi.org/10.2298/TSCI1001183M
23. Farhadi, M., Sedighi, K., Korayem, A.M.: Effect of wall proximity of two staggered triangular cylinders on the transport process in a channel. Engineering Science and Technology, an International Journal. **19**, 1177–1189 (2016)
24. Ashjaee, M., Arzaghi, M., Jarrahi, M., Yousefi, T.: Experimental and numerical study of free convection on an isothermal downward cone. Exp. Heat Transfer. **20**, 307–322 (2007). https://doi.org/10.1080/08916150701418286

Free Convection in a Square Enclosure from Two Submerged Cylinders of Different Aspect Ratio in Shear-Thinning Fluids

Roshan Kumar, Yogendra Nath Prajapati, and Ashok Kumar Baranwal

Abstract The present study focuses on heat transfer due to natural convection between horizontally positioned (side-by-side) hot and cold cylinders kept in a cold non-Newtonian fluid in a square enclosure. Computation fluid dynamics (COMSOL package) was used to solve the differential equations of continuity, momentum, and energy over a wide range of $10 \leq \text{Gr} \leq 10^4$, power-law index $0.2 \leq n \leq 1$, Prandtl number $0.7 \leq \text{Pr} \leq 100$, and an aspect ratio of $0.16 \leq R_1/L \leq 0.24$. The flow and heat characteristics have been reported in terms of streamlines and isothermal contours, respectively. The results of average Nusselt number show a positive dependence on the Prandtl number and Grashof number. No changes in local Nusselt number were observed with Prandtl number by varying Grashof number up to 100, indicating conduction is predominant for heat transfer. For a higher Grashof number (>100), the average Nusselt number increases with increasing the Prandtl number. The average Nusselt number also increases with increasing the aspect ratio for all dimensionless numbers used in the study. Effect of Prandtl number is more at lower value of aspect ratio $R_1/L = 0.16$ with respect to that of at higher values while heat transfer is more at higher value of aspect ratio $R_1/L = 0.24$.

Keywords Enclosure · Cylinders · Grashof number · Prandtl number

Nomenclature

c_p	Thermal heat capacity of fluid [$\text{JKg}^{-1}\ \text{K}^{-1}$]
d	Diameter of cylinder [m]
g	Acceleration due to gravity [ms^{-2}]
h	Heat transfer coefficient [$\text{Wm}^{-2}\text{K}^{1}$]
k	Thermal conductivity of fluid [$\text{Wm}^{-1}\ \text{K}^{-1}$]
L	Length of square enclosure [m]

R. Kumar · Y. N. Prajapati · A. K. Baranwal (✉)
Department of Chemical Engineering, BIT Mesra Ranchi, Ranchi, Jharkhand 835215, India
e-mail: akb@bitmesra.ac.in

R. P. Bharti and K. M. Gangawane (eds.), *Recent Trends in Fluid Dynamics Research*, Lecture Notes in Mechanical Engineering,
https://doi.org/10.1007/978-981-16-6928-6_8

m	Power-law consistency index [Pa s^n]
n	Power-law index –
Nu_L	Local Nusselt number –
Nu_{avg}	Average Nusselt number –
p	Pressure –
Pr	Prandtl number –
R_1	Radius of hot circular cylinder [m]
R_2	Radius of cold circular cylinder [m]
Ra	Rayleigh number –
T	Temperature of fluid [K]
T_c	Temperature of the cold cylinder and enclosure (also used as the reference temperature [K]
T_h	Temperature of the hot cylinder [K]
ΔT	Temperature difference = = $(T_h - T_c)$ [K]
U_r	Reference velocity [ms^{-1}]
U_x, U_y	*x*- And *y*-components of the velocity –
x, *y*	Cartesian coordinates

Greek Symbols

α	Thermal diffusivity [m^2 s^{-1}]
β	Coefficient of volume expansion [K^{-1}]
ε	Rate of strain tensor [s^{-1}]
η	Non-Newtonian viscosity [Pa s]
ρ	Density of fluid [Kg m^{-3}]
ρc	Fluid density at reference temperature [Kg m^{-3}]
θ	Temperature –
τ	Extra stress tensor [Pa]

1 Introduction

Natural convection is driven by the movement of fluids, due to the variation of density, and hence, convenient in heat transfer is microelectronics where forced convection is not commended. The geometric configuration of interest is coupled with the presence of different diameter of bodies embedded within the cavity. Natural convection finds abounding engineering applications in the field of nuclear energy generation, heat exchanger, electronics equipment, chemical reactors, pharmaceutical industries, and processing polymeric melt.

Circular cylinder within the polygonal enclosure has been studied by Saleh et al. [1] which shows the complexity and interdependence between geometrical parameter and dimensionless numbers. Newtonian fluid was filled between the cold polygonal enclosure and hot cylinder, with no viscous dissipation but taking the radiative heat flux into account. Study was performed over $10^2 \le \mathrm{Ra} \le 10^6$, radiation parameter $0 \le R_d \le 1$, and $0.25 \le L/D \le 1.02$. It was observed that average Nusselt number (Nu_{avg}) decreased by increasing the cylindrical diameter and enclosure length ratio, as well as on the number of sides of the polygon. The authors observed two clockwise and two counter-clockwise fields for a triangular enclosure while for pentagon and other polygons having sides more than five showed only one clockwise and one counter-clockwise fields. Rath et al. [2] studied the effects of various geometries of cylinders on average Nusselt number in the range of $10^3 \le \mathrm{Ra} \le 10^6$. At low *Ra*, the heat transfer rate increased with shapes in the order of square, triangular, and circular cylinders, respectively, whereas for $Ra \ge 5 \times 10^5$, the heat loss was found maximum through triangular cylinder. Sahu et al. [3] investigated the effects of aspect ratio (*AR*) in the range of 0.2–0.5 of a triangular cylinder inside a square enclosure for $10^4 \le \mathrm{Ra} \le 10^6$. The reported Nusselt number along the wall varied proportionally to the aspect ratio of the cylinder for the fixed enclosure size. The circulation rate was increased with decreasing cylinder size for a smaller enclosure, whereas the rate was directly dependent on cylinder size for the larger enclosure. Yoon et al. [4] studied the effect of different vertical positions (δ) of the cylinder along the centreline inside a square enclosure for $-0.25 \le \delta \le 0.25$ at $\mathrm{Ra} = 10^7$. The flow behaviors were found to be depended on δ. The transition from steady to unsteady state occurred at two critical positions of $\delta_{C,L}$ and $\delta_{C,U}$ as a lower bound and an upper bound $\delta_{C,L} = 0.05$ and $\delta_{C,U} = 0.18$, respectively. With increasing δ, periodic patterns of single and multiples frequencies were observed. The Nusselt number of the inner cylinder changed periodically with a sinusoidal pattern.

In recent years, studies are being focused on the real-world problems related to heat transfer for the various geometries including multiple cylinders inside an enclosure. Because of multiple cylinders in an enclosure, the overall heat and flow patterns depend on each individual cylinder. Numerical and experimental studies related to free convection from multiple cylinders in circular enclosure are also reported in the literature [5, 6]. Park et al. [7] studied the natural convection in a cold square enclosure with a hot and cold cylinder for $10^3 \le \mathrm{Ra} \le 10^6$. The cylinders were moved vertically along the centreline. At low Rayleigh number, as the cylinders reached closer to the top wall, the obtained isotherms were denser in the direction of movement of cylinders. Park et al. [8] reported a similar study in which both cylinders were hot at $\mathrm{Pr} = 0.7$. The steady or unsteady flow was observed based on the values of the dimensionless vertical distance δ and Ra. For $10^3 \le \mathrm{Ra} \le 10^5$, a steady-state behavior was observed in flow and thermal fields, whereas for $\mathrm{Ra} = 10^6$, the fields became unsteady for $0 \le \delta \le 0.1$. Park et al. [9] studied the natural convection in the range of $10^3 \le \mathrm{Ra} \le 10^6$ and vertical distance between the cylinders $0.3 \le \varepsilon \le 0.7$ from four heated cylinders placed in a rectangular array, moving along the diagonal line inside a cold square enclosure. The steady-state flow and thermal fields changed to unsteady state depending on *Ra* and ε values. Park et al. [10] studied the effects of position and inclination of elliptical cylinder for natural convection in a square enclosure with a vertical array of circular and elliptical cylinders in the range of $10^4 \le$

$Ra \leq 10^6$ at $Pr = 0.7$. The study showed an unsteady-state solution at $Ra = 10^6$ except at an inclination angle (ϕ) of 90° for elliptical cylinder. A transition from unsteady to steady flow regime was observed due to change in flow directions which was result of changing the inclination angle of lower elliptical cylinder. Baranwal and Chhabra [11] explored the effects of Prandtl and Grashof numbers for $0.7 \leq Pr \leq 100$ and Grashof number, $10 \leq Gr \leq 10^5$ on natural convection in an enclosure with cold walls containing a hot and a cold cylinder filled with Newtonian fluid. The results showed that at high Grashof number the effect of Pr number was significant. Yoon et al. [12] studied the effect of the diameter of cylinders in a square enclosure. Two hot cylinders with equal diameter were placed at the upper half and at the lower half of a cold square enclosure. The diameter of the cylinders was varied in the range of 0.05–0.2 L at $10^3 \leq Ra \leq 10^5$. Flow and thermal field were found symmetric about vertical centreline independent of Rayleigh number; however, the fields were asymmetric about horizontal centreline with increase in Rayleigh number. Dominating effects of Rayleigh number on convection were observed with decreasing cylinder radius. The Nusselt number showed a positive dependency on the cylinder radius for all Rayleigh number.

In recent years, much focus is devoted to study the flow and heat transfer for non-Newtonian fluids for various geometries related to real-world problems. A free convection heat transfer from a cylinder immersed in a power-law fluid was investigated numerically in the range of $10 \leq Gr \leq 10^5$, $0.72 \leq Pr \leq 100$, and $0.3 \leq n \leq 1.8$ by Prhashanna and Chhabra [13]. The authors suggested that the heat transfer rate could be increased up to 70–80% in shear-thinning fluids under suitable conditions. Mishra and Chhabra [14] have studied laminar natural convection numerically in power-law fluids in a square enclosure from two differentially heated horizontal cylinders in the range $10^2 \leq Gr \leq 10^4$, $0.70 \leq Pr \leq 100$, and $0.2 \leq n \leq 2$ for the symmetric and asymmetric locations of the cylinders. Vertical alignment and diagonal alignment of the cylinders at various locations were considered in the study. An improved heat transfer was observed in shear-thinning fluids for asymmetric positioning of the cylinders particularly at higher value of Prandtl and Grashof numbers.

To the best of our knowledge, no study has been performed to observe the effects of two cylinders within an enclosure containing non-Newtonian fluid where the radius of one cylinder is varied whereas the radius of other cylinder is constant. The present study focuses the effects of Prandtl number and Grashof number on natural convection from heated cylinder in a cold square enclosure. The coupled mass, momentum, and energy equations have been solved numerically for $0.7 \leq Pr \leq 100$ and $10 \leq Gr \leq 10^4$, and hence $7 \leq (Ra = Gr \times Pr) \leq 10^6$. The fluid used in the study is power-law fluid (shear-thinning fluid), with power-law index, $0.2 \leq n \leq 1$.

2 Problem Description and Governing Equations

The schematic representation of the geometry is shown in Fig. 1a. Two infinitely long (in Z-direction) horizontal cylinders of radius R_1 and R_2 are placed adjacently inside an infinitely long square duct. The cavity is filled with incompressible power-law fluids. As per the problem, the flow pattern in the Z-direction would be approximately same, and velocity component and gradient are assumed negligible, hence this could be assumed to be a 2D problem. As a result, from the front view, only 2 circle and a square are visible and for reference we could use two-dimensional term to describe the problem wherever suitable.

In Fig. 1a, both cylinders are of equal radius 0.2 L represented by solid lines. The radius to length ratio (R_1/L) of the left cylinder has been varied from 0.16 to 0.24. The center-to-center distance of two cylinders is fixed and is equal to 0.5 L. During study, the temperatures of both the cylinders are maintained constant. The cylinder with radius R_1 is maintained at a higher temperature T_h whereas the remaining portions are initially at a relatively lower temperature T_C. Due to difference in temperature between hot cylinder and surrounding fluid, buoyancy-induced, natural convection takes place resulting in generation of plume from the top of heated cylinder. The fluid flow is laminar and steady state. The maximum temperature difference in the fluid is 5 K. The thermo-physical properties (power-law constant, heat capacity, conductivity) of the fluid are considered constant as the temperature difference is very low. However, the change in fluid density is calculated using Bousssinesq approximation as given in Eq. (1):

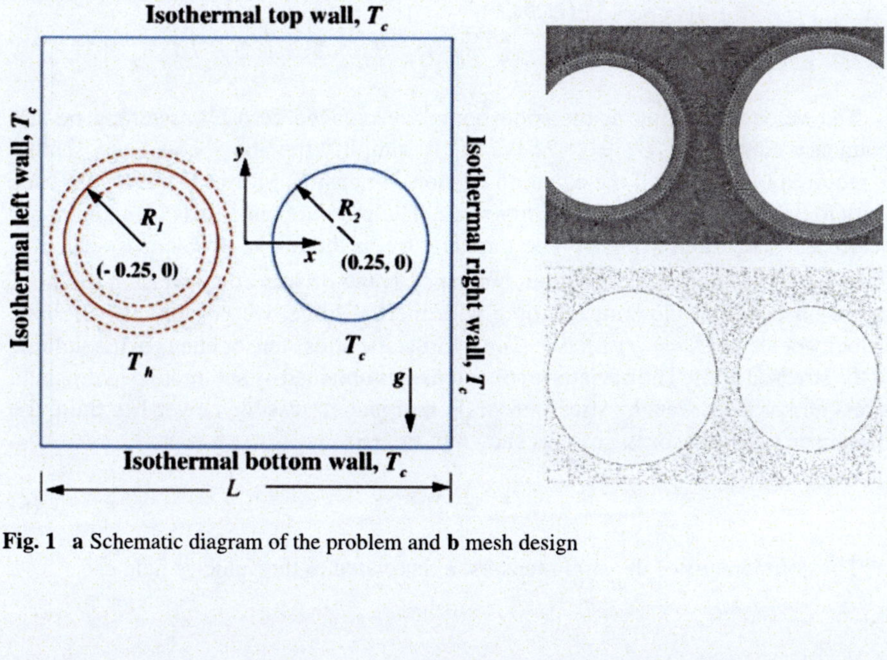

Fig. 1 **a** Schematic diagram of the problem and **b** mesh design

$$\frac{\rho_c - \rho}{\rho_c} = \beta(T - T_c) \tag{1}$$

where ρ_c is density at reference temperature T_c and β is the coefficient of volumetric thermal expansion.

Taking all these assumptions the continuity, momentum and energy equations for power-law fluid are written in following dimensionless forms.

Continuity equation:

$$\frac{\partial U_x}{\partial x} + \frac{\partial U_y}{\partial y} = 0 \tag{2}$$

X-momentum equation:

$$\left(\frac{\partial(U_x U_y)}{\partial x} + \frac{\partial(U_x U_y)}{\partial y}\right) = -\frac{\partial p}{\partial x} + \frac{1}{\sqrt{Gr}}\left(\frac{\partial \tau_{xx}}{\partial x} + \frac{\partial \tau_{yx}}{\partial y}\right) \tag{3}$$

Y-momentum equation:

$$\left(\frac{\partial(U_y U_y)}{\partial x} + \frac{\partial(U_y U_y)}{\partial y}\right) = -\frac{\partial p}{\partial y} + \theta + \frac{1}{\sqrt{Gr}}\left(\frac{\partial \tau_{xy}}{\partial x} + \frac{\partial \tau_{yy}}{\partial y}\right) \tag{4}$$

Thermal energy equation:

$$\left(\frac{\partial(U_x \theta)}{\partial x} + \frac{\partial(U_y \theta)}{\partial y} = \frac{1}{Pr Gr^{\left(\frac{1}{n+1}\right)}}\left(\frac{\partial^2 \theta}{\partial x^2} + \frac{\partial^2 \theta}{\partial y^2}\right)\right) \tag{5}$$

The velocity of fluid at the solid boundary is taken zero by assuming no-slip boundary conditions ($U_x = 0, U_y = 0$). To simplify the above equations, scaling is required to convert all the equations to non-dimensional form. L and U_r are used for length and characteristic velocity scale. The pressure and stress component are scaled as $\rho_c {U_r}^2$ and $m\left(\frac{U_r}{L}\right)^n$. The temperature in the above problem is scaled as $\theta = \frac{(T - T_c)}{(T_h - T_c)}$. In natural convection, buoyancy is the driving force causing the flow, hence on equating the buoyant force with inertial force, we get the characteristic velocity scale as $U_r = \sqrt{gL\beta\Delta T}$. The hydrostatic stress tensor changes the volume of the stressed body. The deviatoric stress tensor is obtained by subtracting hydrostatic stress tensor from Cauchy stress tensor. For an incompressible power-law fluid, the deviatoric stress tensor is given as shown in Eq. (6).

$$\tau_{ij} = 2\eta\varepsilon_{ij} \tag{6}$$

The components of the strain tensors, ε, is related to the velocity field as

$$\varepsilon_{ij} = \frac{1}{2}\left(\frac{\partial U_i}{\partial j} + \frac{\partial U_j}{\partial i}\right) \tag{7}$$

For power-law fluid, the viscosity is given as

$$\eta = m\left(\frac{I_2}{2}\right)^{\left(\frac{n-1}{2}\right)} \tag{8}$$

I_2 is the second invariant of rate of strain tensor, which is defined as

$$I_2 = \sum_i \sum_j \varepsilon_{ij}\varepsilon_{ji} \tag{9}$$

To study the heat transfer more effectively, Nusselt number is derived at every point of cylinder and enclosure. The local Nusselt number is represented at any point as:

$$Nu = \frac{ha}{K} = -\left(\frac{\partial \theta}{\partial n_s}\right)_{\text{surface}}$$

where a is the characteristic length ($a = L$) or ($a = d$) for the square or cylinder, respectively. n_s is unit normal vector to the surface. To calculate the average Nusselt number, it is integrated over the surface

$$\overline{Nu} = \frac{1}{A}\left(\int Nuds\right)_{\text{surface}}$$

3 Numerical Methodology and Choice of Numerical Parameters

All the governing Eqs. (3–6) have been solved using finite element method for the mentioned boundary conditions using COMSOL Multiphysics. In COMSOL Multiphysics, user-controlled, extremely fine triangular meshes are generated for the flow domain under study. The generated meshes are fine enough to capture the sharp temperature and velocity gradient within the thin boundary layer. Figure 1b shows a typical distribution of mesh for the present study. The steady, 2*D*, laminar module coupled with the heat transfer module is solved with PARDISO (parallel direct sparse solver interface) solver. A relative convergence criterion of 10^{-5} was chosen for mass, momentum, and energy equations. To consider the gravitational effects and Boussinesq approximation, a volume force is added in the momentum Eq. 4. Parametric sweep is used for calculating the multiple Prandtl number for the same geometry. In

Table 1 Grid test data at Gr $= 10^4$, Pr $= 100$, and $n = 0.2$

Hot cylinder			
Grid number	G1	G2	G3
$R_1 / L = 0.16$			
Number of elements	23,150	40,414	64,996
Nu_{avg}	5.5577	5.7106	5.7988
$R_1 / L = 0.24$			
Number of elements	20,124	40,910	76,812
Nu_{avg}	11.2323	11.4394	11.5426

the present study, the stable values of Nusselt number are calculated up to four decimal places. The problem statement is solved for each value of governing parameter (Pr, Gr, n, R_1/L). The grid independence test is performed to show grid independency for the average Nusselt number. Table 1 shows the effects of number of grid elements in geometric progression on the calculated Nusselt number. Table 1 shows approximately 1.9% and 1.1% change in Nusselt number with changing the number of grid elements from G1 to G2, and from G2 to G3, respectively. No significant change is observed in Nusselt number for changing the number of grid elements from G2 to G3; therefore, 40,000 grid elements, (approximately same as G2), have been chosen for further study. Figure 2 shows the local Nusselt number along the surface of the hot cylinder. The local Nusselt number for three grids deviates significantly in the range of $46° \leq \theta \leq 232°$. At $\theta = 130°$, the deviation of Nusselt number between G2 and G3 is lower as compared to deviation between G1 and G2.

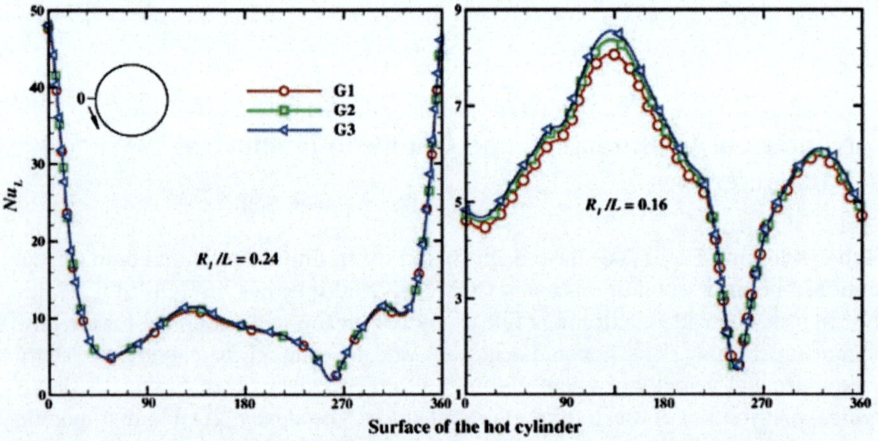

Fig. 2 Grid test at $n = 0.2$, Gr $= 10^4$, and Pr $= 100$

4 Results and Discussion

In this study, the phenomenon of natural convection is elucidated for kinematic parameters (Gr, Pr), geometric parameter (R_1/L), and fluid properties (n). With the help of streamlines and isotherms, the flow and thermal behavior, respectively, of the fluid has been predicted for different size of inner cylinder. Local and average Nusselt numbers are calculated on the surface of hot cylinder which relate all the kinematic parameters, geometric parameters, and fluid properties. Prior to generating data from created COMSOL Multiphysics settings, it was first validated with existing literature data.

4.1 Validation of Results

For accuracy of the results, it is necessary to reproduce the available literature studies. Park et al. [7] reported the effect of the location of the two cylinders differently heated in a cold square enclosure on heat transfer when the cylinders were shifted vertically. Figure 3 shows the variation of surface average Nusselt number at the enclosure (Nu_{av_en}) with different vertical positions (δ) of cylinders. The data obtained using COMSOL Multiphysics settings in the present study is well agreed (<1% deviation) with the data reported by Park et al. [7] at Ra = 10^5 and 10^6.

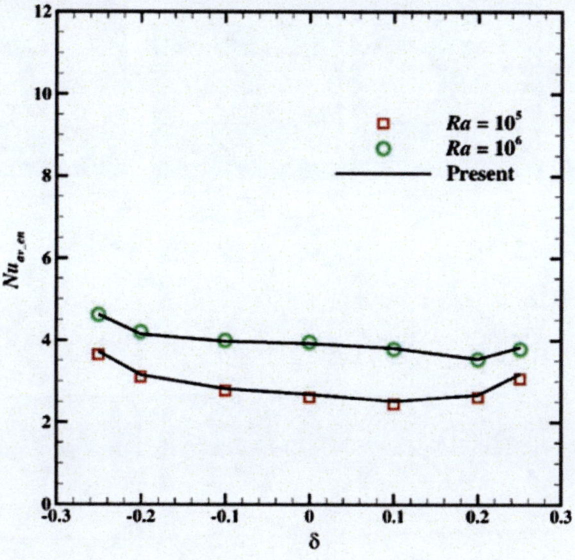

Fig. 3 Validation of average Nusselt number with the study performed by Park et al. [9]

4.2 Streamline and Isothermal Contours

The isothermal contours and streamlines are conventionally used to understand the heat transfer and fluid flow behavior. Figure 4 shows isotherms and streamlines for different values of Grashof number, aspect ratio of hot cylinder, and shear-thinning liquid at different power-law index. Figure 4a shows the isotherm contours and streamlines for an aspect ratio of 0.16 and for a power-law index of 0.2. At the lower value of Grashof number (Gr = 10), heat transfer mainly takes place by conduction, and hence, no significant change in isotherm and streamlines is observed with changing in the Prandtl number from 0.7 to 100. For two different cases (Pr = 0.7 and 100 at constant Gr = 10), there is a symmetricity along centreline x-axis passing through the midpoint of square enclosure. Two primary recirculation regions in opposite directions are observed over the cylinders. The larger recirculation region is observed above the cold cylinder. The result shows the formation of additional two secondary regions over the cold cylinder. A certain amount of cold fluid comes into the contact with the surface of the hot cylinder and gets heated causing the fluid flow

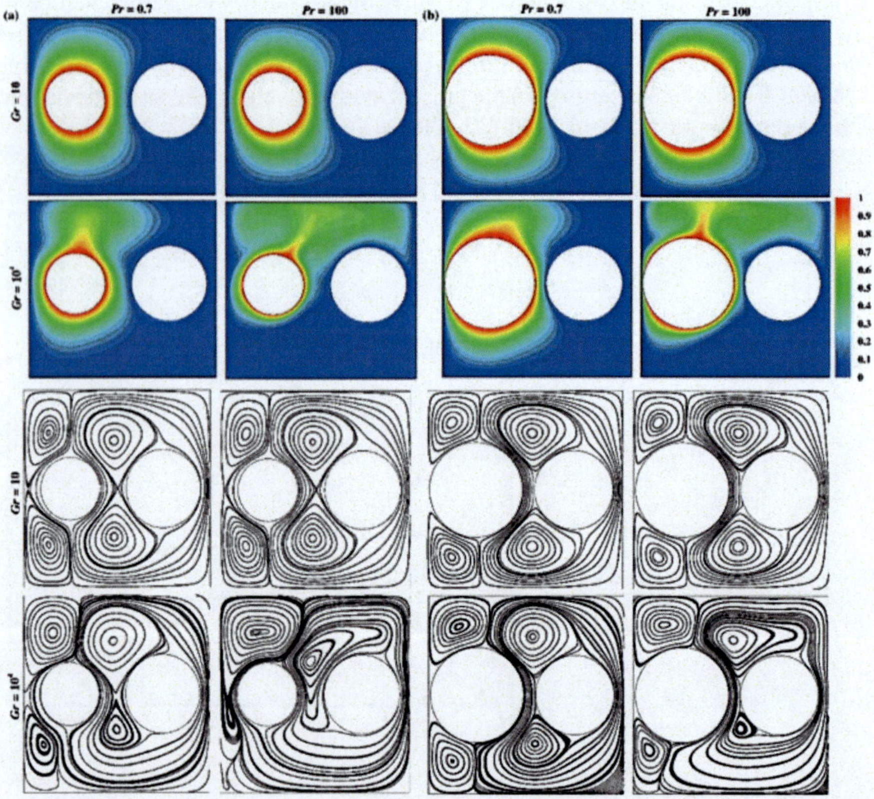

Fig. 4 Isotherm and streamlines at $n = 0.2$ for **a** $R_1/L = 0.16$ and **b** $R_1/L = 0.24$

behavior due to density difference. With increased convection, a large mass of fluid is at a relatively higher temperature. For $R_1/L = 0.16$ as shown in Fig. 4a, more space is available in between the hot cylinder and left wall, which results in the enlargement of primary circulation zone around the cold cylinder. This phenomenon is due to the compression of the circulating zone at left corner near the hot cylinder. At Gr = 10^4, with increasing Prandtl number from 0.7 to 100 (Fig. 4a), secondary circulation zone disappears and reduction in the circulation zone is observed which initially originated at lower left corner of the hot cylinder. In contrary, the two-recirculation zone decreases in the size; however, none of the zones disappear (Fig. 4b). When the fluid comes in the contact with the upper cold isothermal wall of the cavity, it loses its momentum and then it falls in the either direction leading to formation of primary recirculation regions. A stagnant fluid layer is visible over the cold cylinder leading to reduction in heat transfer. The convection comes into picture with increasing Grashof number. However, it also depends on Prandtl number as well as viscosity of the fluid. Figure 4a, b indicates much similarity even though at higher Grashof number (Gr = 10^4) and at Pr = 0.7, which suggests no effects of aspect ratio at these parametric combinations. With increasing Grashof number, a significant distortion is observed in the isotherm and the streamlines which results in decreasing thermal boundary thickness and hence increase in overall heat transfer rate. Plumes are observed in both figures (Fig. 4a, b) at higher Grashof and Prandtl numbers (Gr = 10^4 and Pr = 100). Hence, the symmetric nature of the streamlines is lost. The heat and flow behavior change with the fluid nature. The shear-thinning fluids show significantly higher heat transfer compared to Newtonian fluids because of high shear rate inside boundary layer and less boundary thickness. At lower Grashof number, the isotherm and streamline structures are approximately same with varying power-law index and Prandtl number. With these findings, it is evident that the heat transfer depends on geometric, kinematic, and fluid property parameters. The flow and heat transfer patterns can be controlled and manipulated by using suitable parameters such as Gr, Pr, R_1/L, and n of the fluid.

4.3 Local Nusselt Number Distribution

From the above discussion, a clear picture of heat and flow fields is presented in the form of isotherms and streamlines. The cylinders are under isothermal conditions; however, a normal temperature gradient still exists, which results in the different values of the local heat transfer fluxes along the surface of the hot cylinders. Figure 5 shows the variation in local Nusselt number along the cylinder surface with changing Grashof number (Gr), Prandtl number (Pr), power-law index (n), and geometric parameter (R_1/L ratio). At low Grashof number Gr = 10, the Nusselt number along the surface of cylinder is found to be same and independent of other parameters (i.e., Pr, power index, and aspect ratio). Heat transfer by conduction is the main phenomenon for such observation at low Grashof number. The heat transfer behavior

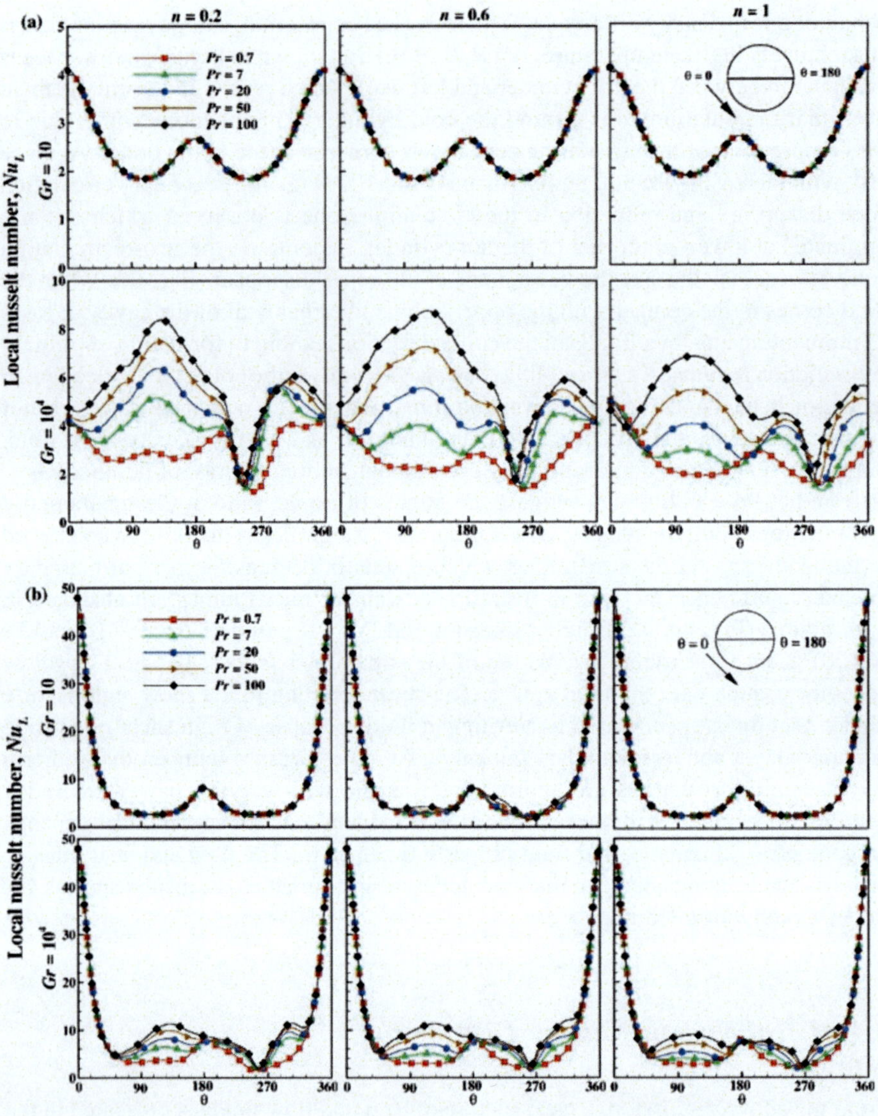

Fig. 5 Local Nusselt number variation for **a** $R_1/L = 0.16$ and **b** $R_1/L = 0.24$

is exactly same in both half of the cylinders (i.e., from $\theta = 0$–180° and $\theta = 180$–360°). Maxima is observed at $\theta = 0°$ and 180°. At $\theta = 0°$, distance between left wall and cylindrical surface is less, and two vortices meet there as evident from the streamline. The Nusselt number is found higher at $\theta = 0°$ due to higher temperature gradient as indicated in isotherm contours. At $\theta = 180°$, the gap between cylinders is large as compared to hot cylinder and the left wall; therefore, the fluid velocity is low and consequently low heat transfer. A constant value of Nusselt number is found at minima (θ = 90 and 270°) which arises due to same proximity from the wall. At relatively higher Grashof number (Gr = 10^4), a significant variation in local Nusselt number is observed for each Prandtl number. However, the peaks are located at approximately a constant θ value. With increasing Prandtl number, higher value of local Nusselt number is observed for each value of power-law index in the θ range of 0–180°. Beyond 180°, the effect of Prandtl number on Nusselt number variation is low. It is observed that the variation in local Nusselt number at high Grashof number (Gr = 10^4) is significant for both cases (R_1/L = 0.16 and 0.24); however, the maximum value is less than 10 for the case of R_1/L = 0.16, whereas the value is higher for the other case (R_1/L = 0.24). Convective transport along with shear-thinning behavior of the fluid is responsible for such phenomena. At θ = 0°, the value of Nusselt number is much higher for R_1/L = 0.24 compared to R_1/L = 0.16.

4.4 Surface Average Nusselt Number

A dimensional analysis indicates that for a fixed geometry, the surface average Nusselt number depends on Grashof number, Prandtl number, and the power-law index. Figure 6 shows the variation in Nusselt number with power-law index for different values of Grashof and Prandtl numbers for two different geometries (R_1/L = 0.16 and 0.24). Figure 6a, b, for low Grashof number (Gr = 10), shows no variation in average Nusselt number for any value of power-law index or Prandtl number due to absence of convective heat transfer. At higher Grashof number, (Gr = 10^3), the value of average Nusselt number remains same at lower values of Prandtl number (Pr = 0.7–7), whereas a significant variation in Nusselt number is observed at higher Prandtl number. On increasing the Prandtl number, the boundary layer thickness decreases resulting in increased heat transfer rate. Hence, average Nusselt number is higher at higher values of Prandtl number (Pr = 20–100). The Nusselt number decreases with increasing power-law index for the low values of Prandtl numbers (Pr = 0.7–7) and at Gr = 10^4. In contrary, for higher values of Prandtl number (Pr = 20–100), first, the Nusselt number increases till a power-law index of 0.4, and then it decreases. Figure 7 shows the variation of average Nusselt number with aspect ratio. This is also in support for the explanation of Fig. 6.

5 Conclusions

This work elaborates the effects of Prandtl number, Grashof number, power-law index, and the aspect ratio of the hot cylinder on the convective heat transfer. The streamline and isotherms are plotted to elucidate the effects of the dimensionless parameters on the average and local Nusselt number in the study. The study is performed for $0.7 \leq \mathrm{Pr} \leq 100$, $10 \leq \mathrm{Gr} \leq 10^4$, $0.2 \leq n \leq 1$, and $0.16 \leq R_1/L \leq 0.24$.

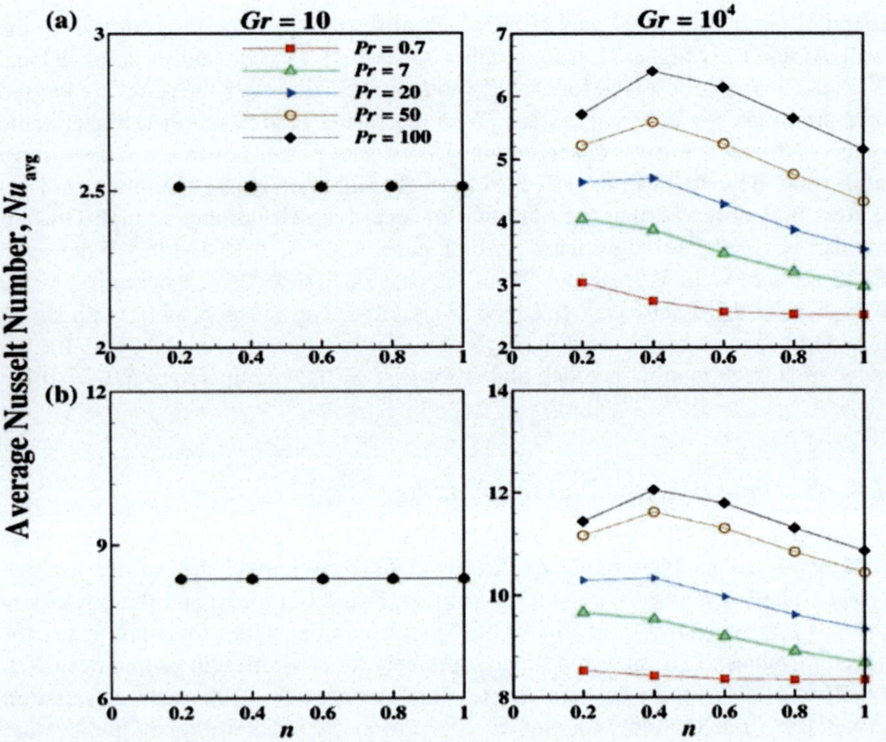

Fig. 6 Average Nusselt number variation with Gr, Pr, and n for **a** $R_1/L = 0.16$ and **b** $R_1/L = 0.24$

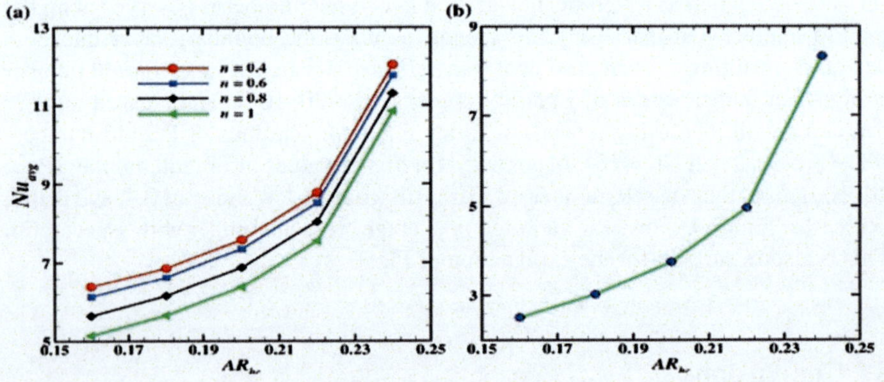

Fig. 7 Average Nusselt number variation with AR_{hc} and n at **a** Gr = $\mathbf{10^4}$, **Pr** = 100 and **b** Gr = 10, Pr = 0.7

The continuity, momentum, and thermal equations have been converted to dimensionless form. The results show that at low Grashof number, conduction is dominant whereas at high Grashof number convection plays a major role in heat transfer. The formation of plumes indicates the convection. The streamline and isotherms are employed to extract the flow behavior and temperature distribution inside the enclosure. In a broad context, it is found that heat transfer increases with decreasing power-law index. The Nusselt number has been increased with increasing the radius of hot cylinder.

References

1. Saleh, H., Alsabery, A.I., Hashim, I.: Natural convection in polygonal enclosures with inner circular cylinder. Adv. Mech. Eng. **7**, 1–10 (2015)
2. Rath, S., Dash, M.K.: Natural convection from horizontal cylinders of different shapes in rectangular enclosure. Heat Mass Transf. Conf. **2017**, 1615–1620 (2017)
3. Sahu, K.B., Singh, R.K.: Analysis of heat transfer and flow due to natural convection in air around heated triangular cylinders of different sizes inside a square enclosure. Procedia Eng. **90**, 550–556 (2014)
4. Yoon, H.S., Ha, M.Y., Kim, B.S., Yu, D.H.: Effect of the position of a circular cylinder in a square enclosure on natural convection at Rayleigh number of 10^7. Phys. Fluids. **21** (2009)
5. Ho, C.J., Cheng, Y.T., Wang, C.C.: Natural convection between two horizontal cylinders inside a circular enclosure subjected to external convection. Int. J. Heat Fluid Flow. **15**, 299–306 (1994)
6. Ho, C.J., Chang, W.S., Wang, C.C.: Natural convection between two horizontal cylinders in an adiabatic circular enclosure. J. Heat Transf. **115**, 158–165 (1993)
7. Park, Y.G., Yoon, H.S., Ha, M.Y.: Natural convection in square enclosure with hot and cold cylinders at different vertical locations. Int. J. Heat Mass Transf. **55**, 7911–7925 (2012)
8. Park, Y.G., Ha, M.Y., Yoon, H.S.: Study on natural convection in a cold square enclosure with a pair of hot horizontal cylinders positioned at different vertical locations. Int. J. Heat Mass Transf. **65**, 696–712 (2013)
9. Park, Y.G., Ha, M.Y., Park, J.: Natural convection in a square enclosure with four circular cylinders positioned at different rectangular locations. Int. J. Heat Mass Transf. **81**, 490–511 (2015)
10. Park, S.H., Seo, Y.M., Ha, M.Y., Park, Y.G.: Natural convection in a square enclosure with different positions and inclination angles of an elliptical cylinder Part I: a vertical array of one elliptical cylinder and one circular cylinder. Int. J. Heat Mass Transf. **126**, 173–183 (2018)
11. Baranwal, A.K., Chhabra, R.P.: Effect of Prandtl number on free convection from two cylinders in a square enclosure. Heat Transf. Eng. **37**, 545–556 (2016)
12. Yoon, H.S., Jung, J.H., Park, Y.G.: Natural convection in a square enclosure with two horizontal cylinders. Numer. Heat Tranf. Part A **62**, 701–721 (2012)
13. Prhashanna, A., Chhabra, R.P.: Laminar natural convection from a horizontal cylinder in power-law fluids. Ind. Eng. Chem. Res. **50**, 2424–2440 (2011)
14. Mishra, L., Chhabra, R.P.: Natural convection in power-law fluids in a square enclosure from two differentially heated horizontal cylinders. Heat Trans. Eng. **39**, 819–842 (2018)

Dynamic Study of Bird Strike on Rigid Plate

Tirth Patel, Atharav Naik, Sankalp Patidar, Gautam Choubey, and Sumit Tripathi

Abstract The issue of bird strike is very paramount in aviation industries and is also a big concern to the safety of passengers. With rapid increase in the number of flights around the globe, the risk of bird strike is also increasing. Typically, bird strike events are more prone when the aircraft is taking off or landing at an airport. In this work, we present the dynamic analysis of bird strike on a rigid plate using smooth particle hydrodynamic (SPH) method in explicit dynamics and AUTODYN software. For impact analysis, the shape of the bird is modeled as a conical frustum of water while aircraft component is modeled as a flat plate made of titanium alloy with a fixed support at the top face of the plate. The SPH model is used to simulate the bird (conical frustum of water) while a Lagrangian plate model is used to simulate aircraft component. Detailed grid convergence analysis was performed by taking different SPH elements for bird model and different element size of structured grid for the plate model. The results are presented in the form of dynamic values of shock pressure, von Mises stress, and absolute displacement of the plate. It is observed that the shock pressure and von Mises stress are maximum at the point of impact, and the maximum displacement occurring because of impact is on the free side of the plate.

Keywords Bird strike · SPH · Impact analysis · Shock pressure

1 Introduction

Bird strikes on an aircraft can cause highly tragic accidents or serious damages to the airplane components and can be a big concern to the safety of passengers on board. The timings of the day and weather conditions around the airport play a major role in it, and huge loss is caused to the aviation industry because of such incidents [1]. The problem of bird strike becomes more relevant with the increase in number of travelers resulting in more aircrafts in the aviation industry. A famous incident happened in

T. Patel · A. Naik · S. Patidar · G. Choubey · S. Tripathi (✉)
Department of Mechanical and Aerospace Engineering, Institute of Infrastructure, Technology, Research and Management, Ahmedabad 380026, India
e-mail: sumittripathi@iitram.ac.in

R. P. Bharti and K. M. Gangawane (eds.), *Recent Trends in Fluid Dynamics Research*, Lecture Notes in Mechanical Engineering,
https://doi.org/10.1007/978-981-16-6928-6_9

January 2009 when a flock of Canadian geese was ingested in the engines of Airbus A320, shortly after takeoff from the New York airport and all of its engines stopped working, the pilot had to land the plane into the Hudson River nearby the airport [2]. Also, in October 1960, world had witnessed the greatest loss of life due to bird strike on Lockheed L-188 Electra, which flew through a flock of starlings (an European bird) during takeoff and all four engines lost all power and the airplane crashed into Boston harbor [3]. After a few major and several minor incidents, the investigations on bird strike cases are extensively studied using numerical methods, as the analysis of such problems using actual testing is infeasible and also confined to experimental restrictions. The main challenges in such study include its occurrence in very short interval of time, and there is a generation of large stress and strains which necessitates selection of a proper numerical techniques. In order to generate a suitable bird model that can be used for bird strike analysis, the physical as well as fundamental things are needed to be examined such as the pressure response of impact when bird comes in contact with the high velocity moving surface. These pressure responses are broadly divided into three phases: initial high-pressure shock, decay of shock pressure decay, and steady pressure rate [4]. Further, the bird consists of mostly water and behaves like liquid when strikes with any surface at high speed; additionally, it will be easy to stimulate by hydrodynamic method [5].

In reality, every bird species is different in terms of their shape, size, mass, and material properties, due to which it is difficult to study impact of one specific type of bird. In the literature, several different types of shapes have been used to simulate birds [5, 6]. In the present study, the bird model is taken as a conical frustum of water. Two computational approaches have been adopted in the present study: Lagrangian method for the plate and smooth particle hydrodynamic (SPH) method for the bird model. The Lagrangian method is a mesh-oriented method in which the solid model is divided into finite elements and material is attached with each element [7]. In the SPH method, there is no presence of mesh, and the model is divided into a finite number of particles instead of points in a mesh [8]. These particles depend on the type of the geometry and their individual sizes and interact with each other as per the initial and boundary condition of the model [9].

2 Flow Modeling and Simulation

2.1 Model Geometry

The geometry of the bird is taken as a conical frustum, having length of 100 mm, base circle diameter 50 mm, and draft angle as 16° as shown in Fig. 1. The plate dimensions are 200 × 200 × 25 mm as shown in Fig. 2. The positioning of bird and plate are shown in Fig. 3.

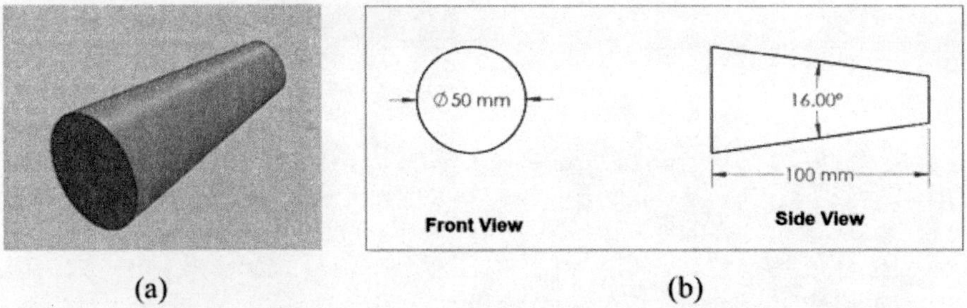

Fig. 1 Geometry of bird model: (**a**) isometric view and (**b**) dimensional details

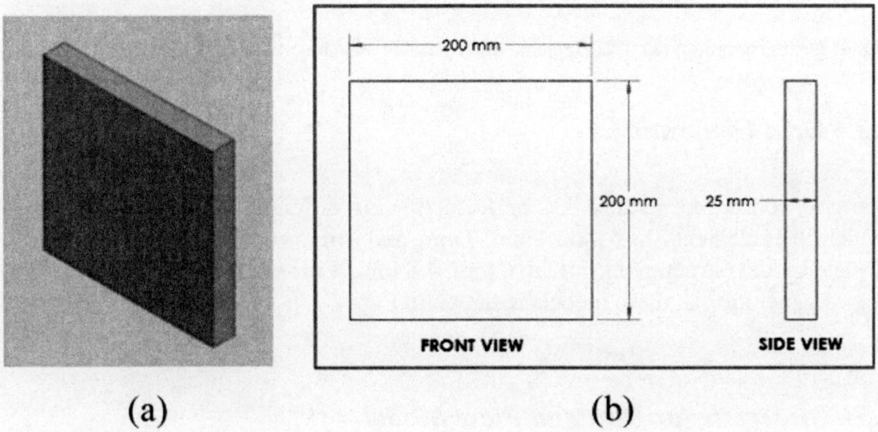

Fig. 2 Geometry of plate model: (**a**) isometric view and (**b**) dimensional details

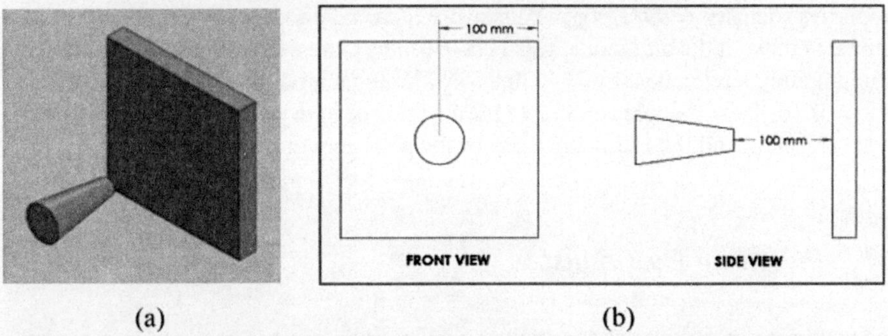

Fig. 3 Positioning of bird and plate models: (**a**) isometric view and (**b**) dimensional details

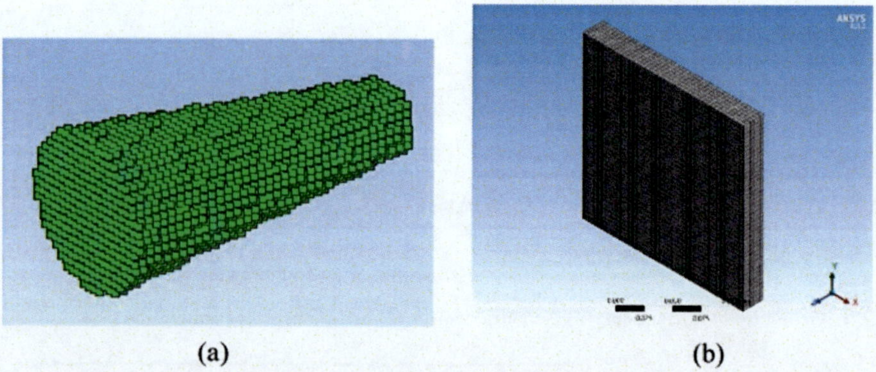

Fig. 4 Grid generation: (**a**) SPH elements in bird model and (**b**) structured grid in plate model

2.2 Grid Generation

The grid convergence study was performed using different element sizes of bird model (SPH elements) as 5 mm, 4 mm, 3 mm, and 2 mm, and different element sizes of plate model (structured grid) are 6 mm, 4.8 mm, 4 mm, and 3.7 mm. A snapshot of grid generation in these models is shown in Fig. 4.

2.3 Materials for Bird and Plate Model

The assigned material for the bird model was water (only liquid) having standard properties (density = 998.2 kg/m^3, viscosity = 10^{-3} Pa-s). As per the geometry, the estimated mass of the bird model was 106.46 g. The plate was assigned the material as titanium alloy having density = 4620 kg/m^3, Young's modulus = 96 GPa, Poisson's ratio = 0.36, shear modulus = 35,294 MPa, yield strength = 930 MPa, bulk modulus = 114.29 GPa, and the estimated mass of the plate model was 4.62 kg.

2.4 Governing Equations

The governing equations are continuity equation, momentum conservation equation, and constitutive relationship of the material [10].

Continuity equation:

$$\frac{D\rho}{Dt} + \rho\nabla.\bar{V} = 0$$

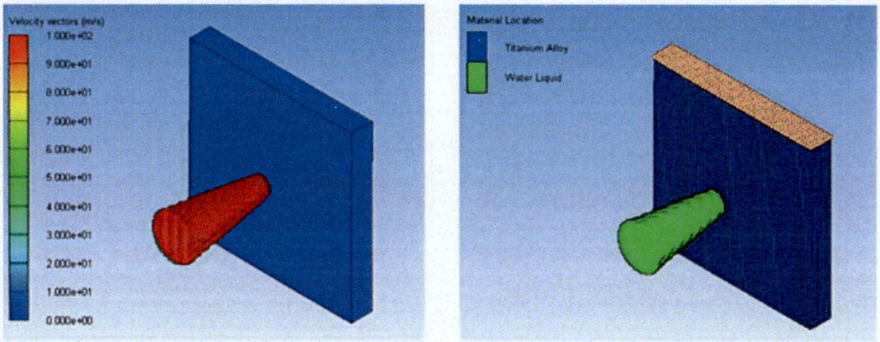

(a) (b)

Fig. 5 Initial conditions: (**a**) initial bird velocity = 100 m/s and (**b**) top face of the plate was fixed

Here, ρ is the density and $\bar{V}$is the velocity vector.

Conservation of momentum:

$$\nabla\sigma + \rho\mathbf{b} = \frac{D}{Dt}\left(\rho\bar{V}\right)$$

Here, ρ is the density, σ is the stress tensor, and **b** is the body force vector per unit mass.

2.5 Initial and Boundary Conditions

The top face of the plate was fixed, and other faces were able to move or deform freely. The bird is set freely to move in the space as per given direction and speed which is toward the plate having a value of 100 m/s. These conditions are shown in Fig. 5. The scale ratio of the system is considered to be 100. The time step was in the range of 0.1 to 0.5 microseconds during the simulation in the AUTODYN solver.

3 Results and Discussion

3.1 Grid Convergence Analysis

In the SPH method, size of an element is given as input, and the output results in the number of elements patched in defined shape. Grid independent study was conducted using 16 combinations of bird and plate element sizes as shown in Table 1.

Table 1 Element sizes in grid-independent studies

Plate refinement (No. of elements)	Bird refinement (No. of SPH elements)
5780	836, 1646, 3833, 13,210
10,584	836, 1646, 3833, 13,210
17,500	836, 1646, 3833, 13,210
21,175	836, 1646, 3833, 13,210

3.2 *Pressure Vs. Time Plots*

Temporal variations of pressure for plate having 21,175 elements and different element sizes of bird model are shown in Fig. 6.

A comparison of Hugoniot pressure and stagnation pressure at the point of contact is shown in Table 2 with the percentage error to that of theoretical values. It can be seen that minimum error is expected somewhere between the bird element sizes of 4 and 5 mm. A good agreement was observed with bird element size of 4.5 mm resulting in Hugoniot pressure and stagnation pressure at the point of contact as

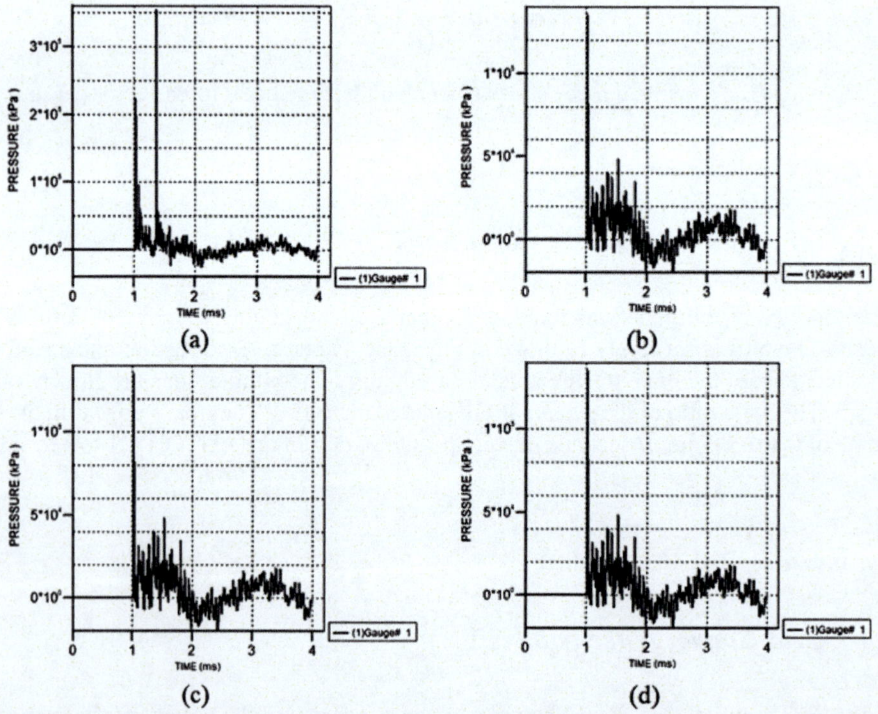

Fig. 6 Pressure Vs. time for plate having 21,175 elements and bird element size of: (**a**) 5 mm; (**b**) 4 mm; (**c**) 3 mm; and (**d**) 2 mm

Table 2 Pressure comparison for bird having velocity 100 m/s

Plate refinement		Bird refinement		theoretical		SPH-FE analysis		Error (%) of P_H
No. of elements	Element size (mm)	No. of elements	Element size (mm)	Hugoniot pressure (P_H) MPa	Stagnation pressure (P_S) MPa	Hugoniot pressure (P_H) MPa	Stagnation pressure (P_S) MPa	
21,175	3.7	836	5	168.3	5	224.6	4	33.452
		1646	4	168.3	5	134.84	3.5	–19.881
		3833	3	168.3	5	123.37	3	–26.696
		13,210	2	168.3	5	123.8	4	–26.441

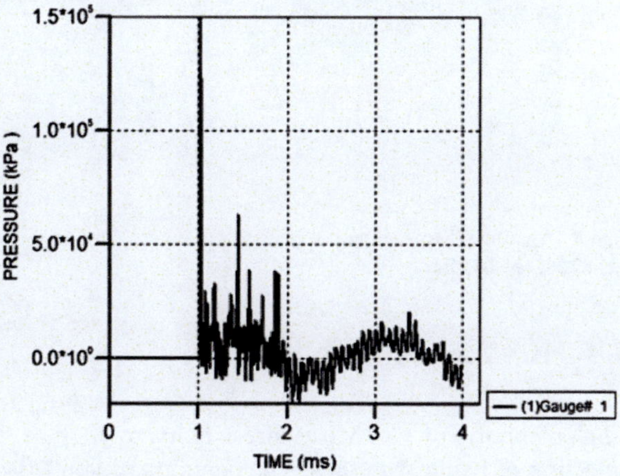

Fig. 7 Pressure Vs. time for grid having bird element size of 4.5 mm and plate element size of 3.7 mm

149.26 MPa and 7 MPa, respectively (as shown in Fig. 7). Error generated in this grid is -11.31% for Hugoniot pressure. The remaining results in the present work are shown for bird element size of 4.5 mm (*i.e.*, 1130 number of elements) and plate element size of 3.7 mm (*i.e.*, 21,175 number of elements).

3.3 *Absolute Displacement*

In this section, the absolute displacement occurred in the system is presented for bird as well as for stationary plate. The total number of cycles generated by the system to perform the analysis is 10284. In Fig. 8, the contours of absolute displacement are shown for a few cycles (2058, 4115, 6172, 8228, 10,284). It is clear that the maximum displacement is on the side which is opposite to the fixed side, and at the end, the maximum absolute displacement at free end is 0.5064 mm.

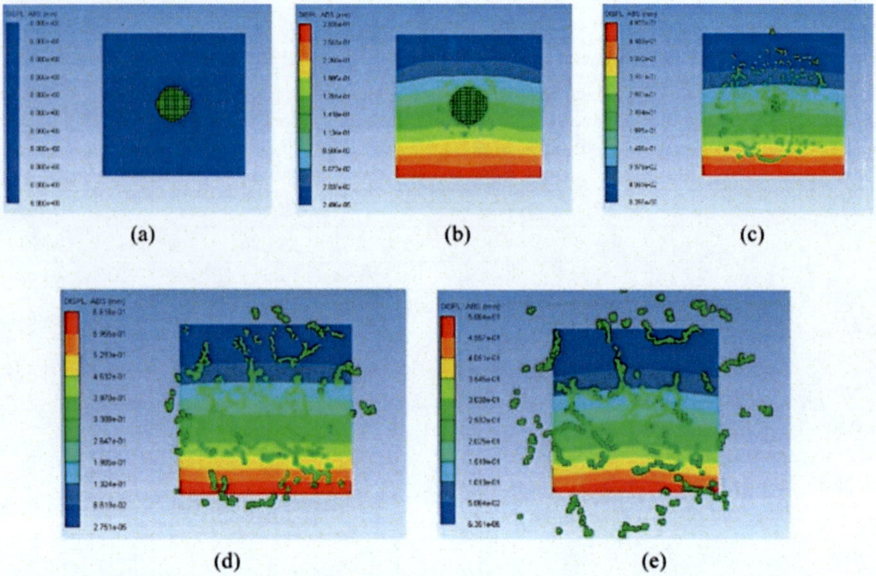

Fig. 8 Absolute displacement of plate at following number of cycles: (**a**) 2058; (**b**) 4115; (**c**) 6172; (**d**) 8228; (**e**) 10,284

3.4 Von Mises Stress

The estimation of von Mises stress helps to analyze the failure of the system and selection of better materials. The contours of von Mises stress for a few cycles are shown in Fig. 9. For all the different cycles of bird strike on the plate, a major stress is generated on the fixed side of the plate, which suggests of having more chances of failure at that side. On the other hand, the free end is free to deform, so the chance of failure of the free side is relatively less; however, the deformation will be more on the free side compared to the fixed one. The temporal variation of von Mises stress at the contact point is shown in Fig. 10 for the studied bird and plate element sizes, which indicates that the von Mises stress is maximum at the time of impact and significantly more than the stress limit of the material.

4 Conclusions

We investigated the effects of collision of a bird model with a plate model. A structured grid was used for the Lagrangian plate model, while a conical frustum shape was used for the SPH bird model which is a rarely used shape for this kind of analysis. A detailed grid convergence analysis was performed in order to get the best element sizes for further analysis. Comparison of error in Hugoniot and stagnation

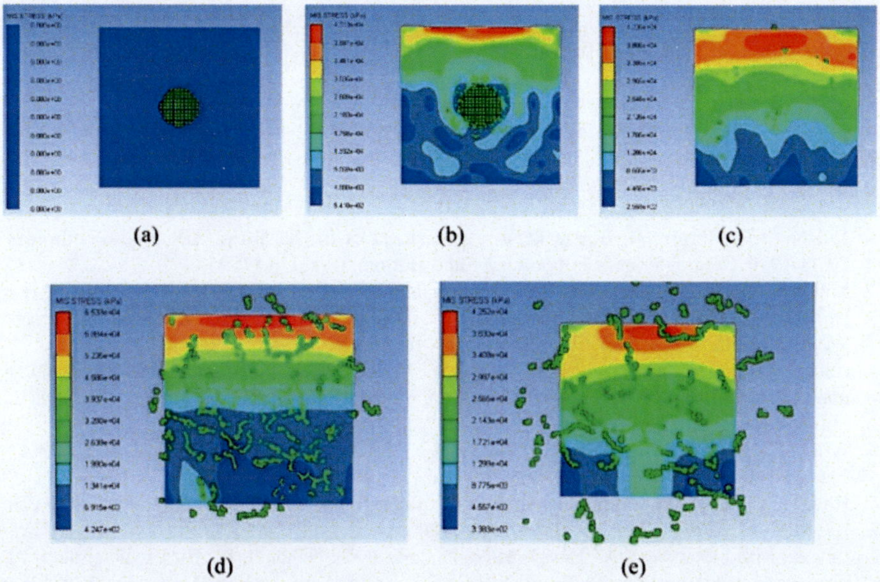

Fig. 9 Von Mises stress on plate at following number of cycles: (**a**) 2058; (**b**) 4115; (**c**) 6172; (**d**) 8228; and (**e**) 10,284

Fig. 10 Von Mises stress Vs. time for grid having bird element size of 4.5 mm and plate element size of 3.7 mm

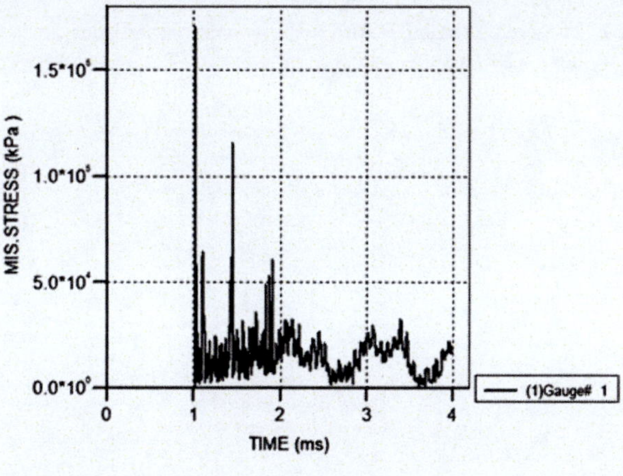

pressures for various grids has been done, and it is concluded that the model having 1130 bird elements and plate having 21,175 elements can be a good choice for the present study as its error is the least among all studied combinations. The results of absolute displacement show that the plate gets very high displacement (0.5064 mm) in comparison to the scale ratio. The high von Mises stress at the support end suggests making a stronger joint at the fixed end. Overall, the obtained results are worthy to be considered for further analysis, and the study can be extended to more parameters like mass and velocity variation impacts.

References

1. Milson, T.P., Horton, N. Birdstrike: An assessment of the hazard on UK civil aerodromes 1976–1990. Central Science Laboratory, Sand Hutton, York, UK (1995).
2. Major bird strike incidents. The Daily Telegraph. 17 June 2011.
3. Dolbeer, R.A.: Birds and aircraft—fighting for airspace in ever more crowded skies. Human-Wildlife Conflicts **3**(2), 165–166 (2009)
4. Aslam, M. A., Rayhan, S. B., ke, Z., Yu, W. J. Ballistic gelatin Lagrange Mooney-Rivlin material model as a substitute of bird in finite element bird strike case studies. Lat. Am. J. Solids Struct. 17(6) (2020).
5. Wilbeck, J.S., Rand, J.L.: The development of a substitute bird model. J. Eng. Power. **103**(4), 725–730 (1981)
6. Hou, N., Li, Y., Liu, J.: Numerical simulation of bird impact on hollow blades of titanium Fan Assembly. J. Aerosp. Eng. **32**(4), 04019044 (2019)
7. Lavoie, M.A., Gakwaya, A., Ensan, M.N., Zimcik, D.G., Nandlall, D.: Bird's substitute tests results and evaluation of available numerical methods. Int. J. Impact Eng **36**(10–11), 1276–1287 (2009)
8. Monaghan, J.J.: An introduction to SPH. Comput. Phys. Commun. **48**, 89–96 (1988)
9. Kalam A., Sd., Vijaya Kumar, R., Ranga Janardhana, G. SPH high velocity impact analysis-influence of bird shape on rigid flat plate. Materials Today: Proceedings, 4(2), 2564–2572 (2017).
10. El-Sayed., Ahmed, F. Bird strike in aviation: statistics, analysis and management. John Wiley & Sons, (2019).

CFD Simulation and Experimental Investigation of the Hydrodynamic Behavior of a Gas–liquid–solid Fluidized Bed

Hara Mohan Jena and Pedina Sibakrishna

Abstract Gas–liquid–solid fluidized beds find numerous applications in industrial processes. Accurate prediction of the hydrodynamic behavior helps operate and design a gas–liquid–solid fluidized bed successfully. CFD is a powerful tool for predicting the gas–liquid–solid fluidized bed's hydrodynamic behavior. In this current work, a three-phase fluidized bed's hydrodynamic behavior is simulated by applying the Eulerian-Eulerian granular multiphase model with a two-dimensional (2D) transient model and validated with the experimental outcomes. Hydrodynamic characteristics, bed expansion or bed voidage, bed pressure drop, and gas holdup are simulated and validated. It has been noticed that the expanded bed height increases with a rise in liquid velocity. At low liquid velocities, the bed height obtained from experiment and simulation differs; in experiment, a bed contraction is observed which is more prominent at higher gas velocities. Above all, the experiment and simulation's expanded bed height value agrees within 10% in beyond the minimum fluidization condition. The CFD simulation results reveal that the gas velocity causes a decline in bed pressure drop, validating the experimental findings. At lower liquid velocity values, the gas holdup values are very close, and at higher liquid velocity range, the agreement is within 20%. The strong concurrence among CFD simulation and experimental values for the current operating parameters demonstrates that the Eulerian-Eulerian multiphase granular flow method can predict a gas–liquid–solid fluidized bed's overall performance.

Keywords Eulerian granular multiphase model · Granular flow approach · Minimum fluidization · Transient model

1 Introduction

CFD is a powerful method for predicting fluid dynamics in different systems, which enables proper system design. It investigates heat and mass transfer processes and

H. M. Jena (✉) · P. Sibakrishna
Department of Chemical Engineering, National Institute of Technology Rourkela, Rourkela 769008, Odisha, India

R. P. Bharti and K. M. Gangawane (eds.), *Recent Trends in Fluid Dynamics Research*, Lecture Notes in Mechanical Engineering,
https://doi.org/10.1007/978-981-16-6928-6_10

fluid flow behavior. For resolving the governing equations for problems concerning heat and mass transfer and fluid flow, the most adaptable discrimination approach is the finite volume method (FVM). Its most convincing aspect is that the resulting solution satisfies momentum, mass, species, and energy conservation. The gas–liquid–solid fluidized beds are the most promising and commonly used multiphase systems, with many industrial applications. They are designed generally by applying available correlation for hydrodynamic characteristics [1]. Heard and Richter [2] illustrated developing a gas–liquid–solid hydrodynamic model at a Conference on Computational Fluid Dynamics in Reaction Engineering 1996. These reactors' hydrodynamics are complex and unknown due to complex phenomena, for example, particle vs particle, liquid vs particle, and particle vs bubble interaction. As a result, for comprehending and exact scaling and design of multiphase reactors [3], computational fluid dynamics (CFD) has been encouraged as a helpful method. In comparison with experimental work, it offers ample information. The multiple (Euler-Euler-Euler) system with the population balance model (PBM) as well as the models of gas–liquid mass transfer was used to predict the behavior of a gas–liquid–solid circulating fluidized bed (GLSCFB) riser by characterizing the phase holdups, mass transfer characteristics, and bubble size distribution [4]. On a complete three-dimensional, unsteady multiple (Euler-Euler-Euler) basis, CFD simulations of three-phase fluidized beds were conducted [5].

CFD simulation was used to investigate particle properties' influence upon the flow characteristics in a three-phase bubble column [6]. On the other hand, the particle density had a more negligible effect on the radial uniformity distributions and hydrodynamic. According to the literature review for three-phase reactors, CFD has recently appeared as a modern prototype for modeling multiphase flow and fluidization. The influence of various parameters on the bed expansion, bed pressure drop, and gas holdup behaviors is poorly recorded in literature. Before applying the system for a particular cause, the fluidized bed reactor's hydrodynamics must be thoroughly characterized to optimize the system's performance.

In this work, an attempt is made to analyze a gas–liquid–solid fluidized bed's hydrodynamics, bed voidage or bed expansion, pressure drop, and gas holdup numerically using CFD and validate with experimental findings. A two-dimensional (2D) transient model with quadrilateral-meshing of a gas–liquid–solid fluidized bed is simulated to characterize hydrodynamic behavior. Commercial CFD package ANSYS Fluent 15.0 using Eulerian-Eulerian multiphase granular flow method is used for the flow characteristics simulation. Experiments with a broad range of operating variables were performed to analyze the co-current gas–liquid–solid fluidized bed's hydrodynamic behavior, including phase hold up, bed expansion (bed voidage), and bed pressure drop. The experimental findings for the gas–liquid–solid phase hydrodynamics have been used to validate the proposed CFD model.

2 Hydrodynamic Flow Model

An Eulerian granular multiphase model is used for simulation, where all the three phases (solid, liquid, and gas) are viewed as continua (i.e., each phase is given an Eulerian treatment), interpenetrating and communicating in the computational domain with each other. All three phases are believed to share the pressure field in the extent of their volume fraction.

2.1 Conservation Equations

Conservation equations of mass and momentum govern the movement of the individual phase.

Equation of Continuity:

$$\frac{\partial}{\partial t}(\varepsilon_k \rho_k) + \nabla(\varepsilon_k \rho_k u_k) = 0 \tag{1}$$

where ε_k, $u_{k,,}$ ρ_k indicate the volume fraction, velocity, and density of individual phase, where $k = L, g, s$, respectively. The gas, liquid, and solid phase volume fraction ε_k meet the criteria mentioned below:

$$\varepsilon_g + \varepsilon_L + \varepsilon_s = 1 \tag{2}$$

Momentum equations:

Gas and Liquid phase:

$$\frac{\partial}{\partial t}(\rho_k \varepsilon_k u_k) + \nabla.(\rho_k \varepsilon_k u_k u_k) = -\varepsilon_k \nabla p + \nabla.\boldsymbol{\tau}_k + \rho_k \varepsilon_k g + F_{i,k} \tag{3}$$

where k represents the liquid and gas phase as L and g, respectively.

Solid phase:

$$\frac{\partial}{\partial t}(\rho_s \varepsilon_s u_s) + \nabla.(\rho_s \varepsilon_s u_s u_s) = -\varepsilon_s \nabla p - \nabla p_s + \nabla.\boldsymbol{\tau}_s + \rho_s \varepsilon_s g + F_{i,s} \tag{4}$$

where P: stands for the pressure contributed by all phases. The RHS side of Eq. (4), i.e., solid phase momentum equation, contains the term ∇Ps, which indicates extra solid pressure because of collision of solids. The terms representing interphase momentum exchange for all three phases are $F_{i,L}$, $F_{i,g}$, and $F_{i,s}$, respectively.

The stress–strain tensors terms $\boldsymbol{\tau}_g$,$\boldsymbol{\tau}_L$, and $\boldsymbol{\tau}_s$ are for gas, liquid, and solid phase, respectively, as represented by a common equation and are defined as;

$$\tau_k = \varepsilon_k \mu_k \left(\nabla u_k + \nabla u_k^T\right) + \varepsilon_k \left(\lambda_k - \frac{2}{3}\mu_k\right) \nabla . u_k \mathbf{I} \tag{5}$$

where k represents the gas, liquid, and solid phases as g, L, s correspondingly.

μ_k indicates the shear viscosity, and λ_k indicates the bulk viscosity of all the three phases.

$\mathbf{I}$ represents identity matrix.

2.2 *Interphase Exchange of Momentum*

The expressions $\boldsymbol{F}_i$, interphase exchange of momentum in the momentum equations comprises of drag force, the lift force, the additional mass force, etc., which are the interaction forces among different phases, combined linearly and are usually expressed as

$$F_i = F_D + F_L + F_{VM} \tag{6}$$

The influence of different interfacial forces has been studied [7]. It was discovered that the additional mass force is lesser than the drag force and the impact of it is only visible when significant frequency variations in the slip velocity exist and that in a bubbly flow. Fluent does not provide the virtual or added mass force by default. Taking into account, there is no solid experimental proof of the magnitude and direction of lift force, the lift force can be overlooked [8], and ignoring the lift force will also steer to adequate resemblances with experimental results [9, 10]. In comparison with the drag force, the lift force is negligible. As a result, in the current simulation using CFD, the interphase momentum exchange is only defined considering the drag force.

Cohesion, friction, pressure, and other effects induce the interphase drag force and are subject to the conditions $F_{D,jk} = -F_{D,kj\ j}$ and $F_{D,jj} = 0$ where the combination of various phases is denoted by subscripts j and k. Then, the expression for $F_{D,jk}$ is represented as:

$$F_{D,jk} = K_{jk}\left(u_j - u_k\right) \tag{7}$$

where interphase momentum exchange coefficient is expressed as K_{jk} $(= K_{kj})$.

In this current work, the dispersed phases are solid, and the gas phase and the liquid phase are continuous. The following expressions address interphase drag force with the phases.

Interphase drag force involving liquid and solid

$\boldsymbol{F}_{D,Ls}$ (interphase drag force involving liquid and solid) is defined by

$$F_{D,Ls} = K_{Ls}(u_L - u_s) \tag{8}$$

Gidaspow drag model (1994) [11] (a union of the Ergun equation with Wen and Yu model) involving the liquid and the solid phases represents the interphase exchange coefficient K_{Ls} and is given by:

$$K_{Ls} = \frac{150\varepsilon_s^2 \mu_L}{\varepsilon_L d_p^2} + \frac{1.75\varepsilon_s \rho_L |u_L - u_s|}{\varepsilon_L d_p}, \quad (\varepsilon_L \leq 0.8) \tag{9}$$

$$K_{Ls} = \frac{3}{4} C_D \varepsilon_s \varepsilon_L \rho_L \frac{(u_L - u_s)}{d_p} f(\varepsilon_L), \quad (\varepsilon_L > 0.8) \tag{10}$$

where Wen and Yu (1966) [12] suggested C_D as the drag coefficient and is represented as

$$C_D = 0.44, \quad \mathrm{Re}_p \geq 1000 \tag{11}$$

$$C_D = \frac{24}{\varepsilon_L \mathrm{Re}_p}\left(1 + 0.15(\varepsilon_L \mathrm{Re}_p)^{0.687}\right), \quad \mathrm{Re}_p \leq 1000 \tag{12}$$

where Reynolds number for the particle is explained as

$$\mathrm{Re}_p = \frac{\rho_L d_p |u_L - u_s|}{\mu_L} \tag{13}$$

$$f(\varepsilon_L) = \varepsilon_L^{-2.65} \tag{14}$$

Interphase drag force involving gas and liquid

$F_{D,gL}$(interphase drag force between gas and liquid) is specified by

$$F_{D,Lg} = K_{Lg}\left(u_L - u_g\right) \tag{15}$$

K_{Lg}: the liquid and the gas phase exchange coefficient. Bubbles formation is thought to arise in the secondary phase (gas phase). For the bubbly flow of the mixture of liquid and gas, the general expression for exchange coefficient is given by:

$$K_{Lg} = \frac{\varepsilon_L \varepsilon_g \rho_g f}{\tau_g} \tag{16}$$

where f: described as the drag function for various exchange-coefficient models distinctively.

τ_g: relaxation time of the particulate (bubble), described as;

$$\tau_g = \frac{\rho_g d_b^2}{18\mu_L} \tag{17}$$

where d_b: the bubbles diameter. Schiller and Naumann (1935) [13] proposed a drag model that is suitable for all pairs of fluid phases and is represented by

$$f = \frac{C_D \mathrm{Re}}{24} \tag{18}$$

where

$$C_D = 0.44, \quad \text{for}\, \mathrm{Re}_b \geq 1000 \tag{19}$$

$$C_D = \frac{24}{\mathrm{Re}_p}\left(1 + 0.15(\mathrm{Re}_b)^{0.687}\right), \quad \text{for}\, \mathrm{Re}_b \leq 1000 \tag{20}$$

the bubble Reynolds number (Re_b) is:

$$\mathrm{Re}_b = \frac{\rho_L |u_L - u_g| d_b}{\mu_L} \tag{21}$$

Interphase drag force between gas and solid

Since particles close to bubbles appear as the bubbles, the momentum exchange gas and solid dispersed phases must be considered for CFD simulation of multiphase flows [14, 15]. The drag force between bubbles and solid particles can be modeled similarly to the drag force between dispersed and continuous phases in our simulation since the two dispersed phases are considered continuous [15–17]. The expression for the interphase drag force involving gas and solid ($F_{D,gs}$) is

$$F_{D,gs} = K_{gs}\left(u_g - u_s\right) \tag{22}$$

$$K_{gs} = \frac{150\varepsilon_s^2 \mu_g}{\varepsilon_g d_p^2} + \frac{1.75\varepsilon_s \rho_g \left|u_g - u_s\right|}{\varepsilon_g d_p}, \quad (\varepsilon_g \leq 0.8) \tag{23}$$

$$K_{gs} = \frac{3}{4} C_D \varepsilon_s \varepsilon_g \rho_g \frac{(u_g - u_s)}{d_p}, \quad (\varepsilon_g > 0.8) \tag{24}$$

$$C_D = 0.44, \quad \mathrm{Re}_p \geq 1000 \tag{25}$$

$$C_D = \frac{24}{\varepsilon_g \mathrm{Re}_p}\left(1 + 0.15(\varepsilon_g \mathrm{Re}_p)^{0.687}\right), \quad \mathrm{Re}_p \leq 1000 \tag{26}$$

where the Reynolds number for a particle is described as

$$\mathrm{Re}_p = \frac{\rho_g d_p |u_g - u_s|}{\mu_g} \tag{27}$$

2.3 Pressure Due to Solids

Normal stresses arising from particle–particle interactions induce the solid phase pressure gradient and become more prominent as the solid phase fraction achieves the highest packing. In this current work, the KTGF model (where a granular temperature and Maxwellian velocity distribution for particles is imported) has been applied and shown in the solid pressure and viscosity expressions. The solids pressure comprises of a term considering particle collisions and a kinetic term:

$$p_s = 2\rho_s(1 + e_{ss})\varepsilon_s^2 g_{0,ss}\Theta_s + \varepsilon_s\rho_s\Theta_s \tag{28}$$

where e_{ss}: the restitution coefficient for particle collisions, 0.9 (the default value in Fluent, used in this work).

$g_{0,ss}$: A correction factor is used for changing the collisions probability among grains for dense solid granular phase (radial distribution function) and is represented as;

$$g_0 = \left[1 - \left(\frac{\varepsilon_s}{\varepsilon_{s,\max}}\right)^{\frac{1}{3}}\right]^{-1} \tag{29}$$

and Θ_s: granular temperature. The granular temperature is worked out from the transport equation and proportionate to the kinetic energy due to fluctuating particle motion and is defined by;

$$\frac{3}{2}\left[\frac{\partial}{\partial t}(\rho_s\varepsilon_s\Theta_s) + \nabla.(\rho_s\varepsilon_s u_s\Theta_s)\right] = (-p_s\mathbf{I} + \boldsymbol{\tau}_s) : \nabla u_s + \nabla.\left(k_{\Theta_s}\nabla\Theta_s\right) - \gamma_{\Theta_s} + \varphi_{ks} \tag{30}$$

where $k_{\Theta_s}\nabla\Theta_s$: the diffusion energy.

$(-p_s\mathbf{I} + \boldsymbol{\tau}_s) : \nabla u_s =$ the energy released by stress tensor of solid.

k_{Θ_s}: the diffusion coefficient and is provided by;

$$k_{\Theta_s} = \frac{15 d_s\rho_s\varepsilon_s\sqrt{\Theta_s\pi}}{4(41-\eta)}\left[1 + \frac{12}{5}\eta^2(4\eta - 3)\varepsilon_s g_{0,ss} + \frac{16}{5\pi}(41 - 33\eta)\eta\varepsilon_s g_{0,ss}\right] \tag{31}$$

where $\eta = \frac{1}{2}(1 + e_{ss})$.

γ_{Θ_s} = the collision dissipation energy (energy dissipation rate within the solids phase on account of particles collisions). This term is derived by Lun et al. (1984) [18], and the expression obtained as;

$$\gamma_{\Theta_s} = \frac{12\left(1 - e_{ss}^2\right) g_{0,ss}}{d_s \sqrt{\pi}} \rho_s \varepsilon_s^2 \Theta_s^{3/2} \tag{32}$$

$\varphi_{ks} = -3K_{ks}\Theta_s$ = the kth fluid phase and the solid phase energy exchange.

2.4 Closure Laws for Turbulence

This experiment uses the *k*-ε dispersed turbulence model because it is a valid and potent model for continuous phase as primary and the remaining secondary phases as dispersed. Applying the *k*-ε dispersed turbulence model, the subsequent equations are utilized to measure continuous phase's turbulence. The eddy viscosity model is employed for the computation of averaged fluctuating parameters. For the liquid phase (continuous), the Reynolds stress tensor is represented below:

$$\tau''_L = -\frac{2}{3}(\rho_L k_L + \rho_L \mu_{t,L} \nabla.\mathbf{U}_L)\mathbf{I} + \rho_L \mu_{t,L}\left(\nabla \mathbf{U}_L + \nabla \mathbf{U}_L^T\right) \tag{33}$$

where $\mathbf{U}_L$: the phase-weighted velocity.

$\mu_{t,L}$: The turbulent viscosity, expressed in connection with the turbulent kinetic energy and explained as:

$$\mu_{t,L} = \rho_L C_\mu \frac{k_L^2}{\varepsilon_L} \tag{34}$$

where $C_\mu = 0.09$, ε_L: the dissipation rate.

The turbulence dissipation rate ε_L and turbulence kinetic energy k_L values are acquired straightly from the various transport equations, which are defined by:

$$\begin{aligned} &\frac{\partial}{\partial t}(\rho_L \varepsilon_L \boldsymbol{\varepsilon}_L) + \nabla.(\rho_L \varepsilon_L \mathbf{U}_L \boldsymbol{\varepsilon}_L) \\ &= \nabla.\left(\varepsilon_L \frac{\mu_{t,L}}{\sigma_\varepsilon} \nabla \boldsymbol{\varepsilon}_L\right) + \varepsilon_L \frac{\boldsymbol{\varepsilon}_L}{k_L}(C_{1\varepsilon} G_{k,L} - C_{2\varepsilon} \rho_l \boldsymbol{\varepsilon}_L) + \rho_L \varepsilon_L \Pi_{\varepsilon_L} \end{aligned} \tag{35}$$

$$\begin{aligned} &\frac{\partial}{\partial t}(\rho_L \varepsilon_L k_L) + \nabla.(\rho_L \varepsilon_L \mathbf{U}_L k_L) \\ &= \nabla.\left(\varepsilon_L \frac{\mu_{t,L}}{\sigma_k} \nabla k_L\right) + \varepsilon_L G_{k,L} + \rho_l \varepsilon_L \boldsymbol{\varepsilon}_L + \rho_L \varepsilon_L \Pi_{KL} \end{aligned} \tag{36}$$

Here,$G_{k,L}$: the turbulent kinetic energy production.

Π_{ε_L} and Π_{k_L}: the impact of the continuous liquid phase on the dispersed phases(solid and gas).

The term Π_{k_L} is obtained from the continuous phase's instantaneous equation and has the following formula,

$$\Pi_{k_L} = \sum_{j=1}^{J} \frac{K_{jL}}{\varepsilon_L \rho_L} \left(k_{jL} - 2k_L + u_{jL}.u_{dr}\right) \tag{37}$$

where J: number of secondary phases:

k_{jL}: the velocity covariance of the dispersed phase j and the continuous phase (liquid),
u_{dr}: the drift velocity.
u_{jL}: the relative velocity,
Π_{ε_L} is represented as,

$$\Pi_{\varepsilon_L} = C_{3\varepsilon} \frac{\varepsilon_L}{k_L} \Pi_{k_L} \tag{38}$$

where $C_{3\varepsilon} = 1.2$

The multiphase flow turbulent drag term $(K_{jk}(u_j - u_k))$ is designed for the dispersed (solid and gas) phase and the liquid phase (continuous) and represented as:

$$K_{jL}\left(u_j - u_L\right) = K_{jL}\left(\mathbf{U}_j - \mathbf{U}_L\right) - K_{jL} u_{dr} \tag{39}$$

The drift velocity is explained as:

$$u_{dr} = \frac{D_j}{\sigma_{jL}\varepsilon_j} \nabla \varepsilon_j - \frac{D_L}{\sigma_{jL}\varepsilon_L} \nabla \varepsilon_L \tag{40}$$

Here σ_{jL}: dispersion Prandtl number (0.75), D_j and D_L: diffusivities.

3 Experimentation

The experimental arrangement used is depicted in Fig. 1. Table 1 shows the scope of the experiment. The flow of air and water is in an upward direction and co-current. A precise quantity of material is weighed and fed into the column, which is then regulated to a given initial static bed height. Using a calibrated rotameter, liquid (water) is transported at the desired flow rate to the fluidizer. The air sparger is then

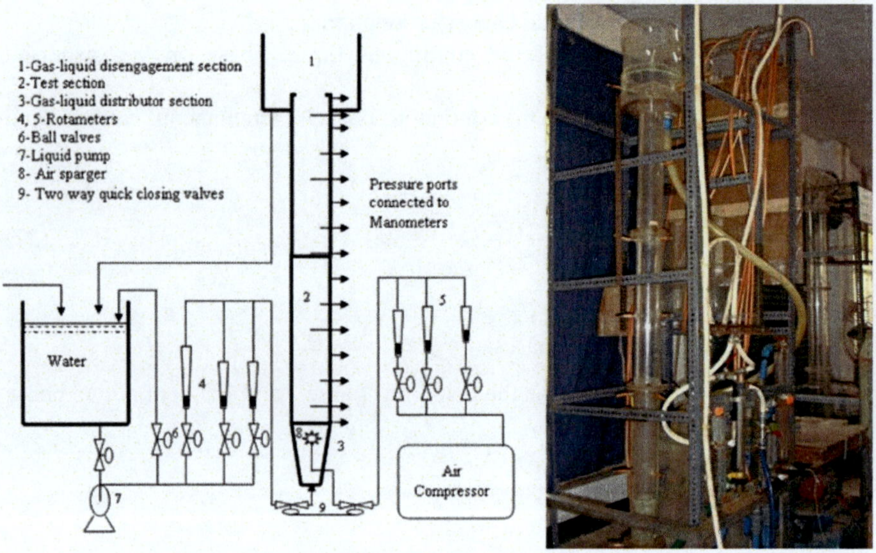

Fig. 1 Schematic and camera-friendly view of the experimental unit

Table 1 Explanation of the system applied in the experiment and simulation

Column diameter:	0.1 m	Gas phase (air), 30 °C	
Column height:	1.88 m	Density:	1.166 kg/m^3
Solid phase (glass beads):		Viscosity:	1.794 × 10^{-5} Pas
Particle size:	2.18 mm	Superficial gas velocity, U_g:	0 to 0.1019 m/s
Particle density:	2470 kg/m^3	Liquid phase (water), 30^{0}C	
Initial static bed height, H_s:	0.171 m, 0.213 m	Density:	995.7 kg/m^3
Static bed voidage:	0.41	Viscosity:	7.98 × 10^{-4} Pas
Bed inventory:	1.965 kg, 2.450 kg	Superficial liquid velocity, U_l:	0.004246 to 0.1746 m/s

used to introduce air into the column at the desired flow rate. It is given around five minutes to achieve a steady state. The expanded bed heights and the pressure drop readings are then recorded. All experiments are conducted at a temperature of (30 ± 5) °C. The gas holdup is measured by closing the quickly closing valves (9, Fig. 1 (Schematic).) in the air and water lines at the same time. To measure the two-phase fractional gas holdup, freeboard trials with a broad range of liquid and gas flow are performed first, using the equation:

$$\varepsilon_g = \frac{H - H_l}{H} \tag{41}$$

where H_l: liquid height in the column following gas runoff (m) and H: test section total height (m).

For the fluidization experimentation with particles, the measurement of the gas holdup is done similarly. By deducting the gas holdup in the two-phase area from the three-phase zone, the three-phase zone's gas holdup is measured. The two-phase area was located above the expanded bed. At various static bed heights, the experiment is replicated for various gas and liquid velocities.

4 Numerical Methodology

The commercialized CFD software package ANSYS 15.0 is utilized to resolve the model equations mentioned above. For this simulation, a Plexiglas column (cylindrical shape, diameter: 0.1 m, height: 1.88 m) has been used as the fluidized bed reactor, as revealed in Fig. 1. The numerical mesh used in this simulation is shown in Fig. 2. The finite volume method is used for discretizing the governing equations, and a high-resolution discretization method is applied for spatial discretization, which ensures precision and stability. A backward Euler scheme of second order is applied to discretize the governing equations concerning time. In ANSYS 15.0, sophisticated algebraic multigrid solver (AMG) technology is applied to solve the discretized equations.

4.1 Geometry and Mesh

Figure 2 shows a two-dimensional computational geometry developed using the commercial software ANSYS 15.0 and a top-down approach (face primitive—rectangle), similar to that of the fluidization column. Following the geometry development, using map-structured quadrilateral elements, a uniform mesh has been created with a height:width = 1:1. A total of 7520 cells (size of 0.005 × 0.005 m) has been employed in the calculation.

Fig. 2 2D meshing of the fluidized column

4.2 Initial and Boundary Conditions

Appropriate boundary conditions for the computational domain must be enforced to achieve a well-defined system of equations. A consistent liquid and gas velocities are the inlet boundary conditions, and the pressure boundary condition (which is adjusted to 1.013×10^5 Pa) is the outlet boundary condition. For the gas phase and the solid phase, free-slip boundary conditions and the liquid phase, no-slip boundary conditions are maintained as wall boundary conditions.

The standard wall function method has been used for dealing with the more significant viscous effect and larger velocity gradient close to the wall. The solid volume fraction has been applied as the initial condition (i.e., 0.59 of the column's static bed height). The inventory determines the volume fraction of gas in the freeboard area and at the inlet. The initial and boundary conditions are revealed in Table 1.

4.3 Solution

For pressure–velocity coupling, the Phase Coupled SIMPLE method [19] has been used. The volume fraction equations are discretized by a first-order upwind scheme. In contrast, the turbulence kinetic energy, momentum equation, and turbulence dissipation rate are discretized by the second-order upwind scheme. The time step range is set to 0.001 s. The convergence criterion focuses on observing the mass flow residual for all numerical simulations, with $1.0e^{-04}$ as the converged value.

Pressure = 0.3, momentum = 0.2, body forces = 1, density = 1, volume fraction = 0.5, turbulent viscosity = 1, turbulent dissipation rate = 0.8, turbulent kinetic energy = 0.8, and granular temperature = 0.2 have been used as under relaxation factors for various flow quantities. The simulations have been run until the acquisition of a quasi-steady state of the system, which means that the average of flow variables does not depend on time, which is investigated by observing the phase volume fractions or expanded bed height. The time variation of the bed profile is shown in Fig. 3. The bed profile is nearly identical between 18 and 30 s of simulation time, as shown in this figure. The simulations lasted 60 s, and the analysis was based on the averages of the last 20 s. The averaged quantities are determined after the achievement of a fully developed quasi-steady state in terms of axial, and radial direction and time.

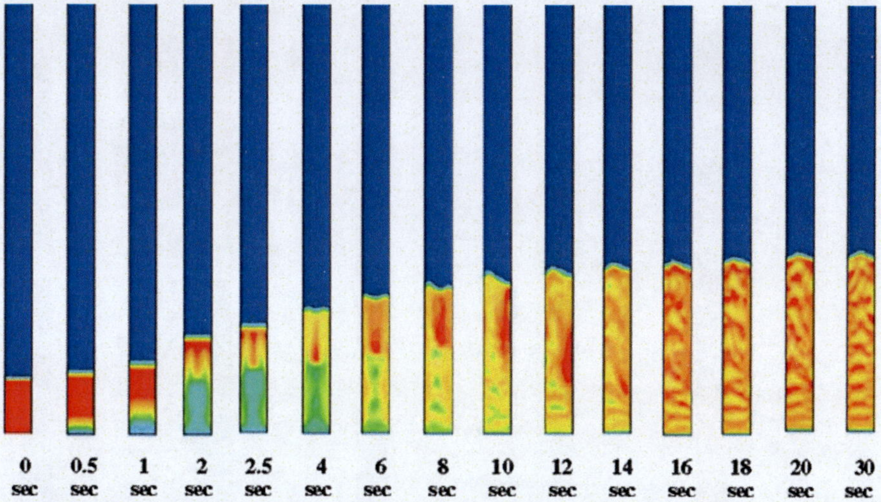

Fig. 3 For 0.213 m initial static bed height (H_s) and at an air velocity of 0.0125 m/s and water velocity of 0.12 m/s, volume fraction contours of glass beads (2.18 mm) with reference to the time

5 Results and Discussions

After attaining the quasi-steady state, the solid, liquid, and gas phase volume fraction contours in the column are attained at 0.0125 m/s air velocity, 0.12 m/s water velocity for 0.213 m initial static bed height, and 2.18 mm diameter glass beads and revealed in Fig. 4. The volume fraction value analogous to the color is given by the color scale shown on each contour's left side. The glass bead contours show that the bed is in a fluidized state. The water contour reveals that the liquid volume fraction in the fluidizing section is lesser than that of the two-phase area above it. The gas holdup in the bed's fluidizing section is shown by air contour and is substantially higher than in the two-phase area above.

From the CFD simulation with inlet air velocity 0.0125 and 0.12 m/s as inlet water velocity, it is obtained that the liquid phase axial velocity is distributed radially and is shown in the XY plot (Fig. 5). This means that the flow in the bed is fully developed. Water has an maximum velocity of around 0.14 m/s. The XY plot for the magnitude of gas phase velocity is depicted in Fig. 6. A fully developed flow with the minimum axial velocity at the wall and the highest axial velocity at the center line has been observed in this plot. Velocity of air bubbles is not zero at the wall because of the gas's free-slip boundary condition used in simulation. Air has an highest velocity of around 0.48 m/s at center.

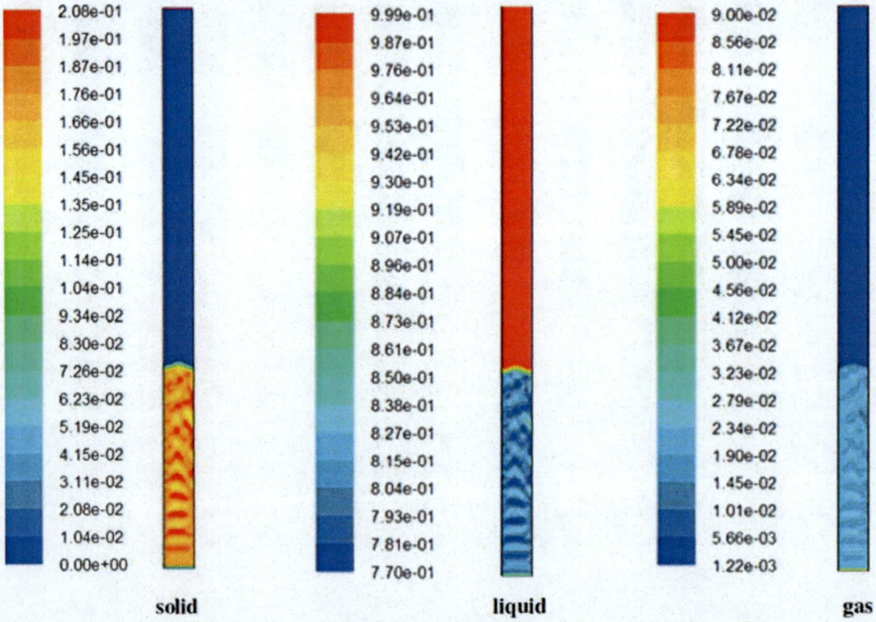

Fig. 4 For 0.213 m the initial static bed height and at 0.0125 m/s air velocity and 0.12 m/s water velocity, volume fraction contours of solid, liquid, and gas

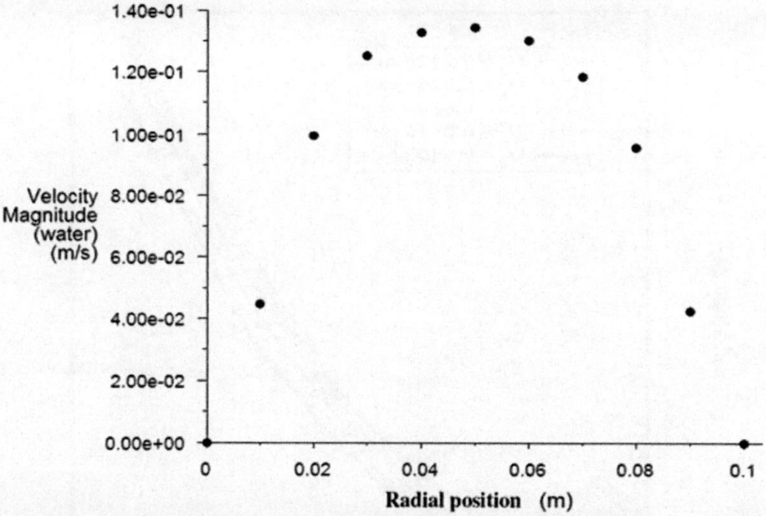

Fig. 5 XY plot for the liquid phase (water) velocity

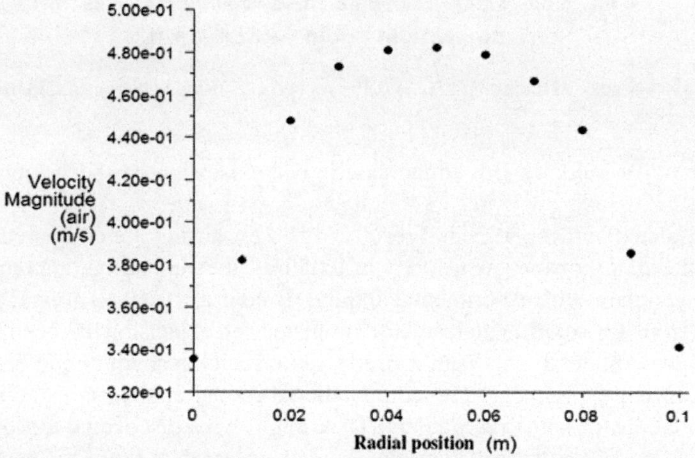

Fig. 6 XY plot for the air velocity

5.1 Bed Expansion (Bed Voidage)

The expanded bed height rises with increased liquid velocity in a three-phase system with constant gas velocity, increasing the bed voidage. Similarly, at constant gas velocity, the CFD simulation outcome also indicates a rise in bed expansion with liquid velocity. The bed expansion vs. liquid inlet velocity plot attained at various gas inlet velocity values is shown in Fig. 7. The increased gas velocity causes a slight decrement in the expanded bed height because of the contraction of the bed

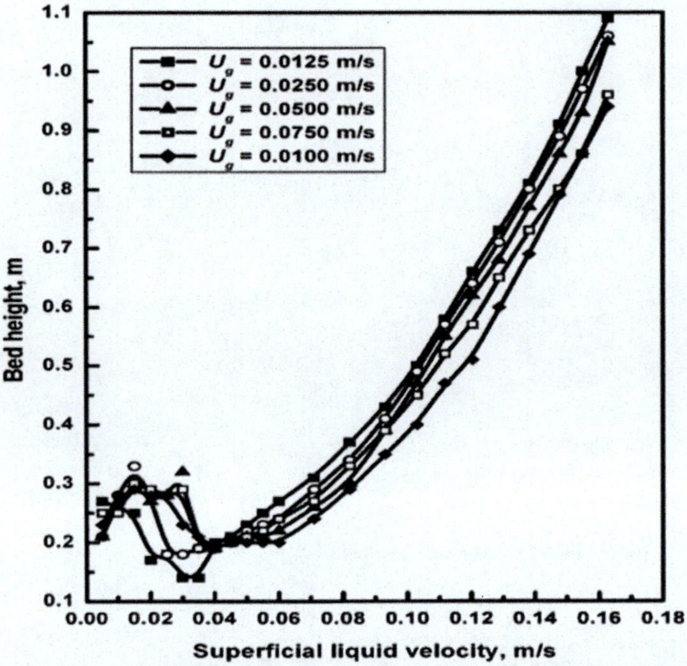

Fig. 7 For glass beads (2.18 mm) and $H_s = 0.171$ m, bed expansion behavior of CFD simulation

with a rise in gas velocity. But an increase in bed expansion has been found in our experiment.

The simulated and experimental results of bed expansion are compared and are depicted in Fig. 8. For the gas velocity of 0.05 m/s, the simulation outcomes show excellent agreement with experimental findings. In contrast, for 0.10 m/s gas velocity, it deviates from the experimental, and the simulated bed expansion has been less than that of the experimental one. With a rise in gas velocity, a small rise in bed height is noticed in the experiment. The computational model applied is established by recommendations of some researchers. They might have discovered a drop in bed height as gas velocity increased. Particles diameter is less than 1 mm and experiences actual bed contraction. For conditions beyond minimum fluidization, the increased bed height value from both experimental and simulation agrees inside the range of 10% in the bed expansion regime. Figures 7 and 8 show that the bed height is higher at lower liquid velocity, and then it declines and further rises as the liquid velocity rises. The larger gas bubbles indicate agitation in the bed that raises a few particles, showing the appearance of fluidization. At low liquid velocities, larger bubbles appear, and bubble size declines with the rise in the velocity of the liquid. The agitated bed height was not taken into consideration when calculating the bed height in experimentation. The agitated bed is seen in the CFD simulation result. In

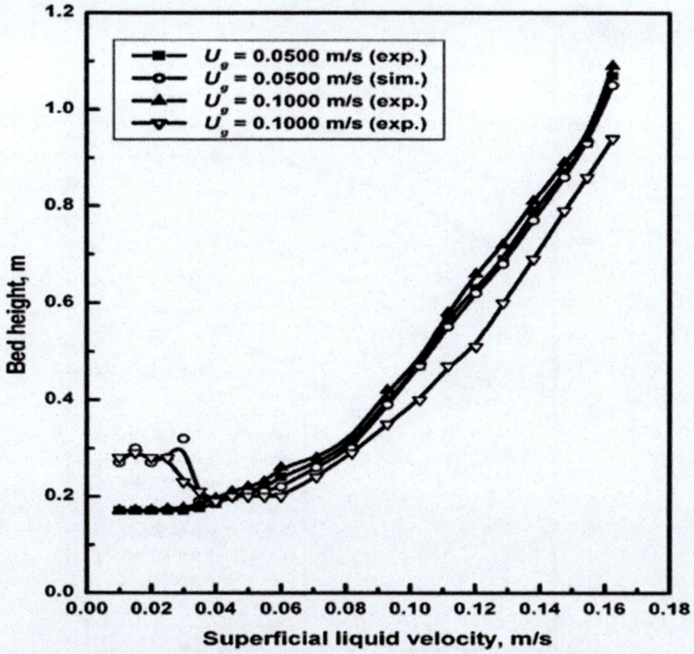

Fig. 8 Evaluation of bed height attained from experimental as well as CFD simulation

the vicinity of the minimum fluidization condition, the agitated bed disappeared in the CFD results.

5.2 *Bed Pressure Drop*

The variation of bed pressure drop with liquid velocity at various inlet air velocity values obtained from simulation is shown in Fig. 9. Rise in pressure drop with a rise in water velocity is observed. Rise in pressure drop with liquid velocity when the air velocity is low ($U_g = 0.0125$ m/s) is not significant. This is for the reason that the gas volume fraction does not change much with the liquid velocity and is low at low air velocity. However, the gas holdup is larger at elevated gas velocity and declines (and the liquid holdup rises) as the liquid velocity rises, which results in a rise in the pressure drop. The relationship between superficial gas velocity and bed pressure drop is revealed in Fig. 10. The influence velocity of liquid on bed pressure drop is also depicted in this plot. An assessment of the closeness of experimental and CFD simulation results of bed pressure drop values is illustrated in Fig. 11, a good agreement is observed.

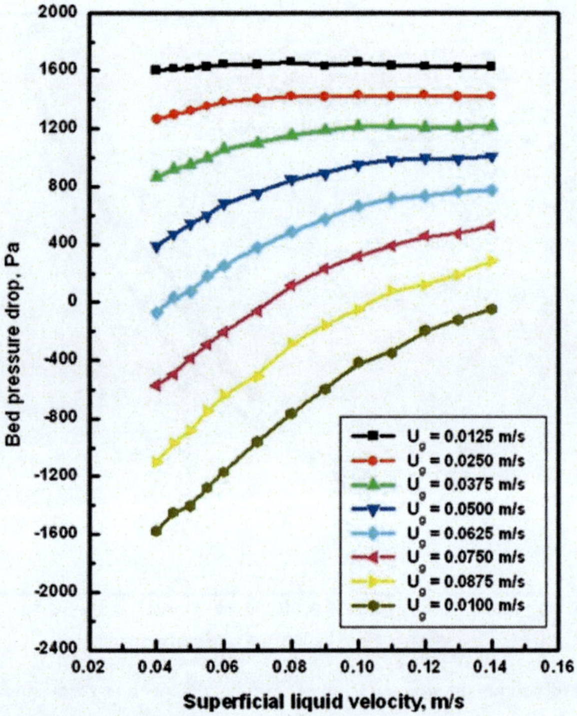

Fig. 9 For 2.18 mm glass beads and at $H_s = 0.171$ m, pressure drop change with the liquid velocity at distinct gas velocity values

5.3 *Gas Holdup*

Figure 12 depicts the effect of variation of superficial liquid velocity on gas holdup obtained from CFD simulation. The rise in liquid velocity induces a decline in the gas holdup, as seen in this plot. CFD simulation shows a steady decline in the gas holdup. Figure 13 shows that with increase in the gas velocity, the gas holdup increases in a monotonic manner. Validation of the gas holdup values with the experiment ones of those obtained from CFD simulation is shown in Fig. 14. Closeness between the two is observed at low superficial water velocities, but they differ moderately at very high water velocities. As previously stated, increasing the liquid velocity results in a nearly constant decrease in the gas holdup.

6 Conclusions

The CFD simulation results for hydrodynamics of solid phase in terms of bed expansion and hydrodynamics of gas and liquid phase concerning phase velocities and

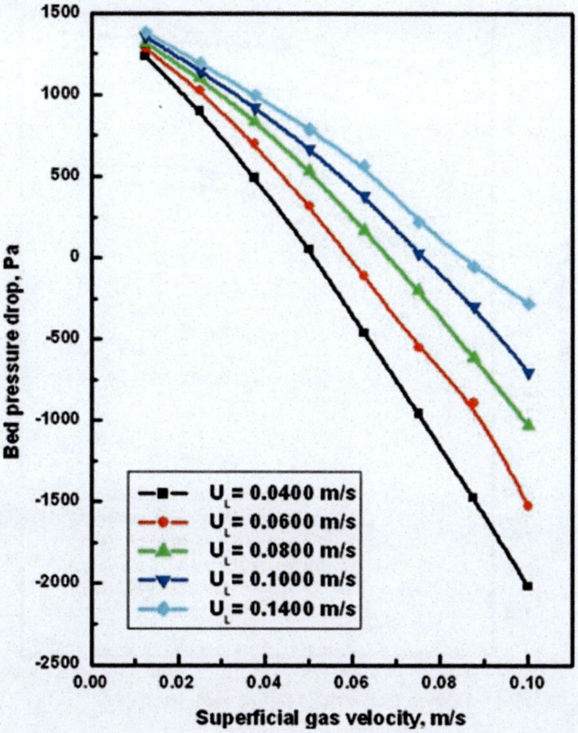

Fig. 10 For 2.18 mm glass beads and at $H_s = 0.171$ m, pressure drop change with the gas velocity at distinct liquid velocity values

holdup showed strong agreement with experimental ones. The bed pressure drop also agrees well. The bed pressure drop and bed expansion values signify that the CFD simulation's drag model accurately reflected the gas–liquid–solid phenomena. The change of bed expansion with gas velocity has only a partial disagreement between the CFD simulation and the experimental results. The experimental results revealed that bed expansion increases with the velocity of the gas, while the CFD simulation showed that bed expansion decreases slightly. For all of the operating conditions, the CFD simulation revealed a decisive solid circulation fashion consistent with previous researchers' findings. The gas holdup (0.1294) in the bed has been estimated with 0.04 m/s liquid velocity and 0.10 m/s gas velocity. At lower liquid velocity ranges, the gas holdup value is very similar, while at higher liquid velocity ranges, the values varied by less than 20%. The validation shows that the Eulerian-Eulerian multiphase granular flow method can predict a gas–liquid–solid fluidized bed's overall performance.

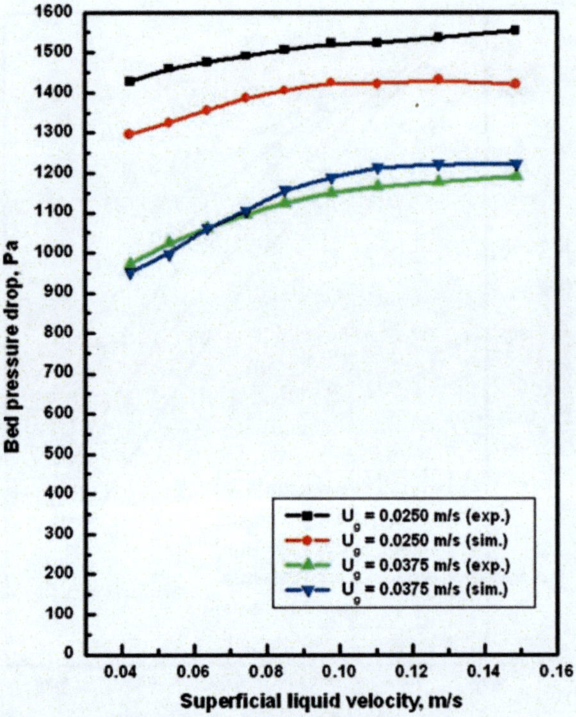

Fig. 11 Bed pressure drop comparison with experimental results

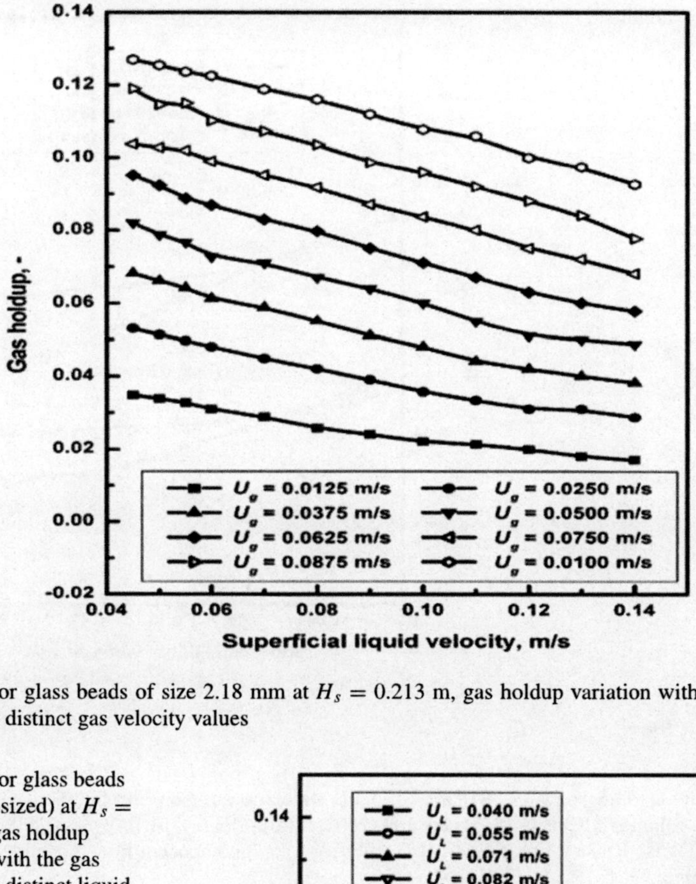

Fig. 12 For glass beads of size 2.18 mm at $H_s = 0.213$ m, gas holdup variation with the liquid velocity at distinct gas velocity values

Fig. 13 For glass beads (2.18 mm-sized) at $H_s = 0.171$ m, gas holdup variation with the gas velocity at distinct liquid velocity values

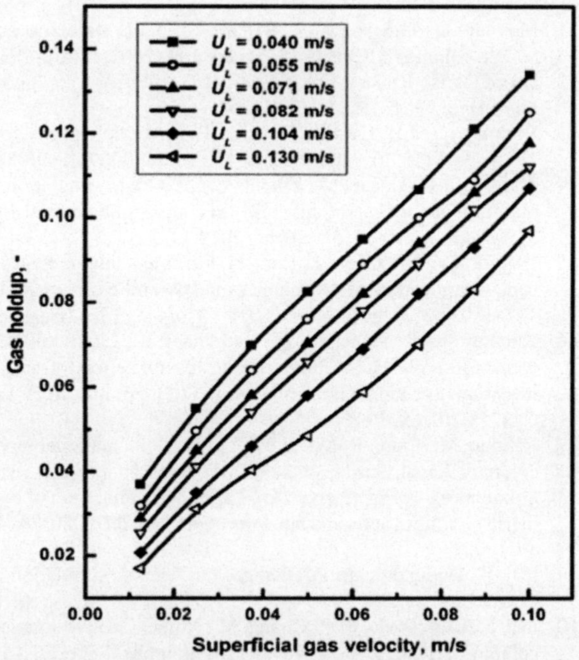

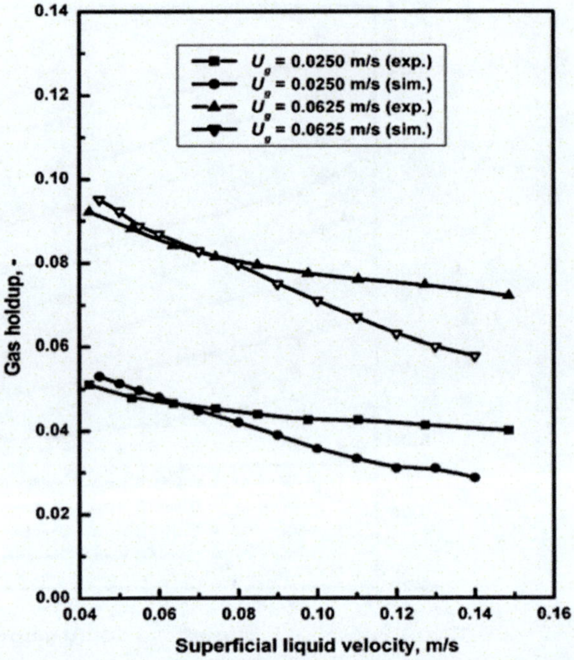

Fig. 14 Gas holdup comparison with experimental

References

1. Matonis, D., Gidaspow, D., Bahary, M.: CFD simulation of flow and turbulence in a slurry bubble column. AIChE J. **48**, 1413–1429 (2002). https://doi.org/10.1002/aic.690480706
2. Heard, W.B., Richter, G.R.: Numerical mutiphase, multicomponent flow modeling. Comput. Fluid Dyn. Chem. React. Eng. 13–18 (1996)
3. Dudukovic, M.P., Larachi, F., Mills, P.L.: Multiphase reactors—revisited. Chem. Eng. Sci. **54**, 1975–1995 (1999). https://doi.org/10.1016/S0009-2509(98)00367-4
4. Zhou, X., Ma, Y., Liu, M., Zhang, Y.: CFD-PBM simulations on hydrodynamics and gas-liquid mass transfer in a gas-liquid-solid circulating fluidized bed. Powder Technol. **362**, 57–74 (2020). https://doi.org/10.1016/j.powtec.2019.11.060
5. Hamidipour, M., Chen, J., Larachi, F.: CFD study on hydrodynamics in three-phase fluidized beds—application of turbulence models and experimental validation. Chem. Eng. Sci. **78**, 167–180 (2012). https://doi.org/10.1016/j.ces.2012.05.016
6. Khongprom, P., Wanchan, W., Kamkham, K., Limtrakul, S.: CFD simulation of the hydrodynamics in three phase fluidized bed reactor: effect of particle properties. In: 2019 research, invention, and innovation congress (RI2C). pp. 1–6. IEEE (2019). https://doi.org/10.1109/RI2C48728.2019.8999951
7. Rafique, M., Chen, P., Duduković, M.P.: Computational modeling of gas-liquid flow in bubble columns. Rev. Chem. Eng. **20** (2004). https://doi.org/10.1515/REVCE.2004.20.3-4.225
8. Sokolichin, A., Eigenberger, G., Lapin, A.: Simulation of buoyancy driven bubbly flow: established simplifications and open questions. AIChE J. **50**, 24–45 (2004). https://doi.org/10.1002/aic.10003
9. Pan, Y., Dudukovic, M.P., Chang, M.: Dynamic simulation of bubbly flow in bubble columns. Chem. Eng. Sci. **54**, 2481–2489 (1999). https://doi.org/10.1016/S0009-2509(98)00453-9
10. Pan, Y., Dudukovic, M.P., Chang, M.: Numerical investigation of gas-driven flow in 2-D bubble columns. AIChE J. **46**, 434–449 (2000). https://doi.org/10.1002/aic.690460303

11. Gidaspow, D.: Multiphase Flow and Fluidization: Continuum and Kinetic Theory Descriptions. Elsevier (1994). https://doi.org/10.1016/C2009-0-21244-X
12. Wen, C.Y.: Mechanics of fluidization. Chem. Eng. Prog. Symp. Ser. 100–111 (1966)
13. Schiller, L.: A drag coefficient correlation. Zeit. Ver. Deutsch. Ing. **77**, 318–320 (1933)
14. Mitra-Majumdar, D., Farouk, B., Shah, Y.T.: Hydrodynamic modeling of three-phase flows through a vertical column. Chem. Eng. Sci. **52**, 4485–4497 (1997). https://doi.org/10.1016/S0009-2509(97)00293-5
15. Schallenberg, J., Enß, J.H., Hempel, D.C.: The important role of local dispersed phase hold-ups for the calculation of three-phase bubble columns. Chem. Eng. Sci. **60**, 6027–6033 (2005). https://doi.org/10.1016/j.ces.2005.02.017
16. Padial, N.T., VanderHeyden, W.B., Rauenzahn, R.M., Yarbro, S.L.: Three-dimensional simulation of a three-phase draft-tube bubble column. Chem. Eng. Sci. **55**, 3261–3273 (2000). https://doi.org/10.1016/S0009-2509(99)00587-4
17. Wang, F., Mao, Z.-S., Wang, Y., Yang, C.: Measurement of phase holdups in liquid–liquid–solid three-phase stirred tanks and CFD simulation. Chem. Eng. Sci. **61**, 7535–7550 (2006). https://doi.org/10.1016/j.ces.2006.08.046
18. Lun, C.K.K., Savage, S.B., Jeffrey, D.J., Chepurniy, N.: Kinetic theories for granular flow: inelastic particles in Couette flow and slightly inelastic particles in a general flowfield. J. Fluid Mech. **140**, 223–256 (1984). https://doi.org/10.1017/S0022112084000586
19. Patankar, S.V.: Numerical Heat Transfer and Fluid Flow. CRC Press (2018).https://doi.org/10.1201/9781482234213

Effect of Contact Angle on Droplet Generation in a T-Junction Microfluidic System

Akepogu Venkateshwarlu and Ram Prakash Bharti

Abstract Microfluidics is an emerging field in the current research world due to its potential applications. A numerical study has been performed to understand the role of contact angle on the droplet generation in a T-shaped cross-flow microfluidic system using the level set method (LSM) and finite element method (FEM). The two-phase fluid system is assumed to be an immiscible, incompressible, and equal density. The present results are compared with published experimental and numerical results, and it is found that there is a good agreement between them. The two-phase flow and droplet generation behavior as a function of contact angle, Reynolds number, and the capillary number of the continuous phase has been studied for a wide range of conditions. The droplet is not able to maintain a proper shape for the contact angle below 90°, i.e., when the surface is hydrophilic. Hence, both the fore and aft radius of the droplet are concave. However, for the higher contact angle ($>90°$), a precise, distinguished droplet is formed, and it is shaped convex on the fore and aft. The size of the droplet is found to be proportional to the contact angle when the surface is superhydrophobic. It is also observed that the droplet length is decreasing by increasing the capillary number of the continuous phase as a function of Reynolds number. The inertial term has a significant effect on the droplet dynamics. The breakup mechanism of the droplet has been elucidated by plotting the pressure evolution at different locations near the T-junction.

Keywords Microfluidics · Contact angle · Droplet generation · Level set method · Finite element method

---

A. Venkateshwarlu · R. P. Bharti (✉)
Complex Fluid Dynamics and Microfluidics (CFDM) Lab, Department of Chemical Engineering, Indian Institute of Technology Roorkee, Roorkee 247667, India
e-mail: rpbharti@iitr.ac.in

R. P. Bharti and K. M. Gangawane (eds.), *Recent Trends in Fluid Dynamics Research*, Lecture Notes in Mechanical Engineering,
https://doi.org/10.1007/978-981-16-6928-6_11

1 Introduction

The study of the droplet generation and dynamics in a microchannel is drawing much attention in the current research domain due to its numerous applications in many fields like material science, pharmaceutical, food, biochemical, crystallization of proteins, synthesis of nanoparticles [1–4]. Various geometrical configurations like flow-focussing, cross-flow, and parallel flow are generally used for the droplet generation in the microchannels. Among others, T-type cross-flow geometry is widely used for the droplet generation mainly due to its ease of operation and control of the flow conditions. In T-type geometry, the continuous phase (CP) flows through the horizontal channel and the dispersed phase (DP) in the vertical (i.e., T-leg) of the channel. The dispersed phase invades into the main channel and interacts with the continuous phase. This enhances the pressure in the upstream of CP, and the developed pressure subsequently assists DP to flow toward the outlet. The droplet formation depends upon the local pressure evolution in the channel during the process. However, experimental measurements of the pressure fluctuations with high accuracy during the droplet formation are extremely difficult and expensive to install the sensors at a specific location in the microfluidic device [5, 6]. The numerical simulation is an alternative way to overcome those challenges and meet demands timely. While significant knowledge is available on the hydrodynamics of droplet generation, understanding the local pressure fluctuations and their sensitivity to the droplet formation is still infancy [4, 7–9]. Therefore, it is essential to carry out a detailed study of the pressure evolution in the microfluidic devices to ease the control and manipulation of the droplet formation. Uncovered findings are reported, in the present numerical study, for a wide range of contact angle (θ) as it plays an important role for the droplet to attain a specific shape and size. For instance, the droplet takes a round/spherical shape when $\theta = 180°$. The forces responsible for the droplet formation are explained in terms of dimensionless numbers, namely capillary number ($\text{Ca} = \mu_c u_c/\sigma$) and the Reynolds number ($\text{Re} = u_c \rho_c w_c/\mu_c$).

2 Problem Statement

Consider the incompressible, laminar, and two-dimensional flow of two immiscible fluids through a T-type microfluidic device, as seen in Fig. 1.

The device consists of the vertical channel (width w_d μm and length L_s μm) perpendicular to the main channel (width w_c μm and length L_c μm) located at upstream distance (L_u) and the downstream distance (L_d). The continuous and dispersed phases (CP & DP) are injected at flow rate (in μL/s) of Q_c and Q_d through inlets of the main and vertical channels. The outlet of the channel is open to ambient. Both Newtonian liquid phases of similar density but different viscosity meet at the junction point and flow downstream toward the outlet. Depending on the flow and fluid properties, the droplet formation may take place in the downstream of the main channel.

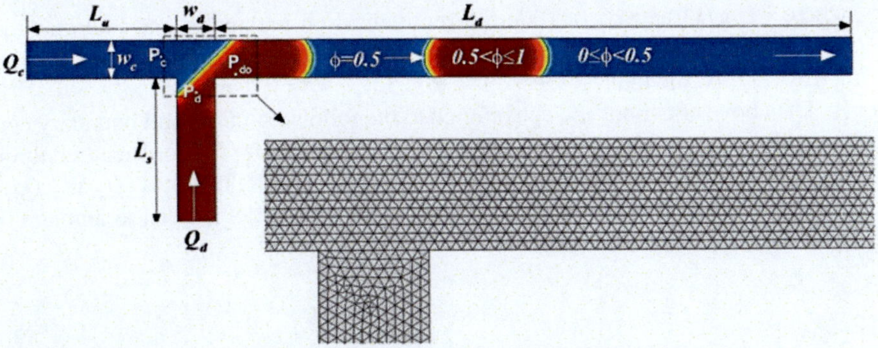

Fig. 1 Schematics of T-junction microchannel

2.1 Governing Equations and Boundary Conditions

Considered flow physics is represented by the following governing equations.

$$\nabla \cdot \boldsymbol{u} = 0 \tag{1}$$

$$\rho(\phi)\left[\frac{\partial \boldsymbol{u}}{\partial t} + \boldsymbol{u} \cdot \nabla \boldsymbol{u}\right] = -\nabla p + \nabla \cdot \boldsymbol{\tau} + \boldsymbol{F_\sigma} \tag{2}$$

where ρ is the density, μ is the dynamic viscosity, $\boldsymbol{u}$ is the velocity vector, $\boldsymbol{\tau}$ is stress field, and p is the pressure field, respectively. The interfacial force ($\boldsymbol{F_\sigma}$) has been adopted from the continuum surface force model (CSFM) [11] as follows.

$$\boldsymbol{F_\sigma} = \sigma\kappa\delta\boldsymbol{n} \tag{3}$$

where interface normal ($\boldsymbol{n}$) is defined as: $\boldsymbol{n} = \nabla\phi/|\nabla\phi|$, the curvature of the interface, $\kappa = 1/R = -\nabla.\boldsymbol{n}$, and δ is the Dirac delta function, respectively. The density (ρ) and viscosity (μ) at any point are given as:

$$\rho(\phi) = \rho_c + (\rho_d - \rho_c)\phi \quad \text{and} \quad \mu(\phi) = \mu_c + (\mu_d - \mu_c)\phi \tag{4}$$

where the subscripts 'c' and 'd' denote the continuous and dispersed phases, respectively.

To understand the droplet dynamics, the evolution of interface between the two immiscible fluids is essential. Therefore, an additional equation is required to track the topological changes and movement of the interface during the flow. Hence, the level set equation [12] has been adopted to capture the interface changes as follows.

$$\frac{\partial \phi}{\partial t} + \boldsymbol{u} \cdot \nabla\phi = \gamma\nabla \cdot [\epsilon_{ls}\nabla\phi - \phi(1-\phi)\boldsymbol{n}] \tag{5}$$

where γ (m/s) and $\epsilon_{\rm ls}$ (m) are the reinitialization parameter and interface thickness control parameter. The level set variable (ϕ) ranges from 0 (for CP) $< \phi <$ 1 (for DP). At the interface, ϕ equals to 0.5.

The above equations are constrained to the following initial and boundary conditions: The pure fluid phases, i.e., $\phi = 0$ (for CP) and 1 (for DP), are entered through the respective inlets during the simulations. The constant flow rates ($Q_{\rm c}$ and $Q_{\rm d}$) are imposed at the inlets of the immiscible phases. The outlet is open to ambient (i.e., $p = 0$ atm). The channel walls are taken to solid (i.e., $\boldsymbol{u} = 0$).

2.2 Solution Approach and Numerical Parameters

The above mathematical model has been solved using the finite element method (FEM)-based COMSOL Multiphysics. The numerical simulations are performed with the optimized [9] geometrical parameters ($w_{\rm c} = w_{\rm d} = 100$ µm, $L_{\rm u} = 9w_{\rm c}$, $L_{\rm d} = 30w_{\rm c}$, $L_{\rm c} = 40w_{\rm c}$, and $L_{\rm s} = 9w_{\rm c}$) for the following physical parameters: contact angle as $30° \leq \theta \leq 180°$, interfacial tension of $\sigma = 5$ mN/m, density of fluids as $\rho_c = \rho_d = 1000$ kg/m^3, flow rate of dispersed phase as $Q_{\rm d} = 0.2$ ml/h, and viscosity of dispersed phase as $\mu_{\rm d} = 0.001$ Pa.s by using the time step (Δt) of 10^{-4} s and non-uniform triangular mesh consisting of 13766 elements with maximum element size of 10 μm. The non-uniform mesh structure is obtained after the mesh independence study [9]. The level set parameters are taken as $\gamma = 1$ m/s and $\epsilon_{\rm ls} = 5$ µm.

3 Results and Discussion

In this work, the numerical results in terms of phase distribution and pressure evolution profiles, droplet length and diameters are presented for the following ranges of conditions: capillary number ($10^{-4} \leq {\rm Ca_c} \leq 1/10$), Reynolds number ($1/10 \leq {\rm Re_c} \leq 1$), and contact angle ($30° \leq \theta \leq 180°$).

The length of the droplet (L) is measured by plotting the level set variable (ϕ) along the length of the channel. The effective droplet diameter ($d_{\rm eff}$) is determined [8, 13, 14] using the following relation.

$$d_{\rm eff} = 2\left[\frac{1}{\pi}\int_{\Omega}(\phi > 0.5)d\Omega\right]^{1/2} \qquad (6)$$

Before presenting new results, the numerical approach is validated in the subsequent section.

3.1 Validation of Results

The present results have been validated with the experimental and numerical results [10] for the effective droplet diameter (d_{eff}) at $Q_{\text{d}} = 0.2$ ml/h and $\theta = 135°$ over the wide range of Q_{c} in Fig. 2. The variation of effective droplet diameter is seen to be consistent and have shown satisfactory agreement with both experimental and numerical results [10]. The comparison shows that level set method (LSM)-based present results based on highly refined mesh have shown higher accuracy over the lattice Boltzmann method (LBM)-based literature results. A thorough validation has been presented in our previous work [9]. The subsequent results presented are believed to have excellent ($\pm 1 - 2\%$) accuracy.

3.2 Instantaneous Phase Composition Distribution

The instantaneous phase composition distribution of immiscible liquids during the droplet formation is shown in Fig. 3 for $\text{Re}_{\text{c}} = 0.1$, $\text{Ca}_{\text{c}} = 10^{-4}$, and $\theta = 30° - 180°$. The formation of the droplet is classified into four stages. In the initial stage, the dispersed phase enters through the vertical channel into the main channel. In the second stage, the dispersed phase opposes the flow of the continuous phase and hence causes the pressure rise in the continuous phase upstream. When the pressure in the upstream of the continuous phase is sufficiently large enough, it starts exerting shear on the dispersed phase and eventually collapses the interface. On moving further, after detachment, the dispersed phase splits into two separate domains. It is

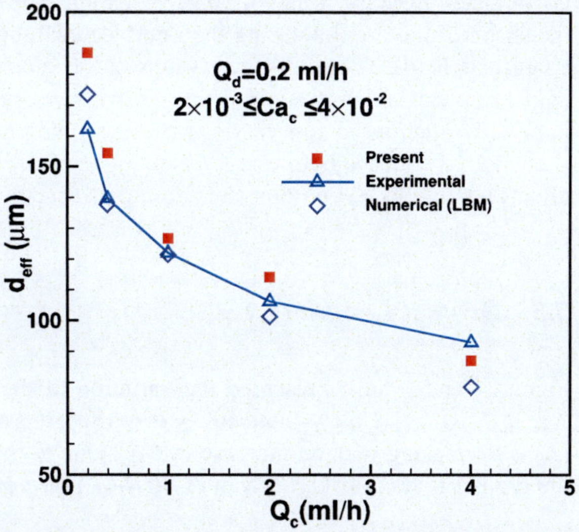

Fig. 2 Comparison of the present and literature [10] results

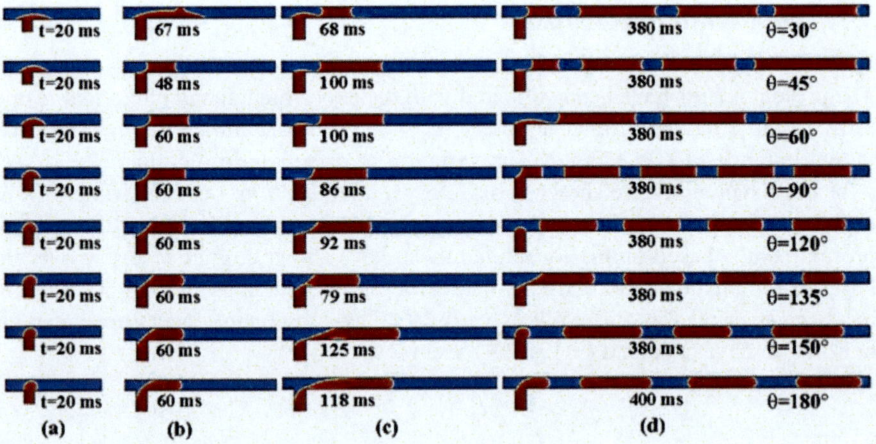

Fig. 3 Phase profiles during the droplet formation for $\mathrm{Re_c} = 0.1$ and $\mathrm{Ca_c} = 10^{-4}$ representing **a** initial stage, **b** growing stage, **c** breakup stage, and **d** channel filled with the droplets

observed that at lower values of $\theta = 30, 45°$, i.e., hydrophilic, the dispersed phase is completely wetting the channel walls. Hence, there is no clear droplet formation. For the larger value of $\theta = 60 - 90°$, the dispersed phase has less contact with the walls and consequently attains a concave shape with sharp edges. For the hydrophobic conditions ($90° < \theta < 150°$), the dispersed phase is forming a clear droplet with minimum contact and escapes from the walls. For the superhydrophobic conditions ($\theta > 150°$), the dispersed phase has negligible contact with walls. Thus, the droplet being formed is a round, clear, and monodispersed. The pressure difference (ΔP) between the inside and outside the droplet, i.e., the dispersed and continuous phase, is calculated from the Young–Laplace equation. The pressure difference that affects the shape of the droplet during the interface evolution is related with the dynamic contact angle (θ). The radius of curvature of the interface depends on the value of the contact angle whether to attain a concave or convex shape. For instance, the interface is initially attaining a convex shape, and as time progresses, it slowly changes its shape into a concave before its breakup. Therefore, the contact angle significantly affects the interface evolution and attaining a specific shape of the droplet.

3.3 *Droplet Length (L)*

Figures 4 and 5 have presented the variation of the droplet length (L) for $0.1 \leq Re_c \leq 1$, $10^{-4} \leq Ca_c \leq 1$ and $60° \leq \theta \leq 180°$. It is observed that the droplet length (L) is decreasing with an increase in Ca_c. Due to the inertial effects, the length of the droplet is less for higher values of Re_c. Therefore, the length of the droplet is

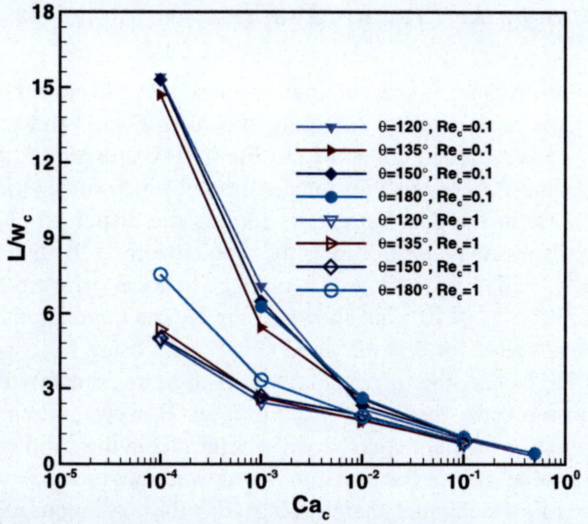

Fig. 4 Droplet length as a function of Ca_c for different Re_c and θ

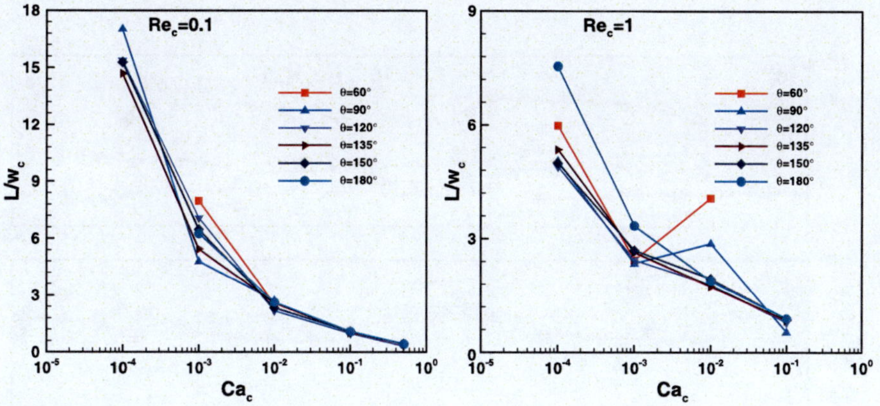

Fig. 5 Droplet length as a function of Ca_c and θ

inversely proportional to Re_c and Ca_c. However, the droplet size is increasing with contact angle (θ) for the same conditions.

For very low values of θ (< 90°), i.e., the hydrophilic conditions, the length of the droplet is randomly changing as it is having more contact with the solid walls, and the curvature is attaining a concave shape with sharp edges (Fig. 5). The droplet is attaining a smooth shape for the hydrophobic conditions as the curvature changes from the concave to convex.

3.4 Evolution of the Pressure Profiles

The evolutions of pressure in the continuous phase (P_c), dispersed phase (P_d), and downstream of the main channel (P_{do}) are shown in Figs. 6 and 7. The locations at which pressure is recorded are shown in Fig. 1. The pressure in the upstream of the continuous phase is responsible for the droplet pinch-off or the breakup. The developed pressure in the upstream starts forcing the dispersed phase toward the outlet after the dispersed phase occupies the whole width of the main channel. This particular stage is called squeezing or necking. The local pressure fluctuations for hydrophilicity ($30° \leq \theta \leq 90°$) are shown in Fig. 6. The fluctuations are insensitive in the continuous phase for $\theta = 30°$, and slowly increasing for $\theta = 45°$, and $60°$, as seen in Fig. 6a. Hence, the dispersed phase is in more contact with the wall and unable to attain a precise shape to form a droplet. However, when $\theta = 90°$, there is a droplet formation as the dispersed phase repels from the solid wall. During the initial stage, a gradual rise in the pressure is followed by sudden shoot-up when the dispersed phase fills the channel, thereby restricting the continuous phase flow. In the final stage, the developed pressure in the upstream of the continuous phase breaks the interface and forms two separate dispersed phase domains.

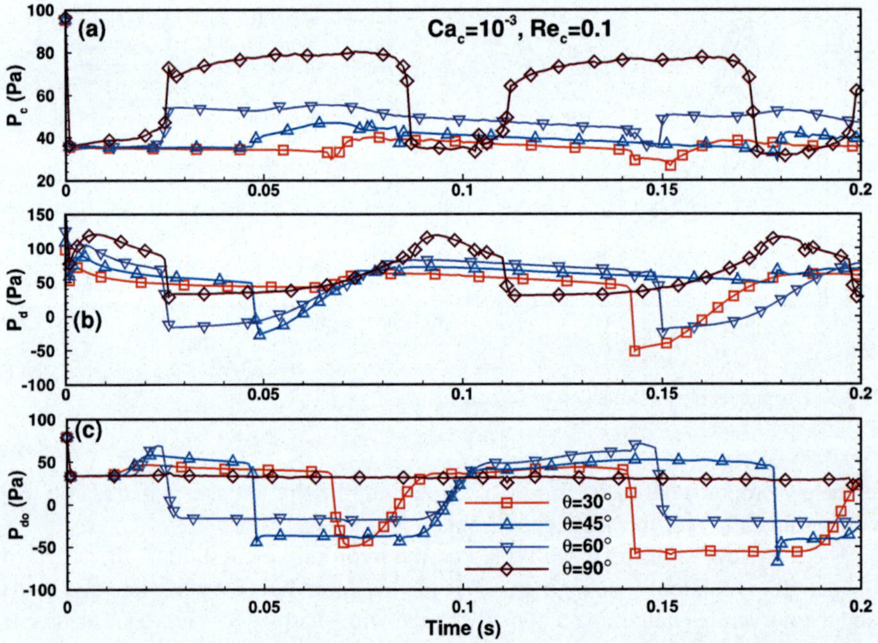

Fig. 6 Evolution of the pressure profiles as a function of contact angle ($30° \leq \theta \leq 90°$) for $Re_c = 0.1$ and $Ca_c = 10^{-3}$

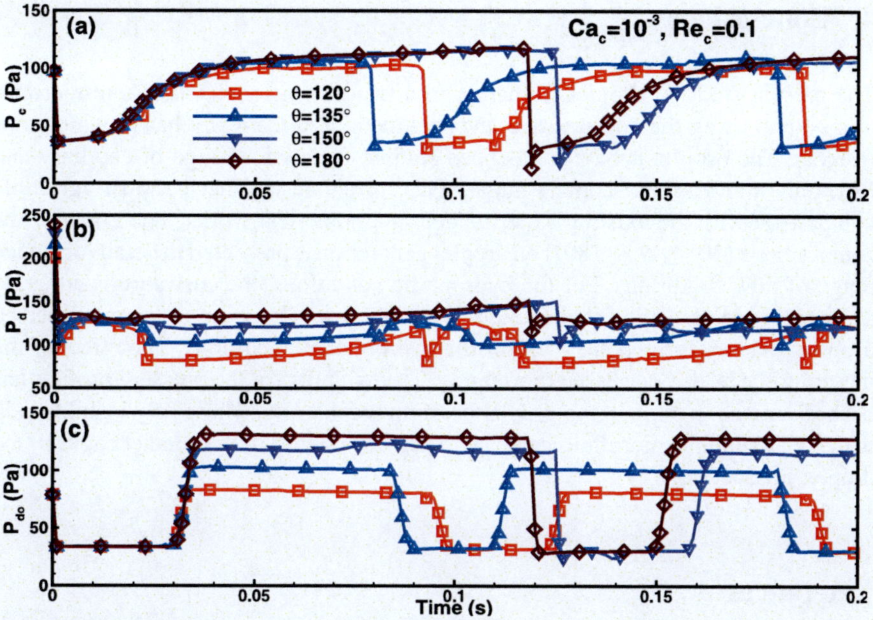

Fig. 7 Evolution of the pressure profiles as a function of contact angle ($120° \leq \theta \leq 180°$) for $\mathrm{Re_c} = 0.1$ and $\mathrm{Ca_c} = 10^{-3}$

Similarly, the sensitivity of the pressure fluctuations to hydrophobicity ($\theta > 90°$) is presented in Fig. 7. In the continuous phase of the main channel upstream, the pressure fluctuations remain the same in the entering stage for all θ (Fig. 7a). However, the dispersed phase starts experiencing the shear from the surrounding fluid once it restricts the flow of the continuous phase and thereby the pressure developed in the upstream of the channel. The squeezing time is less for hydrophobic ($90° < \theta < 150°$) surface and more for superhydrophobic ($150° < \theta < 180°$). It is also observed that the maximum value of the pressure is obtained for the superhydrophobic conditions. After hydrodynamic development, the process of droplet formation repeats and becomes cyclic. The fluctuations of pressure in the dispersed phase (P_d) depicted in Figs. 6b and 7b are small for $\theta = 30°$. However, the fluctuations are in opposite or antiphase with the continuous phase upstream pressure (P_c), and it is noticeable for $\theta = 60°$ and $90°$. The pressure fluctuations in the downstream (P_do) are negligible, as shown in Figs. 6c and 7c.

In summary, the droplet hydrodynamics is strongly dependent on the set of operating conditions (θ, Ca_c for different Re_c). The pressure history can efficiently be utilized for the suitable design and development of the microfluidic device for the on-demand droplet generation.

4 Conclusions

The present study of droplet formation in a two-dimensional T-junction microfluidic system using the conservative level set method numerically has provided new insights. The droplet behavior has been studied for a wide range of capillary and Reynolds numbers and contact angle. The current results have shown agreeable comparison with the existing experimental and numerical works. The effect of the contact angle ($30^\circ \leq \theta \leq 180^\circ$) on droplet generation is presented in detail for a wide range of flow conditions. For the hydrophilic conditions, the curvature is adopting a concave, and thus, the droplet length is not uniform. However, the monodispersed droplets that are formed are seen for the hydrophobic conditions. The effect of the inertial force is also discussed thoroughly. The evolution of the pressure profiles has been discussed to explain the droplet breakup mechanism. These new insights help to design a suitable microfluidic device to ease the control of the contact angle-based droplet production.

References

1. Whitesides, G.M.: The origins and the future of microfluidics. Nature **442**(7101), 368–373 (2006). https://doi.org/10.1038/nature05058
2. Baroud, C.N., Gallaire, F., Dangla, R.: Dynamics of microfluidic droplets. Lab Chip **10**(16), 2032–2045 (2010). https://doi.org/10.1039/C001191F
3. Günther, A., Khan, S.A., Thalmann, M., Trachsel, F., Jensen, K.F.: Transport and reaction in microscale segmented gas-liquid flow. Lab Chip **4**(4), 278–286 (2004). https://doi.org/10.1039/B403982C
4. Garstecki, P., Fuerstman, M.J., Stone, H.A., Whitesides, G.M.: Formation of droplets and bubbles in a microfluidic T-junction: scaling and mechanism of break-up. Lab Chip **6**(3), 437–446 (2006). https://doi.org/10.1039/b510841a
5. Cristini, V., Tan, Y.-C.: Theory and numerical simulation of droplet dynamics in complex flows: a review. Lab Chip **4**(4), 257–264 (2004). https://doi.org/10.1039/b403226h
6. Abate, A.R., Mary, P., Van Steijn, V., Weitz, D.A.: Experimental validation of plugging during drop formation in a T-junction. Lab Chip **12**(8), 1516–1521 (2012). https://doi.org/10.1039/c2lc21263c
7. Li, X.-B., Li, F.-C., Yang, J.-C., Kinoshita, H., Oishi, M., Oshima, M.: Study on the mechanism of droplet formation in T-junction microchannel. Chem. Eng. Sci. **69**(1), 340–351 (2012). https://doi.org/10.1016/j.ces.2011.10.048
8. Wong, V.-L., Loizou, K., Lau, P.-L., Graham, R.S., Hewakandamby, B.N.: Numerical studies of shear-thinning droplet formation in a microfluidic T-junction using two-phase level-SET method. Chem. Eng. Sci. **174**, 157–173 (2017). https://doi.org/10.1016/j.ces.2017.08.027
9. Venkateshwarlu, A., Bharti, R.P.: Effects of Capillary number and flow rate on the hydrodynamics of droplet generation in two-phase cross-flow microfluidic systems. J. Taiwan Inst. Chem. Eng., **129**, 64–79 (2021). https://doi.org/10.1016/j.jtice.2021.07.045
10. van der Graaf, S., Nisisako, T., Schroën, C.G.P.H., van der Sman, R.G.M., Boom, R.M.: Lattice Boltzmann simulations of droplet formation in a T-shaped microchannel. Langmuir **22**(9), 4144–4152 (2006). https://doi.org/10.1021/la052682f
11. Brackbill, J.U., Kothe, D.B., Zemach, C.: A continuum method for modeling surface tension. J. Comput. Phys. **100**(2), 335–354 (1992). https://doi.org/10.1016/0021-9991(92)90240-Y

12. Olsson, E., Kreiss, G.: A conservative level set method for two phase flow. J. Comput. Phys. **210**(1), 225–246 (2005). https://doi.org/10.1016/J.JCP.2005.04.007
13. Liu, H., Zhang, Y.: Lattice Boltzmann simulation of droplet generation in a microfluidic cross-junction. Commun. Comput. Phys. **9**(5), 1235–1256 (2011). https://doi.org/10.4208/cicp.231009.101110s
14. Jamalabadi, M.Y.A., DaqiqShirazi, M., Kosar, A., Shadloo, M.S.: Effect of injection angle, density ratio, and viscosity on droplet formation in a microfluidic T-junction. Theor. Appl. Mech. Lett. **7**(4), 243–251 (2017). https://doi.org/10.1016/j.taml.2017.06.002

Slip Effects in Ionic Liquids Flow Through a Contraction–Expansion Microfluidic Device

Jitendra Dhakar and Ram Prakash Bharti

Abstract In the present study, electroviscous effects in pressure-driven steady laminar flow of electrolytes in a contraction–expansion-type rectangular microfluidic device with surface charge-dependent slip are numerically investigated. The surface charge is assumed to be uniform at the device walls, and liquid is considered as symmetric (1:1) electrolyte solution. The NSNP (Navier–Stokes in conjunction with Nernst–Planck) model equations are solved using finite element method to obtain electrical potential, charge, induced field strength, and pressure drop in the microfluidic devices for various range of dimensionless parameters like Reynolds number ($Re = 0.01$), Schmidt number ($Sc = 1000$), surface charge density ($S = 4$–16), inverse Debye length ($K = 2$–20), and slip length ($B_0 = 0.15$). The current results have shown agreement with the literature for the available limiting conditions. The results show that electroviscous correction factor Y (ratio of apparent to physical viscosity) increases for no-slip ($B_0 = 0$) flow as surface charge density (S) varies from 4 to 16 at fixed K. It has been further observed that the surface charge-dependent slip at the walls plays a vital role on the hydrodynamics. The magnitudes of total electrical potential and pressure drop increase with increasing charge-dependent slip for fixed S and K. Further, the electroviscous correction factor (Y) increases with increasing slip over the range of S from 4 to 16 at fixed K. Finally, the present results are presented in terms of the predictive correlations for their practical use in the design and engineering of microfluidic devices.

Keywords Electroviscous effect · Pressure-driven flow · Surface charge-dependent slip · Ionic liquids · Microfluidic device

J. Dhakar · R. P. Bharti (✉)
Complex Fluid Dynamics and Microfluidics (CFDM) Lab, Department of Chemical Engineering, Indian Institute of Technology Roorkee, Roorkee 247667, India
e-mail: rpbharti@iitr.ac.in

R. P. Bharti and K. M. Gangawane (eds.), *Recent Trends in Fluid Dynamics Research*, Lecture Notes in Mechanical Engineering,
https://doi.org/10.1007/978-981-16-6928-6_12

1 Introduction

Over the years, micro-electromechanical systems (MEMS) have gained increasing attention in the science and engineering fields [1, 2]. Electrokinetic phenomena advance through the interaction of charged channel walls with ionic liquid (electrolytes). The charged walls attract counter-ions and repeal co-ions of electrolyte liquid. In turn, the rearrangement of ions forms an electric double layer (EDL) in which the counter-ions density reduces away from the channel wall. When the pressure-driven flow is imposed in the channel, the flow of ions in a diffusive layer generates a current, known as streaming current. The accumulation of ions in the downstream channel creates a potential difference along the device called streaming potential. It induces a hydrodynamic resistance; i.e., electrostatic force opposes the primary liquid flow. The additional resistance can be accounted for through the apparent viscosity. This behavior is commonly called an electroviscous effect [3].

Various efforts are made to explore the electroviscous effects for uniform geometries that included slit microchannels [4, 5, 6], cylinders [7, 8, 9], rectangular microchannels [10, 11], and elliptical cross section [12]. Relatively less attempts are also noticed for non-uniform geometries that included contraction–expansion slit microchannel [13], cylindrical contraction–expansion microchannel [14], and rectangular contraction–expansion microchannel [15].

No-slip boundary condition is used widely on the walls of microfluidic devices. However, boundary slip conditions and surface charge are also accentuating the microfluidic flow characteristics. An earlier study [16] has concluded a reduction in slip length with increasing surface charge density. Researchers have also explored the combined effects of charge with slip on fluid flow in microfluidic devices, as reviewed elsewhere [17]. The literature lacks such efforts on complex flow geometries, which are widely encountered in practical applications. This paper numerically analyzes the electroviscous effect in a geometrical complex microfluidic device considering surface charge-dependent slip.

2 Mathematical Formulation

Consider the pressure-driven flow of ionic solution through a contraction–expansion microfluidic device at a steady state, as shown in Fig. 1. The lengths of the channel's upstream, downstream, and contraction sections are taken to be equal (i.e., $L_U = L_D = L_C$). The contraction ratio is defined as $d_c = W/W_C$ where W and W_C are the half-widths of the main channels in upstream/downstream and contraction sections, respectively. The fully developed flow (average inflow velocity of $\overline{V}$) is assumed to be incompressible and Newtonian. Further, the liquid contains symmetric ions with equal valences ($z_+ = -z_- = z$), diffusivity ($D_+ = D_- = D$) of ions, and the bulk concentration of each ion species is n_0. The uniform charge density (σ) is taken on

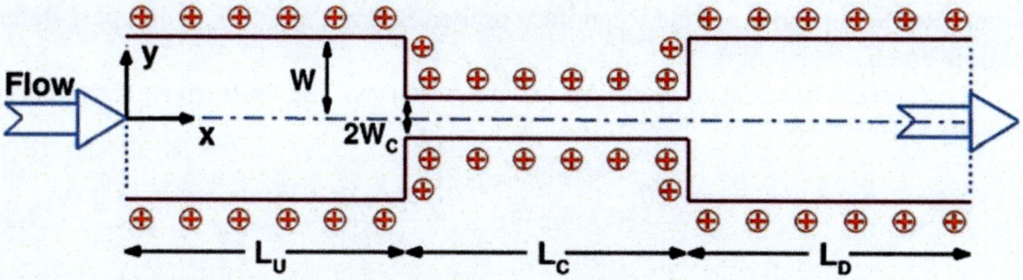

Fig. 1 Schematics of the contraction–expansion microfluidic device

the walls. The density (ρ), viscosity (μ), and dielectric constant (ε) are also spatially uniform.

This physical problem can be mathematically represented by the following coupled set of dimensionless equations.

$$\nabla^2 U = -\frac{1}{2} K^2 (n_+ - n_-) \tag{1}$$

$$\frac{\partial n_-}{\partial t} + \nabla \bullet (\mathbf{v} n_-) = \frac{1}{ReSc} [\nabla^2 n_- - \nabla \bullet (n_- \nabla U)] \tag{2}$$

$$\frac{\partial n_+}{\partial t} + \nabla \bullet (\mathbf{v} n_+) = \frac{1}{ReSc} [\nabla^2 n_+ + \nabla \bullet (n_+ \nabla U)] \tag{3}$$

$$\frac{\partial \mathbf{v}}{\partial t} + \nabla \bullet (\mathbf{v}\mathbf{v}) = -\nabla p + \frac{1}{Re} \nabla \bullet \left[\nabla \mathbf{v} + (\nabla \mathbf{v})^T\right] - \frac{\beta K^2}{(Re)^2} (n_+ - n_-) \nabla U \tag{4}$$

$$\nabla \bullet \mathbf{v} = 0 \tag{5}$$

where U, **v**, n_+ , n_-, and p are the total electrical potential, velocity vector, cation and anion concentration, and pressure. Poisson–Boltzmann Eq. (1) relates the total electrical potential (U) and local charge density ($n_+ - n_-$). Nernst–Planck Eqs. (2) and (3) express conservation of the positive and negative ionic species (n_+ and n_-), respectively. Navier–Stokes Eq. (4) includes extra body force (i.e., electrical) terms in the momentum equation because of the free charges. Equation (5) depicts the mass conservation of an incompressible fluid flow.

Equations (1) to (5) are obtained by using $k_b T/ze$, n_0, $\overline{V}$, W, $W/\overline{V}$ as scaling variables for electrical potential, the number density of anions and cations, velocity, length, and time, respectively. The resulting dimensionless groups are defined as follows.

$$Re = \frac{\rho \overline{V} W}{\mu}, Sc = \frac{\mu}{\rho D}, \beta = \frac{\rho {k_b}^2 T^2 \varepsilon_0 \varepsilon}{2 z^2 e^2 \mu^2}, K^2 = \frac{2 z^2 e^2 n_0 W^2}{\varepsilon_0 \varepsilon k_b T} \tag{6}$$

where Re, Sc, β, and K are the Reynolds number, Schmidt number, liquid parameter, and inverse Debye length.

The relevant boundary conditions are noted below. The uniform surface charge density maintained at the channel is expressed as:

$$\frac{\partial \mathrm{U}}{\partial \mathrm{n}} = S \text{ where } \mathrm{S} = \frac{\mathrm{ze}\sigma \mathrm{W}}{\varepsilon_0 \varepsilon k_b \mathrm{T}} \tag{7}$$

where n represents the outward normal to the wall and S is the dimensionless surface charge density. For ion concentration, zero flux of ions normal to the device wall ($\mathbf{f}_i.\mathrm{n} = 0$) is applied. The wall velocity is imposed as a surface charge-dependent slip velocity condition [18]. It is expressed in dimensionless form as:

$$\mathrm{v_{wall}} = \mathrm{B}\frac{\partial \mathrm{v}}{\partial \mathrm{y}} \text{ where } \mathrm{B} = \frac{b}{W} = \frac{1}{W}\left[\frac{\mathrm{b_0}}{1 + (1/\alpha)(\sigma \mathrm{d}^2/\mathrm{e})^2(\mathrm{l_B}/\mathrm{d}^2)\mathrm{b_0}}\right] \tag{8}$$

where B and b are the dimensionless slip length and the dimensional charge-dependent slip length [16, 19]. The $\alpha \sim 1$, and b_0, l_B and d are the slip length without charge, Bjerrum length, and equilibrium distance of Lennard–Jones potential.

All fields at the inlet are applied from the solution of steady, laminar, fully developed electroviscous flow in a 2D uniform slit. The net axial current [7, 13, 14], i.e., the total current passing across the inlet and outlet, becomes zero at a steady state. Uniform axial potential gradients at the inlet and outlet are considered and chosen to satisfy the net current condition, respectively.

The numerical solution of the above-discussed mathematical model yields velocity, pressure, ion concentrations, and electrical potential fields throughout the computational domain. These fields are further used to deduce the engineering parameters. The subsequent section briefs about the numerical approach and parameters used in this work.

3 Numerical Methodology

The mathematical model shown above is simulated using the finite element method (FEM)-based COMSOL Multiphysics CFD solver. The low Reynolds number (Re = 0.01, based on the average inflow velocity $\overline{\mathrm{V}}$) flow of electrolyte solution in a microfluidic device is accounted for by coupling the laminar flow (spf), electrostatics (es), and transport of dilute species (tds) modules of COMSOL. The iterative solution of the coupled Multiphysics has been obtained by using the PARDISO linear and Newton's nonlinear solvers. The steady-state solution yields the total electrical potential (U), induced electrical field ($E_{induced}$), pressure (P), velocity (v), and the ion concentration (n_+ and n_-) fields.

The following geometrical (Fig. 1) parameters for contraction–expansion microfluidic device are considered in the present work: $L_U = L_D = L_C = 5$ W

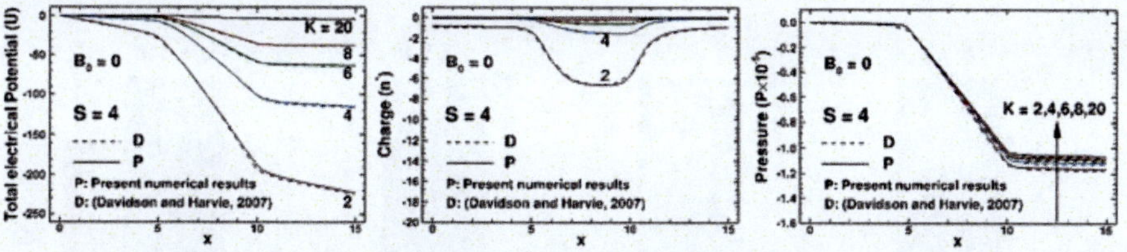

Fig. 2 Validation of the present results for field variables over the centerline of the device with the literature [13] for S = 4 and K = 2 – 20

and $d_c = W/W_C = 0.25$. A grid independence test is carried out for 50, 100, and 150 grid points distributed uniformly over the half-width (W) of the microchannel. The results of total electrical potential, charge, induced electric field, and pressure drop changed insignificantly (i.e., less than 1–2%) with the grid refinement from 100 to 150 over the ranges of conditions. The final calculations are thus performed on a uniform square grid, including corner refinement and boundary layer at the wall with 100 mesh cells per half-width (W).

The present numerical results are validated with the literature [13] in Fig. 2 at S = 4 and K = 2- 20. A close agreement is found between the two results for centerline profiles of total electrical potential, charge, and pressure. It is to be noted here that the literature values are based on the finite volume method (FVM) solver.

4 Results and Discussions

In this work, numerical results are obtained and presented for contraction–expansion slit microfluidic flow of symmetric (1:1) electrolyte liquid (Sc = 1000, $\beta = 2.34 \times 10^{-4}$ based on the water at T = 298 K). The Reynolds number is taken low, Re = 10^{-2} and dimensionless slip length of $B_0 = 0$ (no-slip) to 0.20 in the gap of 0.05. The K and S values are taken as 2 to 20 and 4–16, respectively.

The above ranges of parameters are justified as follows: Reynolds number is taken to be low (Re = 0.01) as the microfluidic flows are generally laminar [13]. K is taken in the range of 2–20 because at lower K (= 2), EDL is thicker (overlapping), and at a higher value of K (= 20), EDL is very thin. So, range of K shows the variation of EDL thickness. S is taken in the range of 4 - 16 as it accounts for practical ranges of zeta potential variation from 50 to 100 mV when K = 2 and 12–50 mV when K = 8 for a uniform microchannel [13]. B_0 is taken in the range of 0–0.20 because the lower value of B_0 (= 0) shows the no-slip condition, and a higher value of B_0 (= 0.20) shows the high slippery surface of the microchannel [20].

Figure 3 shows the total electrical potential (U) distribution for different values of B_0 = 0–0.20 at K = 4 and S = 8. The figure displays an increasing drop in the electrical potential with an increasing slip length (B_0). Further, the normal potential

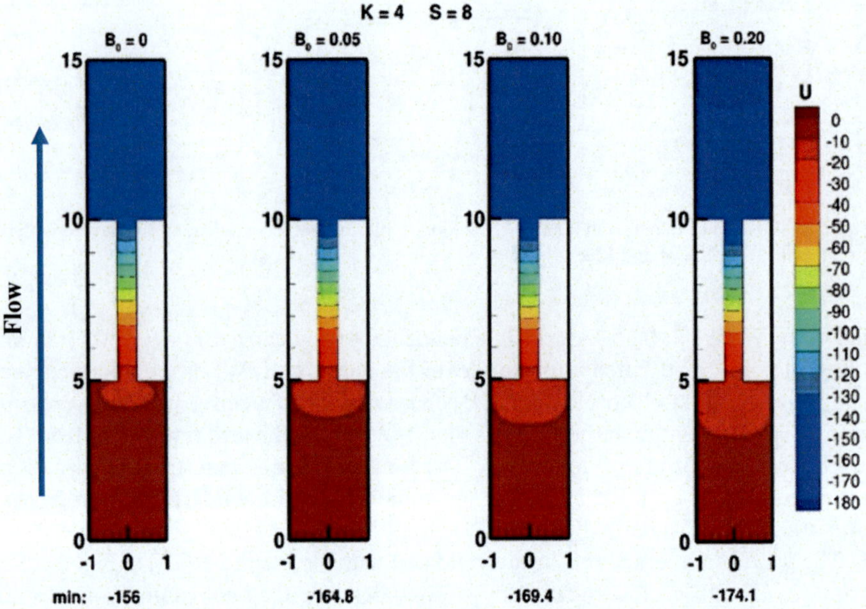

Fig. 3 Dimensionless potential (U) distribution for $B_0 = 0$ to 0.20 at S = 8 and K = 4

gradient at the wall is considered as fixed charge density (Eq. 7). Thereby, the lateral curving of the potential contour is obtained.

Similarly, Fig. 4 shows the dimensionless charge ($n^* = n_+ - n_-$) distribution for different values of slip lengths (B_0) at K = 4 and S = 8. The figure depicts that the charge distribution within the microfluidic device is less affected by the slip length changes from 0 to 0.20. Since the ionic species transfer is highly dependent on the charge capacity of the walls than the convection velocity near the surface, Figure 5 shows the pressure distribution for different values of $B_0 = 0$–0.20 at K = 4 and S = 8. Since the wall slip reduces the wall stress and thereby assists the flow, the pressure drop is seen to decrease with increasing B_0 from 0 to 0.20.

Figure 6a shows the total electrical potential (U) variation across the centerline of microfluidic device for S = 4–16, K = 2–20, and at $B_0 = 0$, 0.15. In the flow direction, total potential decreases due to the advection of negative ions along with the microfluidic device. The gradient is maximum in the contraction section due to increased convective velocity and reduced flow area. The magnitude of total potential increases with increasing the slip length compared to that of the no-slip case. Furthermore, the electrical potential decreases with increasing K, whereas it increases with increasing S, irrespective of the slip length.

Figure 6b presents the pressure variation along the horizontal centerline of the device for S = 4–16, K = 2–20, and at $B_0 = 0$, 0.15. Like electrical potential, the pressure drop is also maximum in the contraction section. The magnitude of the pressure increases with increasing slip length. While the pressure variation is

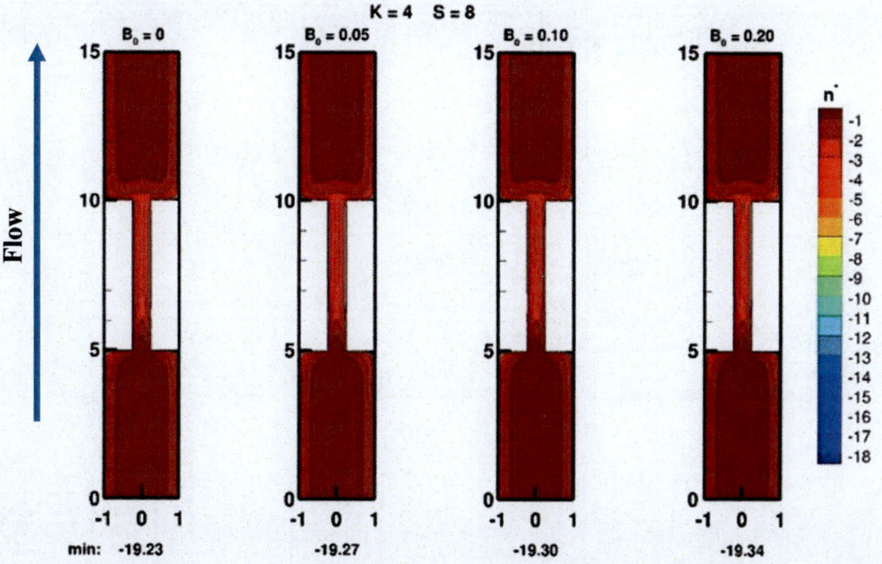

Fig. 4 Dimensionless charge ($n^* = n_+ - n_-$) distribution for $B_0 = 0$ to 0.20 at S = 8 and K = 4

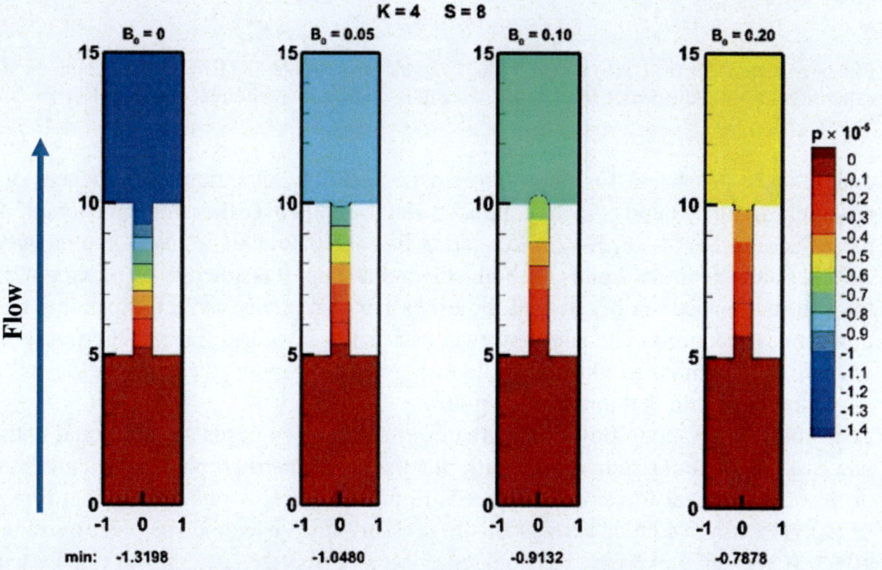

Fig. 5 Dimensionless pressure (p) distribution for $B_0 = 0$–0.20 at S = 8 and K = 4

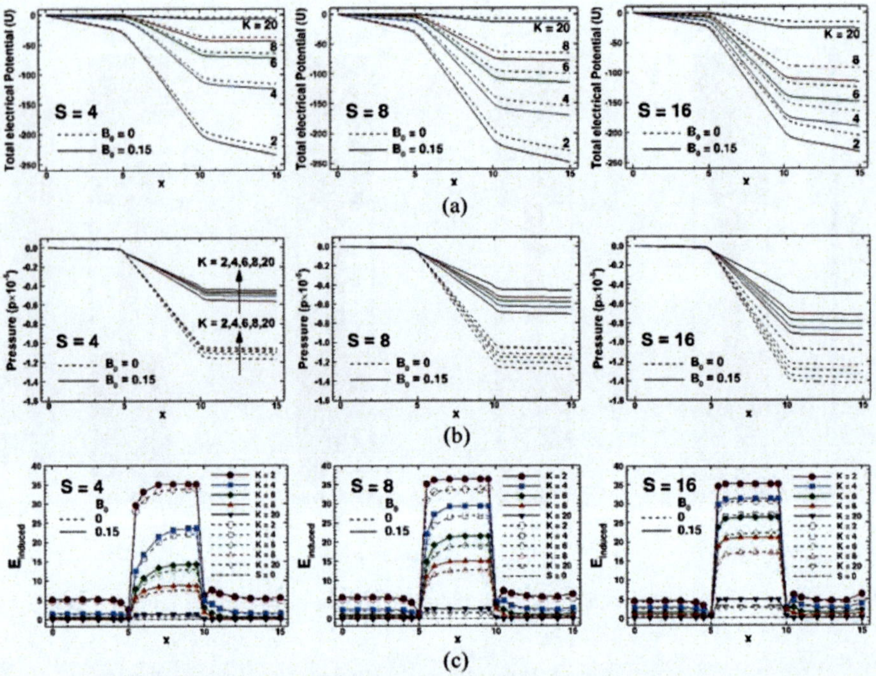

Fig. 6 Variations of (**a**) U, (**b**) p, and (**c**) $E_{induced}$, along the centerline length of the contraction–expansion microfluidic device for S = 4, 8, 16 and $B_0 = 0, 0.15$ at different values of K

qualitatively consistent for both slip and no-slip flow, quantitative influences are notable. Figure 6c depicts locally induced electric field over the centerline length of the device for S = 4–16, K = 2–20, and at $B_0 = 0, 0.15$. Qualitatively, the induced field has shown similar trends for both slip and no slip. It is noted to be maximum in the contraction section because of the excessive charge compared to upstream and downstream sections. The slip, however, assists the flow, and thereby the transport of more anions enhances the charge; in turn, a stronger induced electric field is seen in comparison with that under no-slip flow.

In the electrokinetic flows, the streaming potential imposes an additional resistance to flow. The resistance manifests, at a fixed volumetric flow rate, as a pressure drop with electrical forces (Δp) exceeds to pressure drop without electrical forces (Δp_0). The relative enhancement of the pressure drop is called the electroviscous effect. It is usually expressed as apparent viscosity (μ_{eff}). It (μ_{eff}) is the viscosity in the absence of an electric field needed to achieve the pressure drop (Δp). For low Re flow, the electroviscous correction factor $Y = \mu_{eff}/\mu = \Delta p/\Delta p_0$ as the nonlinear advection term is small in the momentum equation.

Figure 7 depicts Y as a function of K, S, and B_0. Qualitative variations of Y with K and S are similar in both slip and no-slip conditions. The Y increases with increasing S, whereas it is decreasing with increasing K. The thickness of EDL is reduced with

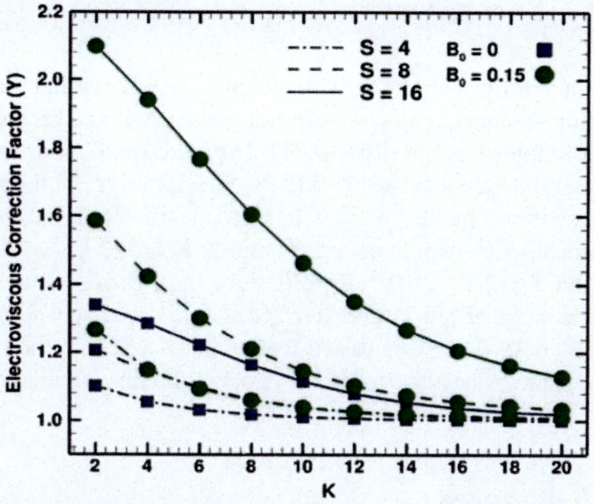

Fig. 7 Electroviscous correction factor (Y)

increasing K, due to which the electroviscous effects weaken with increasing K; however, vice versa is true with increasing S. The electroviscous correction factor is also seen to increase with increasing B_0 **from 0 to 0.15**. The electroviscous correction factor increases by 15–56% for $B_0 = 0.15$ than that for $B_0 = 0$ when S varied from 4 to 16.

Although overall pressure drop for estimation to electroviscous correction factor ($Y = \mu_{eff}/\mu$) can be calculated from a simple model as given below:

$$\Delta p_m = (\Delta p_i + \Delta p_c + \Delta p_o) + \Delta p_e \tag{9}$$

where Δp_i, Δp_o, and Δp_c denote the pressure drops in upstream, downstream, and contraction sections of lengths L_U, L_D, and L_C, respectively. Since these sections are depicting uniform slits, the standard solution of creeping flow through a channel, i.e., $\Delta p = \left(\frac{3}{Re}\right)\left(\frac{L}{W}\right)$ where L and W are the length and width of the uniform section, can be used to obtain for the pressure drop in upstream, contraction, and downstream sections. The excess pressure drop (Δp_e), due to converging and diverging flow areas, is expressed as follows for the present ranges of conditions.

$$\Delta p_e = \frac{156}{\pi d_c^2 Re} \tag{10}$$

where d_c represents the contraction ratio. Such a simple approach [7, 13, 14] for predicting pressure drop, and thereby, electroviscous correction factor enables the use of present results in the design and engineering of relevant microfluidic devices.

5 Conclusions

This study analyzes the slip effects in steady symmetric electrolyte flow in a contraction–expansion slit microfluidic device with surface charge-dependent slip at low Reynolds number (Re = 0.01). The field equations are solved using the finite element method-based solver COMSOL Multiphysics. Numerical results for total electrical potential, charge, pressure, induced electric field, and electroviscous correction factor are discussed for parameters range as K = 2–20, S = 4–16, $B_0 = 0, 0.15$ at Sc = 1000 and $\beta = 2.34 \times 10^{-4}$. Results show the electroviscous correction factor increases in the range of 15–56% for surface charge-dependent slip ($B_0 = 0.15$) than no-slip flow ($B_0 = 0$) when S is varied from 4 to 16. Overall, the surface charge-dependent slip enhances the electroviscous effect in the microfluidic device than the no-slip flow.

References

1. Bhushan, B.: Springer Handbook of Nanotechnology. Springer, Berlin (2010)
2. Li, D.: Encylopedia of Microfluidics and Nanofluidics. Springer Science & Business Media, New York (2008)
3. Hunter, R.J.: Zeta Potential in Colloid Science: Principles and Applications. Academic Press, London (1981)
4. Chun, M.-S., Kwak, H.-W.: Electrokinetic flow and electroviscous effect in a charged slit-like microfluidic channel with non-linear Poisson-Boltzmann field. Korea-Australia Rheol. J. **15**(2), 83–90 (2003)
5. Mala, G.M., et al.: Flow characteristics of water through a microchannel between two parallel plates with electrokinetic effects. Int. J. Heat Fluid Flow. **18**(5), 489–496 (1997). https://doi.org/10.1016/s0142-727x(97)00032-5
6. Mala, G.M., et al.: Heat transfer and fluid flow in microchannels. Int. J. Heat Mass Transf. **40**(13), 3079–3088 (1997)
7. Bharti, R.P., et al.: Electroviscous effects in steady fully developed flow of a power-law liquid through a cylindrical microchannel. Int. J. Heat Fluid Flow. **30**(4), 804–811 (2009). https://doi.org/10.1016/j.ijheatfluidflow.2009.01.012
8. Bowen, W.R., Jenner, F.: Electroviscous Effects in Charged Capillaries. J. Colloid Interface Sci. **173**, 388–395 (1995)
9. Brutin, D., Tadrist, L.: Modeling of surface-fluid electrokinetic coupling on the laminar flow friction factor in microtubes. Microscale Thermophys. Eng. **9**(1), 33–48 (2005). https://doi.org/10.1080/10893950590913314
10. Li, D.: Electro-viscous effects on pressure-driven liquid flow in microchannels. Colloids Surfaces A Physicochem. Eng. Asp. **195**, 35–37 (2001)
11. Ren, L., et al.: Electro-viscous effects on liquid flow in microchannels. J. Colloid Interface Sci. **223**(1), 12–22 (2001). https://doi.org/10.1006/jcis.2000.7262
12. Hsu, J.P., et al.: Electrokinetic flow through an elliptical microchannel: Effects of aspect ratio and electrical boundary conditions. J. Colloid Interface Sci. **248**(1), 176–184 (2002). https://doi.org/10.1006/jcis.2001.8200
13. Davidson, M.R., Harvie, D.J.E.: Electroviscous effects in low Reynolds number liquid flow through a slit-like microfluidic contraction. Chem. Eng. Sci. **62**(16), 4229–4240 (2007). https://doi.org/10.1016/j.ces.2007.05.006
14. Bharti, R.P., et al.: Steady flow of ionic liquid through a cylindrical microfluidic contraction-expansion pipe: Electroviscous effects and pressure drop. Chem. Eng. Sci. **63**(14), 3593–3604 (2008). https://doi.org/10.1016/j.ces.2008.04.029

15. Davidson, M.R., et al.: Electroviscous effects in low Reynolds number flow through a microfluidic contraction with rectangular cross-section. World Acad. Sci. Eng. Technol. **40**(4), 256–260 (2009)
16. Joly, L., et al.: Liquid friction on charged surfaces: From hydrodynamic slippage to electrokinetics. J. Chem. Phys. **125**, 204716 (2006). https://doi.org/10.1063/1.2397677
17. Jing, D., Bhushan, B.: The coupling of surface charge and boundary slip at the solid–liquid interface and their combined effect on fluid drag: a review. J. Colloid Interface Sci. **454**, 152–179 (2015). https://doi.org/10.1016/j.jcis.2015.05.015
18. Jing, D., Bhushan, B.: Electroviscous effect on fluid drag in a microchannel with large zeta potential. Beilstein J. Nanotechnol. **6**(1), 2207–2216 (2015). https://doi.org/10.3762/bjnano.6.226
19. Jing, D., Bhushan, B.: Effect of boundary slip and surface charge on the pressure-driven flow. J. Colloid Interface Sci. **392**(1), 15–26 (2013). https://doi.org/10.1016/j.jcis.2012.09.082
20. Buren, M., et al.: Electroviscous effect and electrokinetic energy conversion in time periodic pressure-driven flow through a parallel-plate nanochannel with surface charge-dependent slip. J. Phys. D. Appl. Phys. **51**, 20 (2018). https://doi.org/10.1088/1361-6463/aabc73

Effect of Shear Rate on Non-Newtonian Droplet Generation in T-junction Microfluidic System

Pradeep Dhondi, Akepogu Venkateshwarlu, and Ram Prakash Bharti

Abstract Mono-dispersed droplet generation in T-junction microchannel is carried out by using the level set method. The dispersed phase fluid is non-Newtonian, and the continuous phase is assumed to be a Newtonian fluid. The flow and phase-field equations are solved for various shear rates ($1 \leq \dot{\gamma} \leq 10^3$) using the finite element method. The power-law fluid model is employed to study the dependence of viscosity of the dispersed phase on droplet generation. The influence of the controlling parameters such as flowrates ratio and shear rate ratios on droplet dynamics is systematically studied. The transient flow in the microchannel has revealed the exciting dynamics of droplet formation with the shear-dependent dispersed phase. The flow regimes have been classified into droplet and non-droplet zones by the power-law index (n) at different shear rates. It is observed that there is a droplet formation at lower values of flowrate ratio ($Q_r < 1$) for shear-thinning (n < 1) fluids for a specific range of $\dot{\gamma}$. However, the flow is becoming parallel at a higher flowrate ratio ($Q_r > 1$). For a fixed flowrate ratio and shear rate, there is a droplet generation for shear-thinning fluids, and whereas the flow is becoming parallel in the case of shear thickening fluids.

Keywords Two-phase flow · Non-Newtonian fluid · the level set method · Droplet generation

1 Introduction

Microfluidics has become a promising area in the current research world due to its versatile applications in various engineering fields like protein crystallization, fermentation industry, drug delivery systems, and the encapsulation of cells [1–3]. Many studies have been carried out to develop efficient and stable methods for

P. Dhondi · A. Venkateshwarlu · R. P. Bharti (✉)
Complex Fluid Dynamics and Microfluidics (CFDM) Lab, Department of Chemical Engineering, Indian Institute of Technology Roorkee, Roorkee 247667, India
e-mail: rpbharti@iitr.ac.in

R. P. Bharti and K. M. Gangawane (eds.), *Recent Trends in Fluid Dynamics Research*, Lecture Notes in Mechanical Engineering,
https://doi.org/10.1007/978-981-16-6928-6_13

conventional single- and multiphase droplet formation and dynamics [4]. Microfluidics study helps to understand the evolutionary dynamics of monodispersed droplets. The forces responsible for the droplet dynamics depend on fluid properties, geometry, and operating conditions. The viscous effects are more critical than inertial at low Reynolds numbers [5, 6]. However, the flow is dominated by surface tension and viscous forces, which play a crucial role in droplet formation at small capillary numbers [7]. The microfluidic device can realize single-step high-yield production of droplets with great flexibility to control the droplet size and anisotropic shape. Each droplet in the microfluidic system acts like a microreactor, allowing species migration or reaction [8]. The droplet size and shape during the interface evolution are controlled and manipulated by changing flow rates of either or both phases [9, 10]. The geometry of channels strongly influences the droplet pattern. The cross-flow, co-flow, and flow-focusing microfluidic devices are widely used geometries. The cross-flow T-type geometries are commonly used due to the ease of droplet generation and uniformity of the formed droplets.

To the best of our understanding, Thorsen et al. [11] are the first to perform the experiments to generate the water droplets in oil in a T-junction microchannel. They reported a stronger relationship between shear stress and interfacial tension in the microfluidic system-controlled droplet production. Many others have studied the formation of droplets using T-junction, explained the formation mechanism of droplets, and proposed a power-law correlation to predict their size with flow rate [12, 13]. There are several current reports on droplet preparation methods. However, more fundamental studies on the underlying mechanism and new phenomena associated with them, especially those involving non-Newtonian fluids, can be significant. Nonlinear rheology of non-Newtonian fluids is well known to have a significant impact on flow dynamics. Abate et al. [14] investigated the formation of monodispersed particles in a non-Newtonian polymer solution using a flow-focusing system. It is clear from their findings that the rheological parameters of non-Newtonian fluids have a significant impact on the droplet formation mechanism.

Further attempts to study the combination of Newtonian and non-Newtonian fluids [15, 16] have reported the effects of the continuous phase rheology and flow rate, interfacial stress, and contact angle on droplet size. The dispersed phase flow rate and rheology, on the other hand, have a more minor but still significant impact on droplet formation. There is still a lack of knowledge of droplet formation and flow in non-Newtonian fluids at the microscale. This work has investigated the two-phase microfluidic flow development of shear-thinning and shear-thickening fluids with the effect of shear rate and flow rate of non-Newtonian fluids and their subsequent evolution. The numerical simulations for the droplet formation in a T-junction microfluidic device filled with a non-Newtonian fluid are performed by a multiphase model based on the level set method.

2 Problem Statement and Numerical Model

Consider T-junction microchannel geometry consisting of two sections: primary and side channels, as shown in Fig. 1. The side channel, of width $w_d = 100\mu$m, is located at upstream length (L_u) of 900μm and downstream length (L_d) of 3000μm from the inlet, and exit of the primary channel, respectively. The total length of the primary channel, of width $w_c = 100\mu$m, is thus taken as $L_c = (L_u + w_d + L_d) = 4000\mu$m. These geometric dimensions are enough to avoid the end effects, as optimized elsewhere [17].

The continuous phase (CP) and dispersed phase (DP) are allowed into the primary and side channels at the flow rate of Q_c and Q_d ml/h, respectively. The relative flow rate is expressed as $Q_r = Q_d/Q_c$. The outlet of the main channel is open to ambient, i.e., pressure is $p = 0$ atm, and no-slip boundary condition is applied on the walls. Both the fluid phases (CP and DP) are assumed to be incompressible and having equal density ($\rho_c = \rho_d$). Furthermore, the continuous phase is Newtonian (i.e., constant viscosity μ_c) and the dispersed phase is non-Newtonian fluid, i.e., $\mu_d = \mu(\dot{\gamma})$. The viscosity of the dispersed phase (μ_d) is defined by the power-law fluid model [18–20] as $\mu_d = m\dot{\gamma}^{n-1}$, where m and n are the fluid consistency index (Pa.s^n), and flow behavior constant, respectively.

The conservation equations for mass, momentum, and interface advection governing the two-phase flow physics are written as follows.

$$\nabla \cdot \mathbf{u} = 0 \tag{1}$$

$$\rho\left(\frac{\partial \mathbf{u}}{\partial t} + \mathbf{u} \cdot \nabla \mathbf{u}\right) = -\nabla p + \nabla \cdot \left[\mu(\dot{\gamma}, \phi)\left(\nabla \mathbf{u} + (\nabla \mathbf{u})^T\right)\right] + \mathbf{F}_{st} \tag{2}$$

$$\frac{\partial \phi}{\partial t} + \mathbf{u} \cdot \nabla \phi = \gamma \nabla \cdot \left(\epsilon_{ls} \nabla \phi - \phi(1 - \phi)\frac{\nabla \phi}{|\nabla \phi|}\right) \tag{3}$$

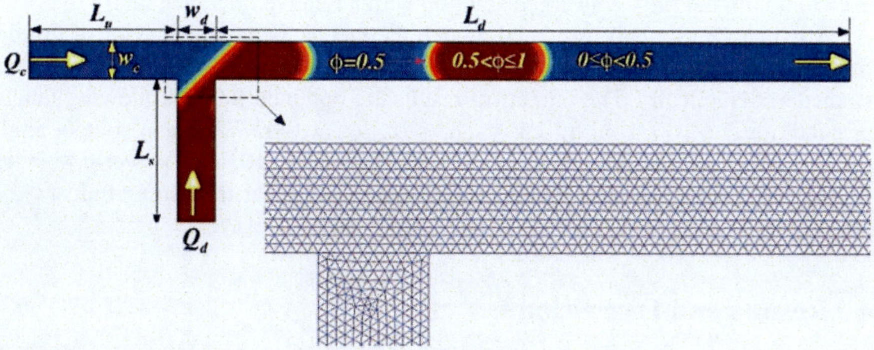

Fig. 1 Schematic T-shaped microchannel with triangular mesh

where ρ, μ, p, $\mathbf{u}$, and ϕ are the density, dynamic viscosity, pressure, velocity vector, and level set variable, respectively. In interface advection (Eq. 3), as per the level set method, level set variable ϕ ranges as $0(CP) \leq \phi \leq 1(DP)$ with an interface at $\phi = 0.5$; and γ and ϵ_{ls} are reinitialization and interface thickness control parameters.

The surface tension force between two immiscible fluids (F_{st}) is given by

$$F_{st} = \sigma \kappa \delta(\phi) n_i \tag{4}$$

where σ is the surface tension coefficient, $\mathbf{n}_i = \frac{\nabla \phi}{|\nabla \phi|}$ is the interface ($\phi = 0.5$) normal, and $\kappa = \frac{1}{R} = -\nabla \cdot \mathbf{n}_i$ is the inverse of the radius of curvature.

The local density and viscosity (ρ, μ) at any point are evaluated as follows.

$$\rho(\phi) = \rho_c + (\rho_d - \rho_c)\phi \tag{5}$$

$$\mu(\phi) = \mu_c + (\mu_d - \mu_c)\phi \tag{6}$$

The continuous phase is denoted by the subscript 'c' and the dispersed phase with 'd'.

3 Methodology and Parameters

In this work, the finite element method (FEM)-based COMSOL Multiphysics CFD solver has been used to obtain the numerical solution of the mathematical model. The 2D, laminar flow with two-phase level set method modules are used to account for the physics. Before starting the simulations, the mesh independence study has been performed [17] to optimize the total number of elements and maximum mesh element size. The optimally refined non-uniform mesh consisting of 13,766 triangular elements is used to discretize the computational domain. The iterative solver has been used to obtain the fully converged solution with a relative tolerance of 10^{-4}.

By using a reasonably validated model, the mechanism and behavior of droplet formation in non-Newtonian liquids under different operating conditions have been studied systematically. The numerical results are obtained for the following ranges of conditions: $Q_c = 1$ ml/h; $0.5 \leq Q_r \leq 2$; $\mu_c = 0.027778$ Pa.s; contact angle $(\theta) = 135°$; $Ca_c = 0.01$; $Re_c = 0.1$; $10^{-5} \leq \dot{\gamma} \leq 10^5$ and power-law index $(0.8 \leq n \leq 1.4)$ representing both shear-thinning and shear-thickening fluids.

4 Results and Discussion

In this section, validation of the present solution approach, the influences of various rheological parameters, such as power-law index (n), shear rate ($\dot{\gamma}$), and flow rate

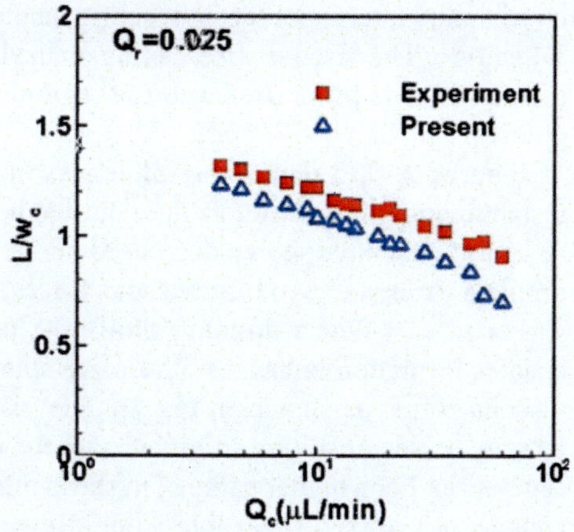

Fig. 2 Comparison of the present (Δ) and experimental [20] (□) work

ratio ($Q_r = Q_d/Q_c$) on the two-phase flow and droplet formation are presented and discussed.

4.1 *Validation of the Numerical Model*

The present numerical solution approach has thoroughly been validated elsewhere [17] for two-phase Newtonian flow through T-junction microfluidic device. The present results for a combination of Newtonian (as DP) and non-Newtonian (as a CP) fluids are validated in Fig. 2 with the experimental [20] results for the following conditions: $Q_r = 0.025$, $\sigma = 3.42$ mN/m, $\mu_d = 49.1$ mPa.s, $m = 32.6$ mPa.s^n, and $n = 0.589$. The continuous phase is assumed to be shear dependent, i.e., non-Newtonian fluid. The length of the droplet is getting decreased with an increase in the flow rate of the continuous phase. It is found that the trend between the experimental and present results to be the same; however, there is a marginal deviation between the results from the two studies. The present results are obtained with 2D flow approximations, whereas the experimental results are with an actual 3D model device. Such deviations may be attributed to various factors, including the dimensionality of the investigated problem, approaches used in the measurements, and precision of the tools applied. Overall, the present results are believed to be accurate within ± 2–3%.

4.2 *Effects of Shear Rate and Power-Law Index*

The influence of the shear rate ($\dot{\gamma}$) and power-law index (n) of the dispersed phase on the droplet formation, size, and frequency are discussed systematically. The droplet formation, in general, depends on the relative strengths of various forces like viscous, inertial, and interfacial tension, acting on the dispersed and continuous phases in the T-junction. The viscous stress acting on the liquid–liquid interface causes the viscous force, which is proportional to the region of the dispersed phase with the velocity gradient.

Figures 2, 3, 5 depict the influences of shear rate and power-law index on the instantaneous phase flow profiles. During the droplet generation, the shear-thickening ($n > 1$) fluids have shown decreased possibility of droplet formation and less stable droplets formed due to high viscous forces.

For $n < 1$ (shear-thinning fluid), the possibility of continuous monodispersed droplet formation enhances, and more stable droplets are formed due to the lower viscous force, as shown in Fig. 3. The viscosity of the dispersed phase increases as n increases, resulting in higher viscous resistance, which prevents rapid droplet separation. For a higher value of n, shear-thinning fluid, the droplet length marginally changes for a fixed shear rate value, Reynolds number ($Re_c = 0.1$), and capillary number ($Ca_c = 0.01$). Surface tension becomes essential in shear-thickening fluids because it resists the breakup of dispersed phase at the T-junction. However, the droplets formed within the channel length are limited for shear-thickening fluids due to increased viscosity and shear force. It can be noted that as the flow regime changes from dripping to jetting, the droplet gets elongated. Formation of the droplet takes place away from the junction point [21], and this is happening due to inertial

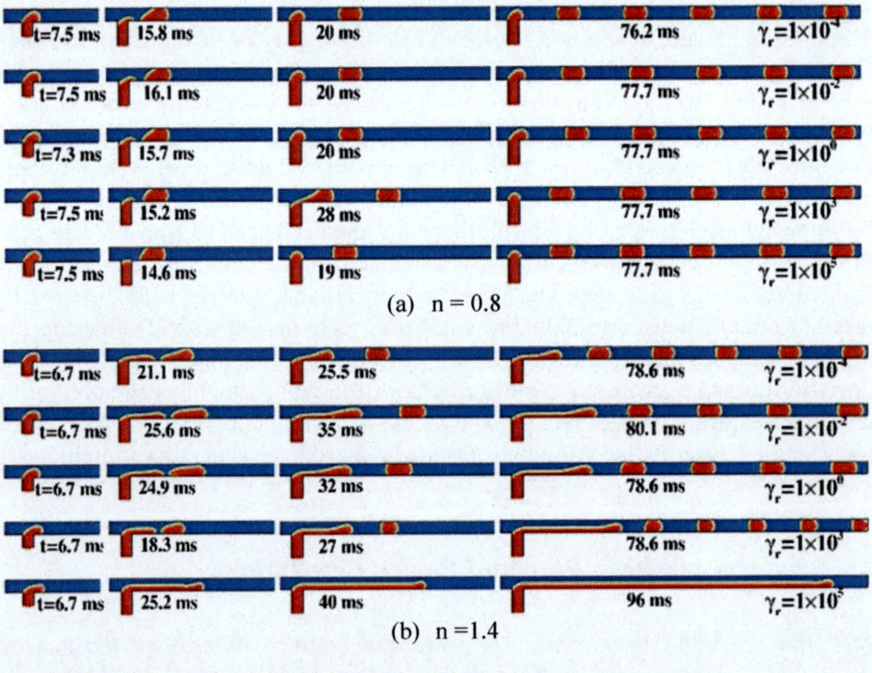

Fig. 3 Effect of shear rate ($\dot{\gamma}$) on droplet formation in T-junction channel for $Ca_c = 0.01$, $Q_r = 0.5$, $Re_c = 0.1$, $Q_c = 1\text{ml/h}$. **a** n = 0.8. **b** n = 1.4

force dominance over the viscous force, and the dispersed phase stretched far from the junction point. The droplet breakup happens due to the shear applied by the continuous phase on the dispersed phase. The shear rate also has a significant effect on the droplet generation, and hence a parametric study performed for a range of the shear rate $\dot{\gamma}$ value $10^0 - 10^5$. For $Q_r = 0.5$ and shear thinning ($n = 0.8$) and varying $\dot{\gamma}$, the change in the droplet length and frequency is insignificant seen in Fig. 3a.

However, in the case of shear-thickening fluids ($n = 1.4$), changing the shear rate ($\dot{\gamma}$) has a significant influence on the droplet length, frequency, and detachment time can be seen in Fig. 3b. It can be observed from the given figures that the droplet formation mechanism for non-Newtonian fluid is the transient flow in the microchannel has revealed the dynamics of droplet formation of the dispersed phase with shear-dependent fluids.

The flow regimes have been classified into droplet and non-droplet zones using the power-law index (n) for different shear values. The flow regime classification thereby establishes the possibility of droplet formation for specific shear rate and power-law index combinations. There is a droplet formation at lower values of flowrate ratio ($Q_r = 0.5$) for shear-thinning fluids for all ranges of the shear rate value. However, the flow is becoming parallel at higher power-law index ($n > 1$) values for a range of shear rate ($10^3 < \dot{\gamma} < 10^5$). There is a droplet generation for power-law index $n < 1$, whereas the flow is becoming parallel in the case of shear-thickening (> 1) fluids. When the power-law index $n = 1.4$, the droplet size getting decreased by increasing the value of the shear rate. As the shear rate increases, the flow becomes parallel due to the inertial force's dominance over the viscous force. On increasing the power index value further, for a specific shear rate, there is a droplet formation, and whereas the flow is becoming parallel for a higher shear rate ($10^3 - 10^5$) values and leaves the channel without getting droplets into a stable form. The flow rate ratio has a significant effect on the droplet breakup time, and it can be observed for Q_r = 0.5–1. The continuous phase generates pressure drop along the channel, and shear stress causes the deformation in the droplets. At the same time, the surface tension resists the deformation.

The breakup time of subsequent droplets in the channel that increases for both shear-thinning and shear-thickening fluids is shown in Fig. 4a and b. Transient flow can be seen as we increase the flow ratio, and also, there is a droplet formation for shear-thinning fluids for all ranges of the shear rate value. But the flow is becoming parallel at higher power index values ($n > 1$) for a range value of shear rate ($10^3 < \dot{\gamma} < 10^5$).

The rate of shear rate ($\dot{\gamma}$) is insignificant for $Q_r = 2$ (Fig. 5) on droplet length, breakup time. However, when the power-law index, $n = 0.8$, there is a droplet formation. After that, the flow becomes a parallel flow. For higher power-law index value n ($n > 1$) and higher flow ratio values, the flow becomes a parallel for particular shear rate values.

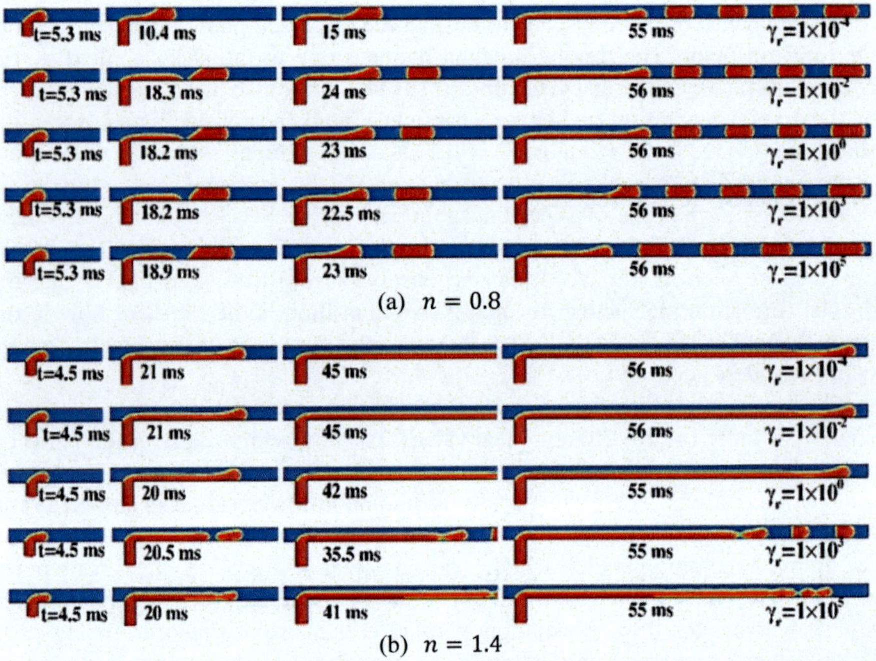

Fig. 4 Effect of shear rate (γ) on droplet formation in T-junction channel for $Ca_c = 0.01$, $Q_r = 1$, $Re_c = 0.1$, $Q_c = 1\text{ml/h}$, **a** $n = 0.8$ **b** $n = 1.4$

4.3 Effect of Shear Rate on the Droplet Size

The graphical results seen in Fig. 6a and b express the effect of shear rate on droplet length and detachment time. The time taken for droplet detachment is less for the lower shear rate in shear-thinning fluids ($n = 0.8$) than the shear-thickening fluids ($n = 1.4$). However, the time taken by the first droplet for the breakup is higher than that of the subsequent droplets in both shear-thinning and shear-thickening fluids.

In the shear rate between $10^{-2} - 10^{0}$ for the shear-thickening fluids, the droplet length is observed to be the same and the curves are overlapped to each other. Similar behavior is seen for the droplet breakup time in Fig. 6a and b. The normalized length of the droplet (L/w_c) is almost constant for shear-thinning fluids. Even the first droplet breakup time is almost the same as that of the subsequent droplets.

A small change can be observed at a higher shear rate value ($\dot{\gamma} = 10^3 - 10^5$) in shear-thinning fluids ($n = 0.8$), as the flow operating condition changes from dripping to transition. The length of the droplet is uniform for the shear rate range between $10^{-4} - 10^3$; however, at the higher shear rates, the droplet length is not uniform as seen in Fig. 7. For a higher flow rate ratio ($Q_r = 2$), the effect of shear rate is insignificant and results in no-droplet formation, as shown in Fig. 5.

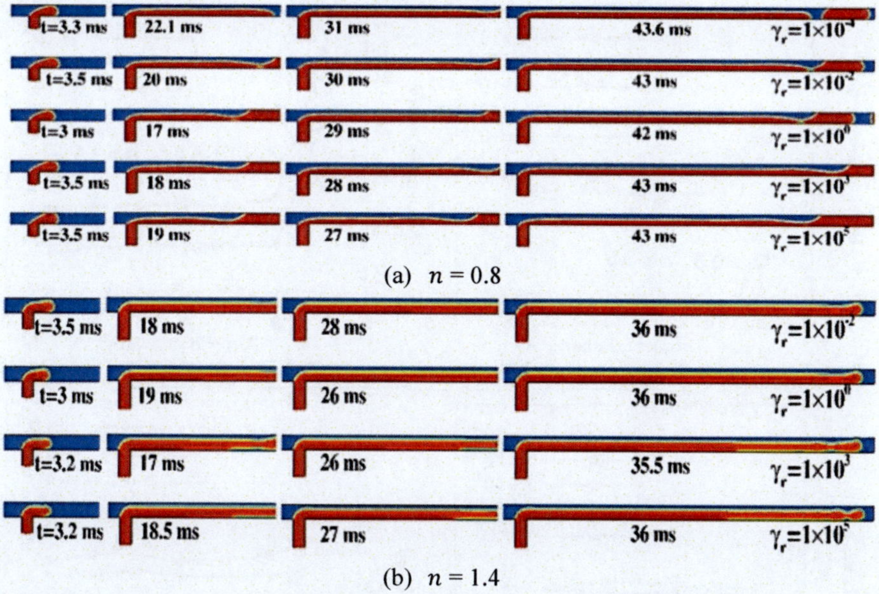

Fig. 5 Effect of shear rate ($\dot{\gamma}$) on droplet formation in T-junction channel for $Ca_c = 0.01$, $Q_r = 2$, $Re_c = 0.1$, $Q_c = 1\text{ml/h}$. **a** $n = 0.8$. **b** $n = 1.4$

5 Conclusions

In this study, two-dimensional laminar two-phase immiscible fluid flow simulations are performed to investigate the microfluidic phenomenon of droplet generation in a T-junction microchannel. The power-law rheological model has been adopted as a simple constitutive model for the non-Newtonian fluid as the dispersed phase. The mathematical model based on the level set method is solved using the finite element method-based COMSOL Multiphysics software. The effects of rheological properties, surface tension, shear rate, and flow rate ratio on droplet characteristics, including droplet size and formation mechanism, are discussed. The possibility of droplet formation reduces with increasing Q_d. The transition flow occurs in the channel when droplet breakup occurs far away from the junction point due to inertial force dominance over the viscous force. At higher shear rate values ($10^0 - 10^3$), the flow rate ratio has a significant impact on droplet size and breakup time for both shear-thickening and shear-thinning fluids. These insights significantly aid in preparing uniform monodispersed droplets of non-Newtonian fluids using microfluidic devices.

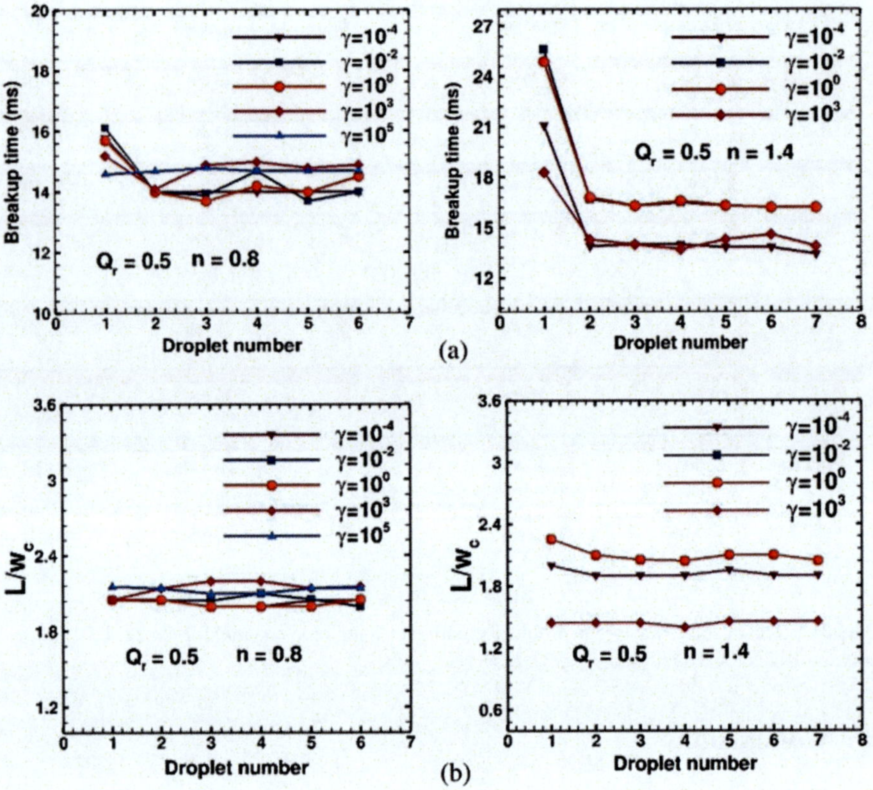

Fig. 6 Effect of the shear rate on the droplet (**a**) breakup time and (**b**) length as a function of droplet number for $Ca_c = 0.01$, $Q_r = 0.5$, $Re_c = 0.1$, and $Q_c = 1\text{ml/h}$

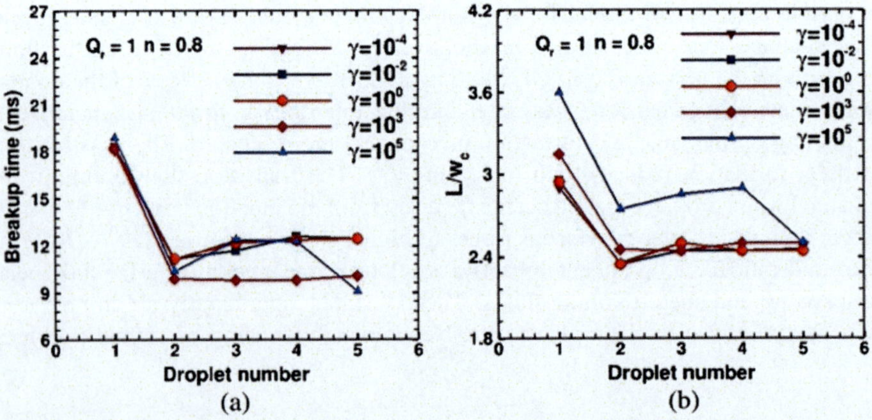

Fig. 7 Effect of the shear rate on (**a**) the droplet breakup time and (**b**) the droplet length as a function of droplet number for $Ca_c = 0.01$, $Q_r = 1$, $Re_c = 0.1$, and $Q_c = 1\text{ml/h}$

References

1. Nguyen, N.-T., Wereley, S.T., House, A.: Fundamentals and Applications of Microfluidics, 2nd edn. Artech House, Boston (2002)
2. Whitesides, G.M.: The origins and the future of microfluidics. Nature **442**, 368–373 (2006). https://doi.org/10.1038/nature05058
3. Ulmeanu, M.: Preparation and characterization of water in oil emulsion via drop break-off. Colloids Surfaces A Physicochem. Eng. Asp. **316**, 119–124 (2008). https://doi.org/10.1016/j.colsurfa.2007.08.038
4. Garstecki, P., Fuerstman, M.J., Stone, H.A., Whitesides, G.M.: Formation of droplets and bubbles in a microfluidic T-junction: Scaling and mechanism of breakup. Lab Chip. **6**, 437–446 (2006). https://doi.org/10.1039/b510841a
5. Granick, S., Zhu, Y., Lee, H.: Slippery questions about complex fluids flowing past solids. Nat. Mater. **2**, 221–227 (2003). https://doi.org/10.1038/nmat854
6. Amiri Delouei, A., Sajjadi, H., Izadi, M., Mohebbi, R.: The simultaneous effects of nanoparticles and ultrasonic vibration on inlet turbulent flow: An experimental study. Appl. Therm. Eng. **146**, 268–277 (2019). https://doi.org/10.1016/j.applthermaleng.2018.09.113
7. Qiu, D., Silva, L., Tonkovich, A.L., Arora, R.: Micro-droplet formation in non-Newtonian fluid in a microchannel. Microfluid. Nanofluidics. **8**, 531–548 (2010). https://doi.org/10.1007/s10404-009-0487-5
8. Song, H., Chen, D.L., Ismagilov, R.F.: Reactions in droplets in microfluidic channels. Angew. Chemie—Int. Ed. **45**, 7336–7356 (2006). https://doi.org/10.1002/anie.200601554
9. Jamalabadi, M.Y.A., DaqiqShirazi, M., Kosar, A., Shadloo, M.S.: Effect of injection angle, density ratio, and viscosity on droplet formation in a microfluidic T-junction. Theor. Appl. Mech. Lett. **7**(4), 243–251 (2017). https://doi.org/10.1016/j.taml.2017.06.002
10. Soh, G.Y., Yeoh, G.H., Timchenko, V.: Numerical investigation on the velocity fields during droplet formation in a microfluidic T-junction. Chem. Eng. Sci. **139**, 99–108 (2016). https://doi.org/10.1016/j.ces.2015.09.025
11. Thorsen, T., Roberts, R.W., Arnold, F.H., Quake, S.R.: Dynamic pattern formation in a vesicle-generating microfluidic device. Phys. Rev. Lett. **86**, 4163–4166 (2001). https://doi.org/10.1103/PhysRevLett.86.4163
12. Gupta, A., Kumar, R.: Flow regime transition at high capillary numbers in a microfluidic T-junction: Viscosity contrast and geometry effect. Phys. Fluids. **22**, 122001 (2010). https://doi.org/10.1063/1.3523483
13. Gupta, A., Sbragaglia, M.: Effects of viscoelasticity on droplet dynamics and breakup in microfluidic T-Junctions: a lattice Boltzmann study. Eur. Phys. J. E. **39**, 1–16 (2016). https://doi.org/10.1140/epje/i2016-16006-9
14. Abate, A.R., Mary, P., Van Steijn, V., Weitz, D.A.: Experimental validation of plugging during drop formation in a T-junction. Lab Chip. **12**, 1516–1521 (2012). https://doi.org/10.1039/c2lc21263c
15. Sontti, S.G., Atta, A.: CFD analysis of microfluidic droplet formation in non–Newtonian liquid. Chem. Eng. J. **330**, 245–261 (2017). https://doi.org/10.1016/j.cej.2017.07.097
16. Wong, V.L., Loizou, K., Lau, P.L., Graham, R.S., Hewakandamby, B.N.: Numerical studies of shear-thinning droplet formation in a microfluidic T-junction using two-phase level-SET method. Chem. Eng. Sci. **174**, 157–173 (2017). https://doi.org/10.1016/j.ces.2017.08.027
17. Venkateshwarlu, A., Bharti, R.P.: Effects of capillary number and flow rates on the hydrodynamics of droplet generation in two-phase cross-flow microfluidic systems. J. Taiwan Inst. Chem. Eng. **129**, 64–79 (2021). https://doi.org/10.1016/j.jtice.2021.07.045
18. Vishal, G., Tomar, J., Bharti, R.P.: Critical parameters for non-Newtonian shear-thickening power-law fluids flow across a channel confined circular cylinder. J. Taiwan Inst. Chem. Eng. **123**, 34–46 (2021). https://doi.org/10.1016/j.jtice.2021.05.025
19. Chhabra, R.P., Richardson, J.F.: Non-Newtonian Flow and Applied Rheology: Engineering Applications. Butterworth-Heinemann, Oxford, UK (2008)

20. Chiarello, E., Gupta, A., Mistura, G., Sbragaglia, M., Pierno, M.: Droplet breakup driven by shear thinning solutions in a microfluidic T-junction. Phys. Rev. Fluids. **2**, 1–13 (2017). https://doi.org/10.1103/PhysRevFluids.2.123602
21. Husny, J., Cooper-White, J.J.: The effect of elasticity on drop creation in T-shaped microchannels. J. Nonnewton. Fluid Mech. **137**, 121–136 (2006). https://doi.org/10.1016/j.jnnfm.2006.03.007

Effects of Inertial Force and Interfacial Tension on Droplet Generation in a T-junction Microfluidic System

Shuvam Samadder, Akepogu Venkateshwarlu, and Ram Prakash Bharti

Abstract In this study, the characteristics of droplet generation in the T-junction microchannel has been studied numerically by varying the Reynolds number of the continuous phase (Re_c) from $10^{-3} - 0.1$, and dispersed phase (Re_d) from $10^{-3} - 10$ for a constant capillary number of the continuous phase (Ca_c). The main focus of our study is to analyze the influence of phase velocity on different flow regimes, droplet size, and droplet detachment time. The results were explored over the range of interfacial surface tension between two phases for 5 and 10 mN/m. It has been found that an immense value of Re_c produces a smaller droplet and increases the droplet detachment time. Also, change in surface tension has a less significant effect on droplet size for a constant low capillary number.

Keywords Microfluidics · Droplet · Inertial force · Capillary number · Reynolds number · CFD

1 Background

1.1 Introduction

Microfluidic devices hold a considerable share in the future point-of-care (POC) and biotechnology market. Microfluidic devices are widely used for chemical reactions [1], food engineering [1], material synthesis [2], and many more. Hence, sufficient knowledge is necessary to tune the flow parameters to precisely control the microfluidic flow behavior. Among others, T-junction is very commonly used to produce microdroplets. It is generally easy to fabricate and operate this geometry for monodisperse droplet generation. The mono-dispersity of the droplet produced is relatively high and efficient at higher flow rates [3]. Earlier studies have made

S. Samadder · A. Venkateshwarlu · R. P. Bharti (✉)
Complex Fluid Dynamics and Microfluidics (CFDM) Lab, Department of Chemical Engineering, Indian Institute of Technology Roorkee, Roorkee 247667, India
e-mail: rpbharti@iitr.ac.in

R. P. Bharti and K. M. Gangawane (eds.), *Recent Trends in Fluid Dynamics Research*, Lecture Notes in Mechanical Engineering,
https://doi.org/10.1007/978-981-16-6928-6_14

limited attempts to explore the effect of inertial forces on droplet size and regime. A relevant prevailing simulation study [4] has analyzed capillary number and flow rate ratio effects on droplet formation characteristics. Mastiani et al. [4] have studied numerically the influence of the continuous phase Reynolds number on gas–liquid system. On the other hand, Jena et al. [5] have focused on the effect of dispersed phase inertia on droplet formation in T-junction microchannel flow. Prevailing studies lack the contribution of both the continuous and dispersed phase Reynolds numbers, which have been studied numerically in detail in this study.

The flow rates of the continuous and dispersed phases have been varied to study the influence of various forces such as inertial, viscous, and interfacial on the droplet generation. The results have been presented as a function of capillary number ($Ca_c = \frac{u_c \mu_c}{\sigma}$) and Reynolds numbers ($Re_c = \frac{u_c \rho_c w_c}{\mu_c}$ and $Re_d = \frac{u_d \rho_d w_d}{\mu_d}$). The subscripts 'c' and 'd' denote the continuous and dispersed phases. The present study uncovers exciting behavior of the droplet size, and flow regime in terms of Re_c and Re_d.

1.2 Objectives

The main objective of the present study is to understand the characteristics of droplet generation in the T-junction microfluidic device and their dependence on the Reynolds number of continuous (Re_c=$10^{-3}-0.1$) and dispersed ($Re_d = 10^{-3} - 10$) phases for a constant capillary number ($Ca_c = 10^{-2}$). Results expressing the effect of flow rates on flow regimes, droplet size, droplet detachment time, and droplet effective diameter. Further, the study was performed by varying the interfacial tension (σ) value as 5 and 10 mN/m to explore their effect on the droplet generation.

2 Modeling

2.1 Problem Statement

Consider an incompressible, non-reacting, Newtonian flow of two immiscible fluids through the T-junction microchannel, as shown in Fig. 1.

In this geometrical flow arrangement, the continuous phase (CP) enters the main horizontal channel (width w_c μm, and length L_c μm) at a flow rate of Q_c μL/s (corresponding Reynolds number, Re_c). The disperse phase (DP) enters at a flow rate of Q_d μL/s (corresponding Reynolds number, Re_d) to the vertical (or T) legged channel (width w_d μm, and length L_s μm) which is intersecting the main channel at an upstream distance (L_u μm) from the inlet of CP and the downstream distance (L_d μm) from the outlet of the main channel. The channel walls are solid and impermeable. Both fluids are assumed to have equal density (i.e., $\rho_c = \rho_d = 1000$ kg/m^3) but variable viscosity (i.e., $\mu_c \neq \mu_d$).

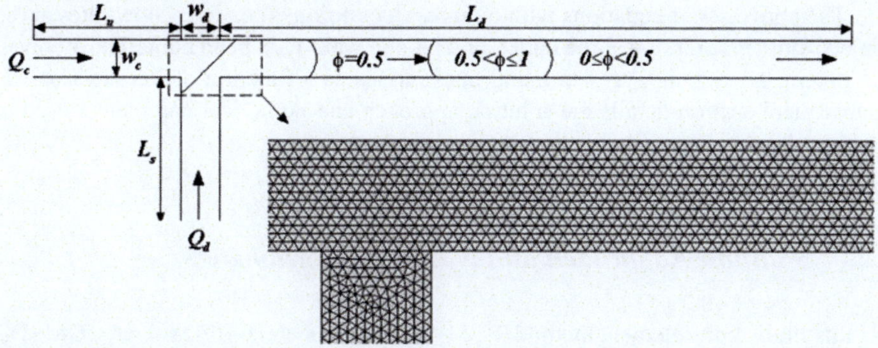

Fig. 1 Schematic geometry of the 2D T-junction microfluidic channel

2.2 Mathematical Model

The mathematical model to represent the above-stated problem is written as follows. Momentum and continuity equations for a two-dimensional, laminar, and unsteady flow are expressed as the following.

$$\rho\frac{\partial \mathbf{u}}{\partial t} + \rho(\mathbf{u}\cdot\nabla\mathbf{u}) = \nabla\cdot\left[-\mathbf{pI} + \mu\left(\nabla\mathbf{u} + (\nabla\mathbf{u})^{\mathbf{T}}\right)\right] + \rho\mathbf{g} + \mathbf{F_{st}} \tag{1}$$

$$\nabla\cdot\mathbf{u} = \mathbf{0} \tag{2}$$

The conservative level set method (CLSM) is adopted to track the interface of the immiscible phases [6–9]. The quantitative fraction of phase (ϕ) varies from 0 (for CP) to 1 (for DP). It is equal to $\phi = 0.5$ at the interface. The phase fraction or level set variable (ϕ) is obtained to represent the interface between two phases from the following level set equation.

$$\frac{\partial\phi}{\partial t} + \mathbf{u}\cdot\nabla\phi = \gamma\nabla\cdot\left(\epsilon_{\mathbf{ls}}\nabla\phi - \phi(\mathbf{1}-\phi)\frac{\nabla\phi}{|\nabla\phi|}\right) \tag{3}$$

where γ is the reinitialization (or stabilization) parameter and ϵ_{ls} is the interface thickness control parameter. The surface tension force ($\mathbf{F}_{\mathrm{st}}$) is given by $\mathbf{F}_{\mathrm{st}} = \sigma\kappa\delta\boldsymbol{n}$**,** where $\boldsymbol{n}$ is interface normal, κ is the interface of curvature, and δ is the Dirac delta function.

The fluid properties like viscosity and density are evaluated as follows.

$$\mu = \mu_c + \phi(\mu_d - \mu_c) \tag{5}$$

$$\rho = \rho_c + \phi(\rho_d - \rho_c) \tag{6}$$

The above set of equations with boundary conditions (i.e., fixed flow rates at the inlets, ambient pressure at the outlet, and no-slip walls) has been numerically solved to obtain the velocity, pressure, and phase fields as a function of Re and Ca. The subsequent section details the solution approach and numerical parameters used in this work.

2.3 Solution Aapproach and Numerical Parameters

In this study, finite element method (FEM)-based commercial CFD solver, COMSOL Multiphysics has been used to simulate the present mathematical model. A two-dimensional (2D), time-dependent study is performed by using the coupled laminar flow and level set modules. The optimally refined [9] free triangular non-uniform mesh (total elements = 13766, degrees of freedom = 53029, and maximum element size = 10 μm) has been used with a custom-defined, obtained after the mesh independence test with different mesh sizes ranging from coarser to more delicate meshing. The parallel direct sparse solver (PARDISO) module has been applied for the iterative computations by using the time step (Δt) = 0.0001 s and relative tolerance (ε) of 0.001.

The following geometrical parameters (**Fig. 1**), free from end effects, have been used in the numerical simulations: $L_u = 900$ μm, $L_d = 3000$ μm, $L_c = 4000$ μm, $L_s = 900$ μm, $w_c = 100$ μm, and $w_d = 100$ μm. The characteristics length is taken as $w_c = 100$ μm. The physical and dimensionless parameters used in this work are noted as follows: First, Re_c is varied in the range as 10^{-3} to 10^{-1} in the 10 folds by fixing $Re_d = 10^{-2}$, and second, Re_d is changed in the range as 10^{-3} to 10 in the 10 folds by fixing $Re_c = 10^{-2}$ to analyze the contributions of inertia exerted by continuous and dispersed phases at the fixed capillary number ($Ca_c = 10^{-2}$). Further, the interfacial tension (σ) is altered as 5 and 10 mN/m in defining the fixed capillary number ($Ca_c = 10^{-2}$). The outlet is open to the ambient ($p = 0$ atm), and the contact angle (θ), suggesting the spread of dispersed phase fluid on a solid wall, is taken as 135° [8–10].

Before presenting new results, the mathematical model and numerical approach have been validated with the limited literature results. Figure 2 presents a comparison of the present numerical results with the experimental and numerical values from the literature [7] for the droplet length (L) versus dispersed phase flow rate (Q_d) at a fixed continuous phase flow rate ($Q_d = 0.028$ μL/s). A close qualitative and a quantitative match between the results obtained from different approaches suggests the reliability and accuracy (within ± 1 - 2%) of the present results discussed hereafter.

3 Results and Discussion

In this section, the present results in terms of the phase flow profiles, droplet length, and detachment time for the ranges mentioned above of conditions have

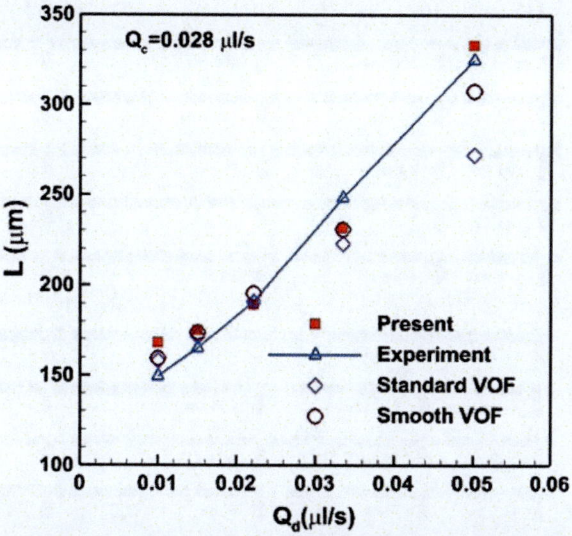

Fig. 2 Comparison of the present results with the literature [11] values

been presented and discussed to analyze the roles of inertial, viscous, and interfacial forces on the droplet generation. The results have been obtained for the following set of conditions: $Re_c = 10^{-3}$, 10^{-2}, and 10^{-1}; $Re_d = 10^{-3}$, 10^{-2}, 10^{-1}, 1, and 10; $Ca_c = 10^{-2}$ for $\sigma = 5$ and 10 mN/m.

3.1 Phase Flow Profiles

Figures 3 and 4 depict the phase composition profiles and different flow regimes for different Re_c and Re_d values at $Ca_c = 10^{-2}$ for $\sigma = 5$, and 10 mN/m, respectively. Figures also show various stages of flow during droplet formation. The dispersed phase is injected through the lateral channel, while the continuous phase is injected into the much-extended continuous channel. The dispersed phase reaches the continuous channel and creates an interface with the continuous phase near the junction where the droplet flow regime develops. Subsequently, the dispersed phase takes the shape of a plug with an effective diameter greater than or equal to the channel width, which eventually occupies the continuous channel. Finally, the delicate balance between inertial, viscous, and interfacial forces results in the droplet generation in varying regimes, as shown in the phase flow profiles.

Phase flow profiles also suggest that the interfacial forces majorly govern the droplet generation rather than inertial forces, only when $(Re_d/Re_c) \approx 1$. When $(Re_d/Re_c) > 1$, the inertial forces due to the dispersed phase come into the picture and cease the droplet generation in the channel. Interestingly, the flow regime for $(Re_d/Re_c) = 1$ transit from squeezing to parallel flow. However, as the ratio (Re_d/Re_c) goes above 1, the regime becomes stratified flow, and ultimately the dispersed phase

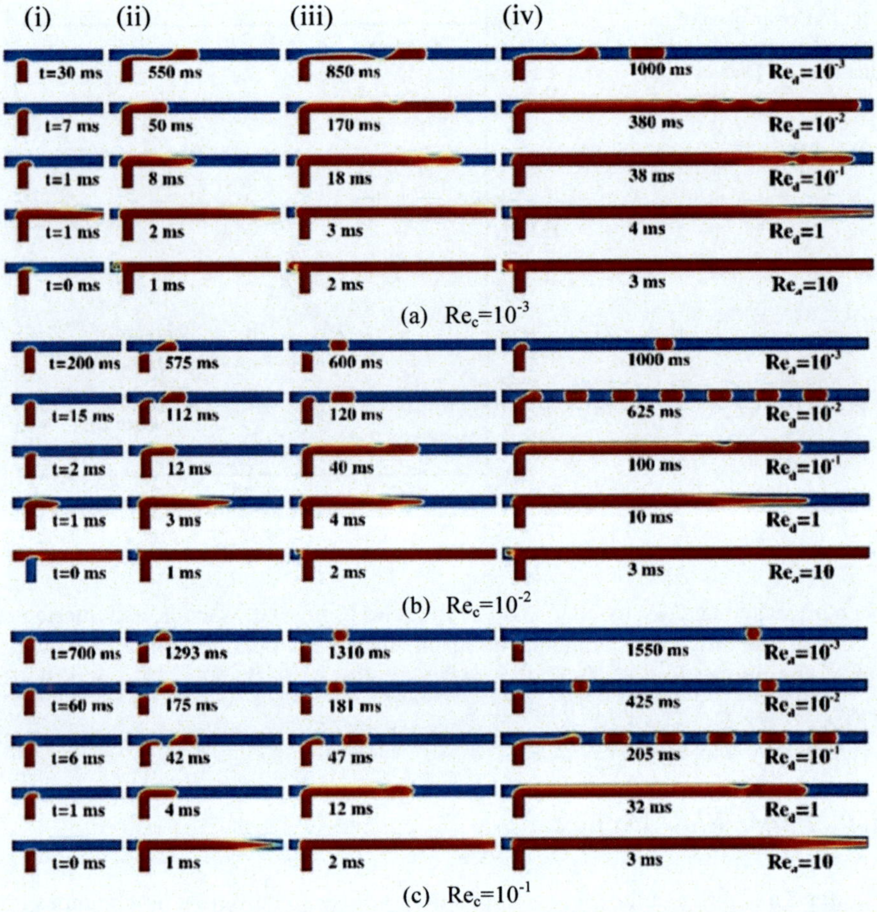

Fig. 3 Instantaneous phase flow profiles as a function of Re_c and Re_d for $Ca_c = 10^{-2}$, $\sigma = 5$ mN/m, and $\theta = 135°$. (i) evolution of dispersed phase, (ii) droplet breakup stage, (iii) stable droplet formation, and (iv) hydrodynamically developed flow filled with droplets/dispersed phase

occupies the downstream channel length. The droplet generation frequency and droplet size increases monotonically with an increase in Re_d until the Reynolds number (Re_d/Re_c) equals 1 for a constant value of Ca_c and Re_c.

Beyond (Re_d/Re_c) > 1, the dispersed phase completely occupies the downstream channel. The inertial force of the dispersed phase dominates over the inertial force of the continuous phase and eventually restricts the flow of the continuous phase in the downstream channel. Therefore, no droplet formation occurs, and the flow regime in the main channel transits from squeezing to parallel flow.

Figure 4 depicts the phase flow profiles for $\sigma = 10$ mN/m, under otherwise for identical conditions of Fig. 3. Qualitatively, the change in interfacial surface tension (σ from 5 to 10 mN/m) negligibly changes the phase flows characteristics as the

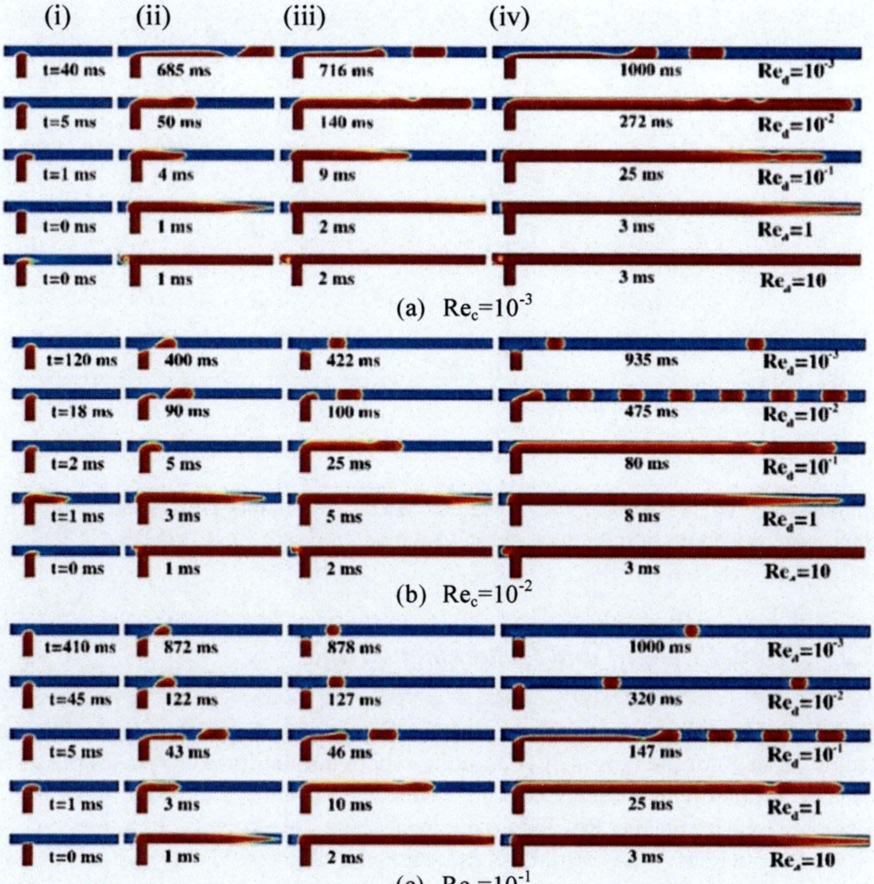

Fig. 4 Instantaneous phase flow profiles as a function of Re_c and Re_d for $Ca_c = 10^{-2}$, $\sigma = 10$ mN/m, and $\theta = 135°$. (i) evolution of dispersed phase, (ii) droplet breakup stage, (iii) stable droplet formation, and (iv) hydrodynamically developed flow filled with droplets/dispersed phase

capillary number is kept constant. Hence, the increased interfacial tension for fixed Ca_c has an insignificant effect on flow kinematics. Some minor notable changes can be observed on droplet detachment time and flow regime with an increase in surface tension. Due to increased surface tension, the flow regime shifts more toward jetting flow from squeezing flow. Also, the droplet detachment length got elongated for a higher surface tension value.

In the liquid–liquid two-phase flow model, droplet generation mechanics is majorly governed by a lower *Re*, which corresponds to a weak inertial effect compared with the viscous effect of the phase flow velocity. When *Re* is less than 1, interfacial and viscous forces are dominant during droplet generation. However

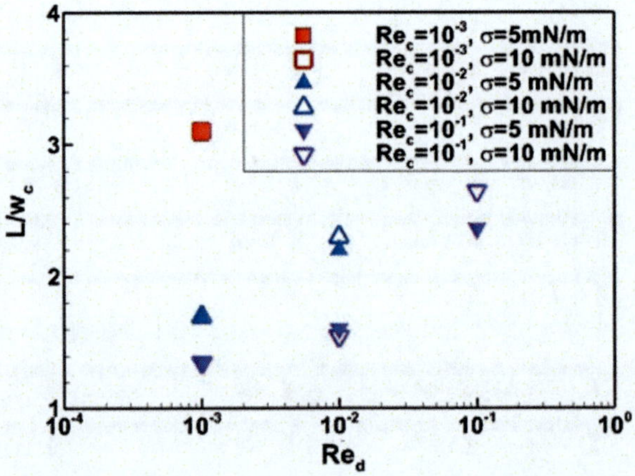

Fig. 5 Effect of Re_c and Re_d on droplet length for $Ca_c = 10^{-2}$

in this study, when the Re_d is revamped from 10^{-3} to 10, the inertial forces become dominant and come into the picture, as visualized through Figs. 3 and 4.

3.2 Droplet Length and Detachment Time

Figure 5 shows the effects of $\mathrm{Re_c}$, $\mathrm{Re_d}$, $\mathrm{Ca_c}$, and σ on the dimensionless droplet length (L/w_c) for the ranges of conditions, where droplet formation takes place. In general, droplet length decreases with increasing $\mathrm{Re_c}$ for a fixed $\mathrm{Re_d}$. However, it increases with increasing $\mathrm{Re_d}$ for a fixed $\mathrm{Re_c}$.

As Re_c increases, the inertial forces increase and dominate over surface tension forces. For a fixed Re_d, the upstream pressure increases with increasing Re_c, which shifts the droplet regime more toward dripping or squeezing. Those kinds of regimes are more inertia/pressure force dominated in the neck of the channel. This is visible when comparing multiple cases of Re_c values keeping Re_d constant. The increasing $\mathrm{Re_d}$ enhances the Laplacian pressure inside the dispersed phase, thereby tending to enlarge the formed droplets, i.e., droplet length increases with $\mathrm{Re_d}$.

The droplet size (or length) can also be correlated with the droplet detachment time. Subsequently, Fig. 6 shows the effects of $\mathrm{Re_c}$, $\mathrm{Re_d}$, $\mathrm{Ca_c}$, and σ on the dimensionless droplet detachment time ($T = t\, w_c/u_c$) over the ranges of conditions.

The detachment time increases with increasing Re_c for fixed Re_d, which is the opposite to the case of a gas–liquid system [4]. It, however, decreases with increasing $\mathrm{Re_d}$ for a fixed $\mathrm{Re_c}$. This behavior can be understood in terms of viscous force. Since Re is the ratio between inertial and viscous force, for a fixed Re_c, the viscous force of the continuous phase is fixed, but the viscous force of the dispersed phase keeps decreasing with increasing Re_d hence the continuous phase needs lesser force, thus less time to form the droplet. On the other hand, for fixed $\mathrm{Re_d}$, it enhances the time for the droplet detachment.

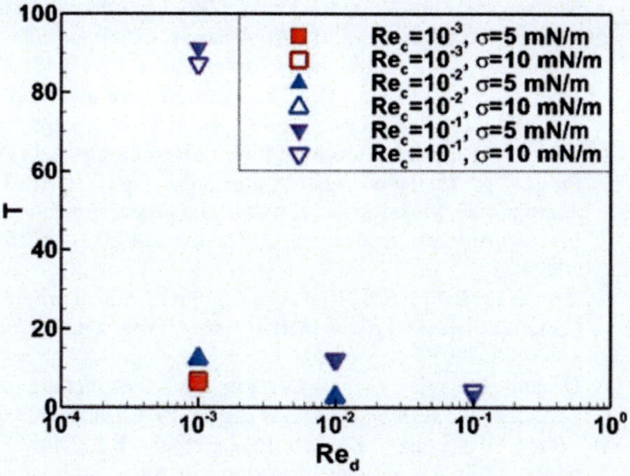

Fig. 6 Effect of Re_c and Re_d on droplet detachment time for $Ca = 10^{-2}$

4 Conclusions

In the present study, the effects of inertial forces on the droplet generation in a T-junction microchannel are explored numerically by using the finite element method-based COMSOL Multiphysics. The mathematical model for the two-phase flow is coupled with the conservative level-set method. The droplet generation nature is studied by varying Reynolds number (*Re*) of both the dispersed and continuous phases. The phase profile results indicate the various stages of droplet formation and flow regimes presented for the broad ranges of conditions ($Re_c = 10^{-3}$ - 10^{-1}; $Re_d = 10^{-3}$ to 10) at fixed $Ca_c = 10^{-2}$. The change in surface tension has shown an insignificant effect on the flow characteristics for a constant Ca_c. Due to increased surface tension, the flow regime shifts more toward parallel from dripping flow. Apart from that, the droplet detachment length got elongated for a higher surface tension value. Droplet formation is mainly seen at a low *Re* which corresponds to the weak inertial effects. Droplet generation is majorly governed by the surface tension forces rather than inertial forces, only when the $Re_d/Re_c \approx 1$. For $(Re_d/Re_c) > 1$, the inertial forces due to the dispersed phase dominate the flow and cease the droplet generation in the channel. However, as $(Re_d/Re_c) = 1$ with increases in Re_d, the droplet generation frequency increases monotonically. The droplet length decreases with increasing Re_c for a fixed Re_d. However, it increases with increases Re_d for a fixed Re_c. The droplet detachment time has shown an inverse trend to the droplet length. These results would help design a suitable microfluidic device for a specific application according to the flow conditions.

References

1. Song, H., Chen, D.L., Ismagilov, R.F.: Reactions in droplets in microfluidic channels. Angew.

Chemie - Int. Ed. **45**, 7336–7356 (2006). https://doi.org/10.1002/anie.200601554
2. Skurtys, O., Aguilera, J.M.: Applications of microfluidic devices in food engineering. Food Biophys. **3**, 1–15 (2008). https://doi.org/10.1007/s11483-007-9043-6
3. Dendukuri, D., Doyle, P.S.: The synthesis and assembly of polymeric microparticles using microfluidics. Adv. Mater. **21**, 4071–4086 (2009). https://doi.org/10.1002/adma.200803386
4. Seemann, R., Brinkmann, M., Pfohl, T., Herminghaus, S.: Droplet based microfluidics. Reports. Prog. Phys. **75**, 016601 (2012). https://doi.org/10.1088/0034-4885/75/1/016601
5. Mastiani, M., Mosavati, B., Kim, M.: Numerical simulation of high inertial liquid-in-gas droplet in a T-junction microchannel. RSC Adv. **7**, 48512–48525 (2017). https://doi.org/10.1039/c7ra09710g
6. Jena, S.K., Bahga, S.S., Kondaraju, S.: Prediction of droplet sizes in a T-junction microchannel: Effect of dispersed phase inertial forces. Phys. Fluids. **33**, 032120 (2021). https://doi.org/10.1063/5.0039913
7. Olsson, E., Kreiss, G.: A conservative level set method for two phase flow. J. Comput. Phys. **210**, 225–246 (2005). https://doi.org/10.1016/J.JCP.2005.04.007
8. Wong, V.L., Loizou, K., Lau, P.L., Graham, R.S., Hewakandamby, B.N.: Numerical studies of shear-thinning droplet formation in a microfluidic T-junction using two-phase level-SET method. Chem. Eng. Sci. **174**, 157–173 (2017). https://doi.org/10.1016/j.ces.2017.08.027
9. Bashir, S., Rees, J.M., Zimmerman, W.B.: Simulations of microfluidic droplet formation using the two-phase level set method. Chem. Eng. Sci. **66**, 4733–4741 (2011). https://doi.org/10.1016/j.ces.2011.06.034
10. Venkateshwarlu, A., Bharti, R.P.: Effects of capillary number and flow rates on the hydrodynamics of droplet generation in two-phase cross-flow microfluidic systems. J. Taiwan Inst. Chem. Eng. 129, 64–79 (2021). https:\\doi.org/10.1016/j.jtice.2021.07.045
11. Van Der Graaf, S., Nisisako, T., Schroën, C.G.P.H., Van Der Sman, R.G.M., Boom, R.M.: Lattice Boltzmann simulations of droplet formation in a T-shaped microchannel. Langmuir **22**, 4144–4152 (2006). https://doi.org/10.1021/la052682f
12. Soh, G.Y., Yeoh, G.H., Timchenko, V.: Numerical investigation on the velocity fields during droplet formation in a microfluidic T-junction. Chem. Eng. Sci. **139**, 99–108 (2016). https://doi.org/10.1016/j.ces.2015.09.025

Drag Reduction of Sphere Using Acrylic and Alkyd Paints: A New Approach

Saroj Kumar Samantaray, Mohammad Hussain, and Basudeb Munshi

Abstract The Acrylic and Alkyd coated spheres of different diameters are used in this study to investigate the variation of coefficient of drag in Newtonian fluid in an inclined cylindrical tube. The diameter and the density of the spheres were kept the same as the uncoated sphere. The drag coefficients of the coated spheres were found comparatively less than the uncoated sphere, especially at a high Reynolds number. The paint medium has influenced the fluid dynamic interaction and resulted in the minimum pressure drop around or across the sphere, which led to less drag coefficient than the uncoated sphere.

Keywords Drag coefficient · Acrylic and Alkyd paint · Drag crisis · Pressure drop

1 Introduction

The polymers and surfactants are added in a fluid medium to reduce the drag acting on the particle. This area of research plays a vital role in many industrial applications like sediment transportation and slurry transportation. The drag reduction is fluidized and packed may reduce the pressure drop to achieve a more economic process. The drag applied on the solid particles was successfully reduced by applying a significantly less polymer (ppm) [1]. Toms [2] reported the first use of polymeric additives to reduce the drag coefficient. The author explained that adding fewer polymers with high molecular weight in a flow mainly turbulent can decrease the friction factor up to 80%. Drag is the function of the friction factor of solution additives by concerning the solvent at the same Reynolds number explained by Lumley [3]. The phenomenon is widely studied in practical industrial applications in engineering due to its benefits. Polyethylene oxide (PEO) also delivers the same impacts on drag reduction. Vlachogiannis et al. [4] studied polymers shows the most effect on drag reduction when the polymers are applied with large structures, i.e., high molecular weights, to singular monomer units. Subsequently, Khanom et al. [5]

S. K. Samantaray · M. Hussain · B. Munshi (✉)
Department of Chemical Engineering, Laboratory of Transport Phenomenon, NIT Rourkela, Odisha 769008, India

R. P. Bharti and K. M. Gangawane (eds.), *Recent Trends in Fluid Dynamics Research*, Lecture Notes in Mechanical Engineering,
https://doi.org/10.1007/978-981-16-6928-6_15

explained the effect of polyacrylamide (PAM) on drag reduction in an area of raw petroleum in channel lines. Tian et al. [6] explained the drag reduction by adding Hydroxypropyl xanthan gum (HXG) depends on the concentration of HXG used. In other cases, it is observed that the high polymer increases the viscosity magnitude of liquid, hence the drag ingresses significantly. Hence, it was advised for determination of the proper ppm before the experiment. The cationic surfactant also delivers the same effect on the reduction of drag. The use of surfactant or polymer in reducing drag depends on the nature of the physical problem and the fluid to be used as a channel medium. Alkyd is an oil-based paint; the viscosity over the whole shear rate range is too low for such a paint because of its typical dispersion rheology. Acrylic polymers are known as acrylics. Acrylate polymers are widely utilized as coatings and thickeners. Acrylics show good resistance to UltraViolet light, stiffening on aging, heat, dry-cleaning solvents, and chemicals. There are applied on automotive fabric and carpets as back coating materials, window drapes, and pile fabrics. Oil-based paints provide good resistance to wear and tear. It shows good properties like less shrinkage, the effect of temperature is more petite, durable is more, holds rust better on steel, and has better adhesive on chalky and dirty surfaces. Still, it is unable to form a better adhesion bond with the surface. Water-based paints can expand and contract with weather conditions. They can withstand a small amount of moisture on a surface. It creates a strong adhesion bond to the surface. It also shows good thermal properties, and drying of a surface is fast.

1.1 Physics Around Spherical Particle

The representation of the particle is given in Fig. 1 the different forces acting on the spherical particle when it is traveling in a fluid medium, and the forces exerted on a particle are drag force (F_D), the net bodyweight of the particle (F_B sin θ), and resistance force (F_R). At equilibrium, the forces are balanced, represented in equation [1].

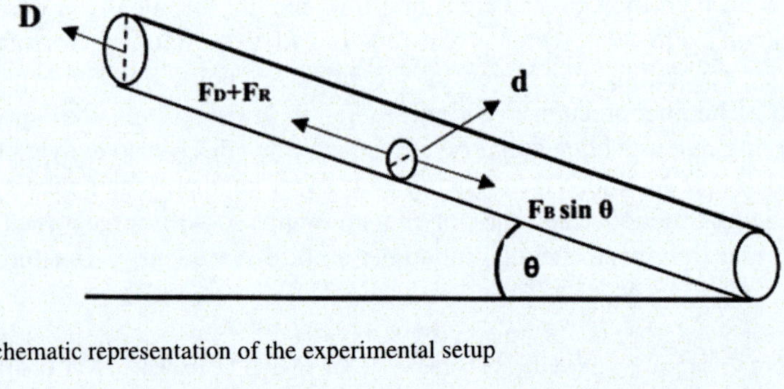

Fig.1 Schematic representation of the experimental setup

$$FB\sin\theta = FD + FR \tag{1}$$

The net drag force observed by the sphere is expressed as a non-dimensional quantity called the coefficient of drag as a function of Reynolds number [7, 8] the drag coefficient, C_D of a particle in an inclined plane defined as, [9, 10]

$$C_D = \frac{4gd}{3V^2}\left(\frac{\rho_S - \rho_F}{\rho_F}\right)\sin\theta \tag{2}$$

For regular plane consideration θ=90^0

$$C_D = \frac{4gd}{3V^2}\left(\frac{\rho_S - \rho_F}{\rho_F}\right) \tag{3}$$

1.2 Contribution of This Work

The thorough literature survey showed that many qualitative works are reported on drag reduction using polymers in flow medium. However, no information is communicated about the effect of paint on drag reduction. Hence, in this work, acrylic (water-based) and alkyd (oil-based) paints are used to coat the sphere, and the comparison among the standard and coated spheres are reported in terms of a drag coefficient.

2 Experimental

2.1 Materials

The steel spheres are used with different diameters ranging from 3.0 mm to 11.40 mm and densities from 1420.95 to 9330.04 kg/m^3. 1.2 m long transparent perplex tube of inner diameter 30.10 mm is used as the flow domain for all spheres. The glycerin solutions with four different concentrations of glycerin and pure form glycerin are utilized as the working fluids. Rheometer (Anton Parr, Germany, RheolabQC) and pycnometer measure the viscosities and densities of the working fluids. The physical properties of working fluids are listed in Table 2. Spheres are coated with a very thin layer of acrylic and alkyd paints manufactured and marketed by Asian paint. The densities and diameters of the sphere are changed with negligible magnitude, and hence, those are neglected in the current study. The properties of the used paints are mentioned in Table 1.

Table 1 Properties of the used paints in the current study

Name	Acrylic paint	Alkyd paint
Product name	Classique premium acrylic interior emulsion	Decora synthetic enamel
Company name	Asian Paints	Asian Paints
Type	Water based	Oil based
Application viscosity	50–60 KU (Krebs Units (KU)) when diluted to 35–40 volume % by water	55–60 KU (Krebs Units (KU)) when diluted to 10–15 volume % by thinner (T101)

Table 2 Physical properties of working fluids at 25 °C

Solution. No	Fluid name	Viscosity (Pa. s)	Density(kg/m^3)
1	100% glycerin	0.7533	1250.93
2	80% glycerin	0.0571	1198.34
3	70% glycerin	0.024	1167.66
4	50% glycerin	0.0059	1144.81
5	35% glycerin	0.003	1121.23

2.2 *Experimental Setup and Procedure*

The experimental setup is supported on a wooden frame. The 1.2 m long transparent tube of diameter 30.10 mm is mounted on a wooden frame, one end is fixed, and the other is allowed to change angles from 10^0 to 90^0 with clamps hooked at different angles. The movable end changes to other angles concerning requirements during the experiment. The working fluid is filled 24 h before the experiment to reach bubble-free conditions and thermal stability. The forceps and a stopwatch are used to release the particle in the fluid domain and note particle settling time. The reading is taken far away from both inlet and outlet of the tube. The experiments are repeated five times, and the mean settling time is recorded to calculate the terminal velocity of the Particle and, finally, the coefficient of drag and Reynolds number.

3 Result and Discussion

The drag and Reynolds numbers coefficient are calculated using fundamental properties of the fluid medium and sphere. The C_D-Re relationships for the used angle of inclinations are reported in Fig. 2 for uncoated, water-based coated, and oil-coated paint. Figure 2a shows a reduction in C_D for painted sphere, and the prominent effect is seen for Alkyd, especially in post-transition and early turbulent zone for 90°, i.e., vertical channel. The same trend was also seen for 70^0 tube inclination (Fig. 2b),

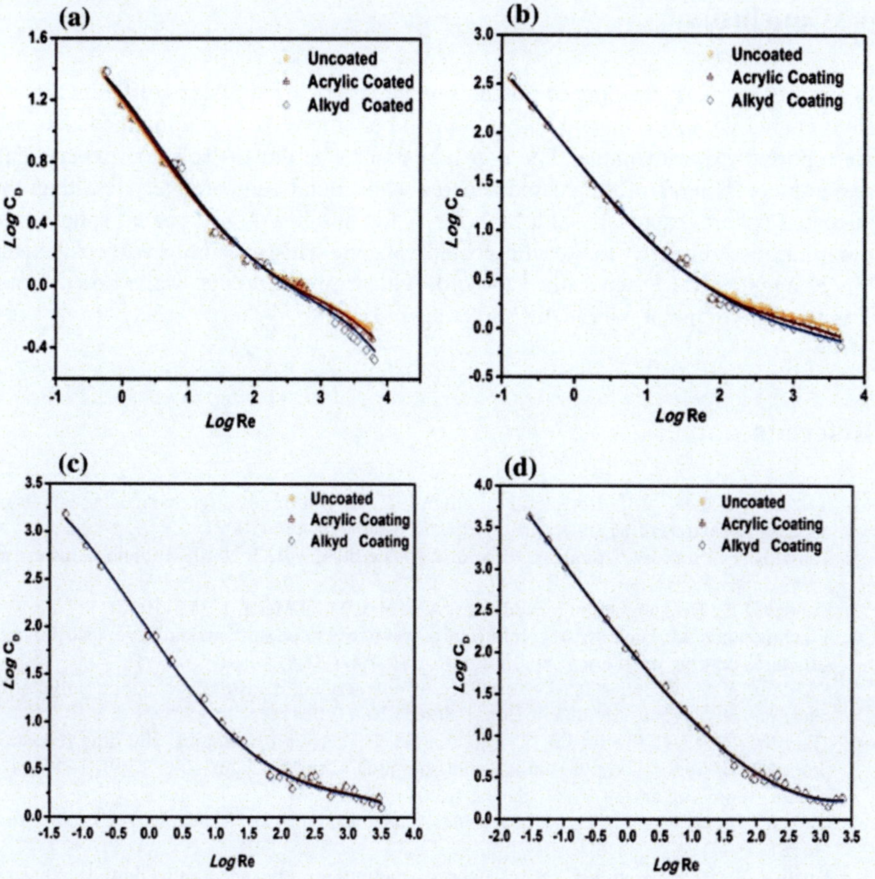

Fig. 2. C_D-Re relationship for different angle of 90°,70°,45° and 30°, respectively

but the effect of paint on C_D minimizes due to extra retardation, i.e., sliding resistance from the wall. Another reason is for the 70^0 tube inclination that the downward body force component, i.e., decreases, which increases the C_D. The above-discussed effect of paint is not visible for 45^0 angles of inclination and below to this. Here, the developed resistance force surpassed the effect of paint. In Fig. 2c and d, the drag coefficients obtained for the Alkyd coated sphere is minimum as compared to the Acrylic coated sphere. This may be reasoned that the binder agent in oil-based paint is Alkyd, which minimizes the wake downstream than the Acrylic paint, where the binder agent is water.

4 Conclusion

In this present work, the drag coefficients of the spherical particles of different dimensions coated with Acrylic and Alkyd paint in both the vertical and inclined channels are reported experimentally. The conclusions can be drawn from the experiment: the drag coefficients of the coated spheres were found comparatively less than the uncoated sphere, especially at a high Reynolds number. Alkyd coated spheres are minimum as compared to Acrylic coated spheres. This may be reasoned that the binder agent in oil-based paint is Alkyd, which minimizes the wake downstream than the Acrylic paint, where the binder agent is water.

References

1. Lyotard, N., Shew, W.L., Bocquet, L., Pinton, J.-F.: Polymer, and surface roughness effects on the drag crisis for falling spheres. Eur. Phys. J. B **60**, 469–476 (2007)
2. Toms, B.A.: Proc. Int. Congress. Rheology Amsterdam, Vol. 2. North-Holland, Amsterdam (1949).
3. Lumley, J.L.: Drag reduction by additives. Ann. Rev. Fluid Mech. **1**, 367 (1969)
4. Vlachogiannis, M., Hanratty, T.J.: Influence of wavy structured surfaces and large scale polymer structures on drag reduction. Exp. Fluids **36**, 685–700 (2004)
5. Khadom, A.A., Abdul, A.A.: Performance of polyacrylamide as drag reduction polymer of crude petroleum flow. Ain Shams Eng. J. **5**(861), 865 (2014)
6. Tian, M., Fang, B., Jin, L., Lu, Y., Qiu, X., Jin, H., Li, K.: Rheological and drag reduction properties of hydroxypropyl xanthan gum solutions. Chin. J. Chem. Eng. **23**(9), 1440–1446 (2015). https://doi.org/10.1016/j.cjche.2015.04.003
7. Clift, R., Grace, J.R., Weber, M.E.: Bubbles, Drops and Particles. Academic Press, New York (1978)
8. Khan, A.R., Richardson, J.F.: the resistance to motion of a solid sphere in a fluid. Chem. Eng. Commun. **62**, 135–150 (1987)
9. Jan, C.D., Chen, J.C.: Movements of a sphere rolling down an inclined plane. J. Hydraulic Res. **35**(5), 689–706 (1997)
10. Chhabra, R.P., Ferreira, J.M.: an analytical study of the motion of a sphere rolling down a smooth inclined plane in an incompressible Newtonian fluid. Powder Technol. **104**, 130–138 (1999)

Low-Frequency Acoustics Assisted Propagating Fires and Related Implications

Saumya Shekhar, Bhushan Thombare, and Vinayak Malhotra

Abstract Fire is one of the finest revelations of mankind. Combustion is basically an imperative portion of lifestyle. Forest fires have become a major concern in terms of human fatalities, financial loss, and environmental destruction as a result of global warming and ever-increasing urban interface regions. To obtain it and to predict potential fire activity, fire modeling is used. From the data estimated by National Fire Prevention Association (NFPA) for the year 2019, losses are: 8.5 million acres of land are lost due to forest fires, 10.7 billion dollars (USD) was estimated to be structured fire loss, 10,700 million dollars (USD) loss of residential, industrial, educational institutions. This shows the complexity of the problem which has been prevented through understanding due to heterogeneous heat and mass transfer. Through actual experimentation, the acoustic effect on propagating fires is investigated in the aid of spread rate. The present work is motivated by the superior standards of fire safety from practical and functional significance as it covers a wide range of engineering and industrial applications. The changes in heat transfer will be reflected in increments or decrements in spread rate. The present work highlights the effect of acoustic frequency and acoustic source distance on the flame spreading in opposed configuration. An experimental setup was upraised comprising of an acoustic source with essential controls, different configurations coupled with external heat sources. The frequency of the sound source is systematically varied keeping the ignition front fixed. Optimal conditions in terms of sound source location, frequency, the orientation of fuel is investigated for applications needing fast-spreading and flame extinction.

Keywords Fire propagation · Forward heat transfer · Compression and rarefaction

S. Shekhar · B. Thombare (✉) · V. Malhotra
Department of Aerospace Engineering, SRM Institute of Science and Technology, Chennai 603203, India
e-mail: bp5463@srmist.edu.in

R. P. Bharti and K. M. Gangawane (eds.), *Recent Trends in Fluid Dynamics Research*, Lecture Notes in Mechanical Engineering,
https://doi.org/10.1007/978-981-16-6928-6_16

1 Introduction

Fires are certainly one of humanity's most remarkable advancements, and it's been a constant source of human evolution. Effective use of fire has resulted in massive progress in the disciplines of technology, industrialization, the practical environment, functional and operating processes in a variety of realms. Although fire is a very normal phenomenon in which almost everyone uses the area of architecture, industrialization, practical environment, functional and operating structures in various realms, it often has the potential to get out of control, posing a risk to human lives and property.

The fire has been the cause of the most devastating incidents, such as massive fires in factories, wildfires, house fires, and aviation and spacecraft accidents, which result in incalculable loss of humanity, wildlife, and wealth, and each year an insane amount of money is spent in science to avoid fires. According to figures provided by the National Fire Prevention Association (NFPA), about 8.5 million acres of land were destroyed due to forest fires, with an annual cost of 10.7 billion dollars (USD). The loss of private, commercial, and educational institutions was estimated to be worth about $10,700 million (USD). Recently, forest fires have occurred in the Amazon River, Australia, Uttarakhand, and California, causing huge accidents and threatening the lives of people, animals, property. As seen from (Figs. 1 and 2) its details about the fire propagation in aviation and rocketry. Similarly, fire is responsible for 6% of aviation mishaps, resulting in a massive death toll and economic damage. The growth and propagation of flames is the primary cause these disasters. Fire proliferation focused on this work. Through this combustion, the development speed of the flame, that is, the moving speed of the flame on the surface of the fuel is

Fig. 1 SpaceX SN4 starship test

Fig. 2 Saudi Arabian flight crash

determined. The established acoustic excitation increases the combustion efficiency of the flame as a result of the flame's instability induced by the introduction of sound waves. Acoustic energy creates localized pressure and velocity fields around the pilot fuel that are influenced by lead.

Disturbances in a thermoacoustic system can arise when acoustic and thermal fluctuations interact constructively, from theory presented by Rayleigh [1]. Understanding the dynamics of acoustic and entropy wave production through the blaze, as well as their propagation in the environment, is therefore critical for predicting and controlling thermoacoustic instabilities. From the conventional work done by Clarke & McChesney [3] suggests that Wave attenuation can occur in the dissociating mixture when the wave drives the non-equilibrium portion of the flow. Elaine et. al. [2] report how scattering frequency emerges when the shape of a sound wave is altered while spreading through a non-equilibrium background. Furthermore, they proposed that only acoustic wave amplification is required when non-equilibrium flow occurs in the background, or when induced by external sources rather than propagating waves. Clarke [4] has shown for the experiment the acoustic wave can be amplified by non-equilibrium background flow. Toong [5] observes from experiments based on diffusion flame and shown evidence of both the suppression and the amplification of sound waves when they interact with a flame. The horizontal spread of the fuel increases the flame width. In the 'DARPA' project which was funded by the US Department of Defense [6], at a pressure of 0.2 to 112 Pa and under the sound frequency from 35 to 150 Hz, a positive correlation was found for a gas burner between the extinction frequency and extinction pressure. During combustion interference, Nair and Sujith [7] observe the existence of multifractality. Furthermore, they characterized the transformation from combustion noise to thermoacoustic instability as a transfer from disorder to order, which is expressed in the pressure fluctuations as a lack of multifractality. Nair et al. [8] also state that the initiation of thermoacoustic instability is accompanied by a condition known as intermittency, which is comprised in the middle of the low amplitude aperiodic oscillations, there are occasional bursts of high amplitude periodic oscillations. Bauerheim et. al. [9] reexamined the quasi1D conservation equations and found that acoustic and entropy perturbations are intertwined. The extinction requirements of Bennewitz et. al. [10] for single droplets from three distinct fuels placed in a closed wave guide by acoustic steady-wave node.

Within in the confines of calculable conceptual obligations, an issue that must be thoroughly investigated is the effect of thermoacoustics on external heat sources for various configurations and orientations. Typically, vital impacts on the engendering wonder and inspired us to research and examine fire propagation. While significant work has been done in the past, some elements are yet to be investigated.

The below are the work's basic objectives:

1. To understand the marvel of fire propagation in the proximity of acoustics.
2. To assess the propagation rate on pilot fuel in the presence of external sources in different configurations and orientations.
3. To become familiar with the most important control parameters.

2 Experimental Setup and Solution Methodology

An experimental setup was developed to investigate the spread of fires. It was achieved in order for a variety of criteria to be varied and analyzed as a result. The assembly consists of—(a) a Matchstick holder, (b) a matchstick as a fuel, and (c) a tone generator (Fig. 3).

For blaze stabilization, the coordinate sticks have 0.5 cm division markings accompanied by 3 one cm markings. Interspace distance between two matchsticks was 0.5 cm. To generate different frequency sounds, a multimedia machine speaker 2.1

Fig. 3 Experimental setup

was used. To create the sound of shifting frequencies, the NCH tone generator application was used. The processes were carried out for acoustic bass, pitch, and frequency spectrum of 500–2500 Hz. To investigate the diverse impacts of acoustic operation, four different configurations of heat energy matchsticks were used: (1) Unilateral (2) Asymmetrical (3) a 'Y' or trilateral configuration, and (4) a '+' or quadrilateral configuration. At a distance of 100 cm, the speaker was held perpendicular to the pilot fuel. The primary goal was to conduct the test in the absence of an acoustic source. In each scenario, the fire spread rate was calculated by noting the time it took to burn 1 cm marking. Speakers were placed at a distance of 100 cm from pilot fuel. Both experiments were guaranteed to be repeatable to the third degree. To document the strange events, photographs and videos were taken.

2.1 Forward Transfer Theory

From 2nd Energy Equation:

Energy Change = Energy produced – Energy lost.

So,

$$\rho_s C_s V(dT/dt) = q_p - q_L \tag{1}$$

$$where,\ q_p = \Delta H_c V C_i A * e_a^{(E}/^{RT)} \tag{2}$$

$$q_L = hA(T - T_a) \tag{3}$$

From Classical Forward Heat Transfer theory, regression rate (r) can be calculated by-

$$r = \frac{\int q_{net}}{\rho_s \tau_s c_s (T_{surface} - T_\infty)} \tag{4}$$

2.2 Measurement

The rate at which the surface burns is called the Flame spread rate which is linearly calculated as:

$$r = \frac{\text{Distance burnt}}{\text{Time taken to burn that region}} \tag{5}$$

In each case where a speaker was used to convey sound at specific frequencies, the fire regression rate was determined. The experiment was carried out at standard room temperature, with valid measurements taken to ensure productivity and consistency in each case. It's important to note that all of the data presented here relates to the third order's repeatability and reproducibility.

3 Result and Discussion

At first, an experiment was performed to evaluate the base case values for the regression rate. It should be noted that the flame propagates rate of pilot fuel was calculated from all aspects. Figure 4 depicts a non-monotone trend at 45 degrees with a median burn rate of 5.142 cm/min. The result was tested using standard heat transfer theory and found to be well matched when compared to Pratik et al. [11]. The experiment was conducted to determine the basic fire regression rate values. Note that the amplitude of the pilot fuel with and without acoustic disturbance was measured.

According to conventional heat transfer theory. This appears as a rise or decline in regression values. Because of the convective dense stream, localized direction, and temperature areas, it forms around the fuel surface. High-temperature smoke generates additional energy and preheats by paralleling thermal flow on the moving surface. High heat transfer from burned to unburned fuel can lead to rapid recovery rates.

Two Effects can be seen for Thermoacoustics influence on fire Propagation-

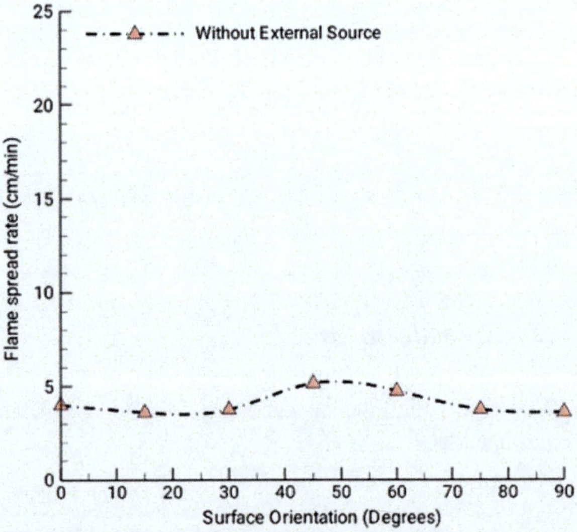

Fig. 4 Single matchstick without acoustic

- **Stabilizing Effect:** The flickering of the flame ceases with this effect, and the flame becomes stable due to acoustics. The combined energy transfer of thermal and acoustic energy is believed to be the same, and pilot fuel has no effect.
- **Destabilizing Effect:** In this case, acoustic impact causes the flame to shine softly, causing the relapse rate to shift. As shown by a rise or decrease in the flame regression rate. The consequence divides the final outcome into three zones.

Heat Sink Zone: In this system, the transfer of heat decreases as the pilot fuel concentrates around. Due to acoustics, the pilot's fuel is not filled with enough oxygen. Consequently, this situation has a lower regression rate than the pilot fuel.

Neutralizing zone: The transfer of thermal heat to that of the pilot fuel in this case is constant and therefore no acoustic effect, i.e., the regression rate is the same than for the pilot fuel base case.

Heat source zone: Total heat transfer is dominant in this area as the heat is being transported by the external impact of acoustics from combusted to unburnt fuel. Consequently, this situation has a higher regression rate than the pilot fuel.

The ratio of the flame spread rate at that particular orientation for a given number of external sources in the context of acoustics to the flame spread rate at that particular orientation for a defined number of external sources without acoustics is the **Fire Stimulation number**. The outcomes are categorized into different regimes. The heat sink zone is labeled FS < 1, the heat source zone is labeled FS > 1, and the neutralizing zone is labeled FS = 1. A non-dimensional number was calculated for each orientation. It was done to verify the configuration's effectiveness and to assess the effects of different positions, orientations, acoustics, and external sources, as well as when no acoustics were present.

$$\text{Fire stimulation number} = \frac{Spread\ rate\ in\ persence\ of\ acoustic\ (cm/min)}{Spread\ rate\ of\ pilot\ fuel\ without\ acoustic\ (cm/min)}$$

A new regime is established based on the FS number.

1. **FS-I:** Potential Heat Sink Effect—In this case, the FS number to FS number ratio is less than 1. (FS > 1) Pilot fuel absorbs energy from external sources in this effect.
2. **FS-II:** Neutralizing Effect—The FS number ratio is 1 in this case. (FS = 1) It isn't affected by either positive or negative heat sources.
3. **FS-III**: Heat Source Effect—The FS number ratio is greater than one in this case. (FS > 1) Pilot fuel is used to supply energy to external sources in this effect.

The following plots (Fig. 5) depict the effect on flame propagation in presence of acoustic bass for pilot fuel when one external source is attached for different orientations. The recurrence frequency was adjusted from 500 to 2500 Hz. The trend is non-monotonic, with a rise of **163.13%** at 90 degrees angle for 1500 Hz frequency, as shown in the plot. The maximum drop-in regression rate for this case was observed at 45-degree orientation and a 2500 Hz frequency of about **35.18%.** These figures

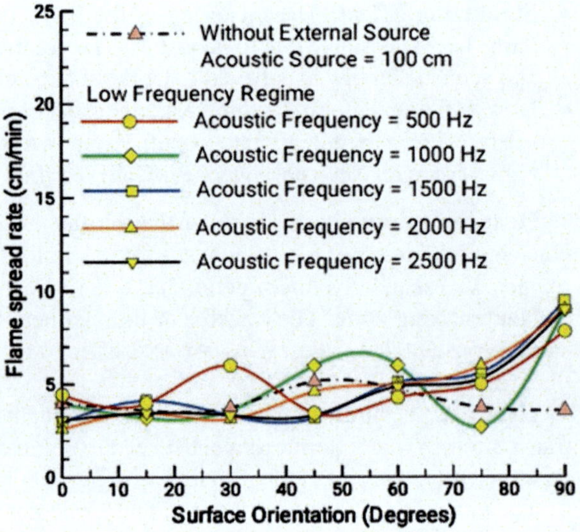

Fig. 5 Pilot fuel in presence of acoustic

were used as a baseline for comparing fire rates for various configurations with external heat sources.

The case of unilateral configuration is depicted in Fig. 6. At 2000 Hz, the highest regression rate was obtained. At 90 degree orientation, there was an increase of about **284.72% (FS > 1)**. The fire extinguished completely before hitting the first 1 cm mark at 500 Hz and 2000 Hz, with a maximum decrease of about **–100% (FS < 1)** at 0 and 15 degrees orientation. It is delivered in a blown-off state. For 0, 15, and

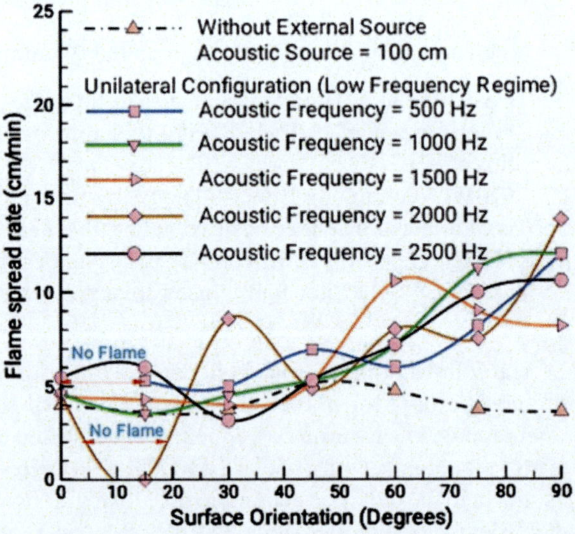

Fig. 6 Unilateral configuration

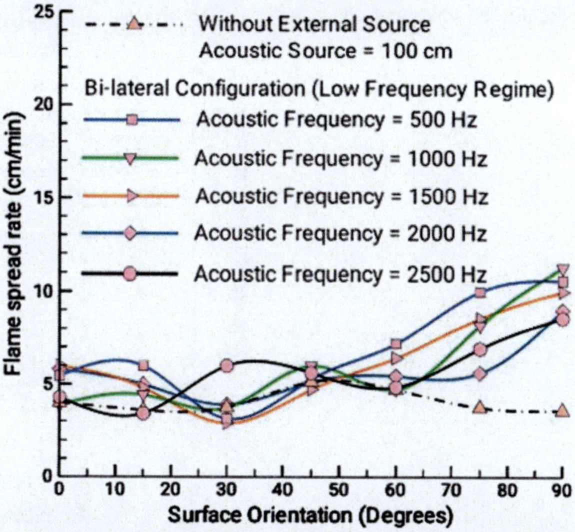

Fig. 7 Bilateral configuration

30 degrees, blow-off was dominant. For the most part, the FS number was less than one because heat transfer in this regime is reduced due to a decrease in the localized temperature around the pilot fuel. It's worth noting that for 500 and 1000 Hz, and a neutralizing effect was observed at a 90 degree angle. Figure 7 represents bilateral configuration. The maximum rise in spread rate was seen at 1000 Hz of **212.5% (FS > 1)** for 90 degree orientation were as a drop-in regression rate was seen at 1500 Hz and 30-degree angle of **38.77% (FS < 1).** Cases of blow-off were seen for 500 Hz and 1000 Hz at 0 to 60 degrees of orientation. The neutralizing effect was seen for 500 Hz to 2000 Hz at 60, 45, and 0 degrees, respectively. There is no acoustic effect in these situations since the heat transfer is constant to that of the pilot fuel, so the regression rate is the same as in the base case of pilot fuel (Table 1).

The variance for the 'Y' configuration is shown in Fig. 8. At 90 degrees, the regression rate was the highest. For 500 Hz, a huge rise of about **256.94% (F S > 1)** was observed. At 15 degrees and 1000 Hz, the maximum decrease is about **–46.45%**

Table 1 FS number variation for unilateral configuration when N = 2

Degree	500	1000	1500	2000	2500
0	0.886	0	0.823	0.606	0.986
15	0.56	0	0.88	1.026	0.724
30	1.101	0.7	0.9615	1.003	0.786
45	0.64	0.705	0.8	0.7266	0.696
60	0.862	0.483	1.293	1.551	1.825
75	1.503	1.428	1.19	0.865	1.904
90	0.666	0.706	0.6	0.8	0.706

Fig. 8 'Y' configuration

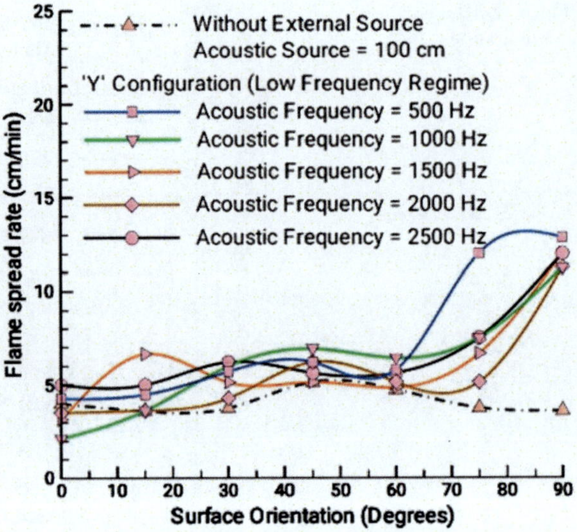

(FS < 1). It is delivered in a blown-off state. For angles 0 to 15 degrees, blow-off was dominant. The heat sink effect was observed in the majority of cases due to thermal effect superiority over acoustic energy. The interesting thing to note is that the neutralizing effect for 1000 Hz was seen for 60–90°. The thermoacoustic effect for the + configuration is shown in Fig. 9. The regression rate for 90 degrees was the highest. At 2500 Hz, there was a huge increase of around **316.66% (FS > 1)**, while the highest decrease was around **4.25% (FS < 1)** at 0 degrees. The fact that no

Fig. 9 ' + ' configuration

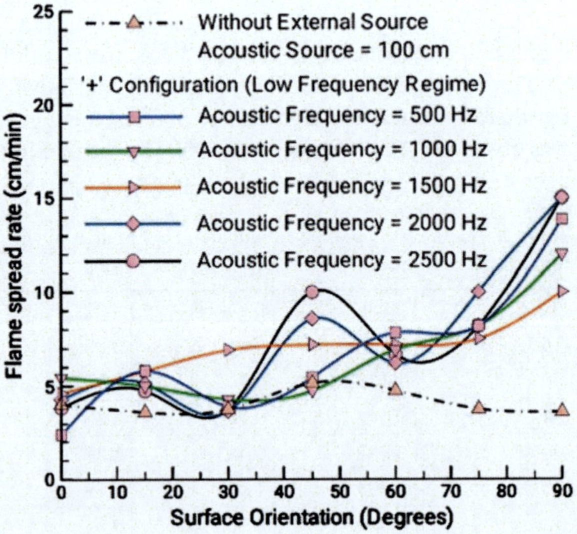

cases of blow-off were found is appealing because of buoyancy, which carries heat from burned to unburned fuel due to external acoustic influence, cumulative heat transfer dominates in this region in cases of heat source effect. At a 45-degree angle, a neutralizing effect was observed for 1000 Hz (Table 2).

The following plots depict the effect on flame propagation in presence of acoustic bass where the number of external sources is fixed at 3 and coupled effect is seen and compared for all configurations at various frequencies. Figure 10 depicts the variance of thermoacoustic energy exchange for different 500 Hz configurations. The significant increase in the regression rate of about **284.72% (FS > 1)** can be seen for the ' + ' configuration at 90 degrees. The maximum drop was around **-16.66% (FS < 1)** for bilateral configuration at 0-degree orientation where the fire was extinguished. The blow-off condition was seen only for zero degrees orientation for unilateral, bilateral, and + configurations. The heat sink effect was observed in the majority of cases due to thermal effect superiority over acoustic energy. It's worth noting that

Table 2 FS number variation for bilateral configuration when N = 1

Degree	500	1000	1500	2000	2500
0	0.75	0.645	1	0.966	0.714
15	0.667	0.845	0.919	0.952	0.652
30	0.771	0.771	0.602	0.805	1.234
45	0.833	1	0.789	0.854	0.937
60	0.754	868	1.179	1	0.892
75	1	1.09	1.142	0.75	0.923
90	0.76	1.187	1.059	0.95	0.905

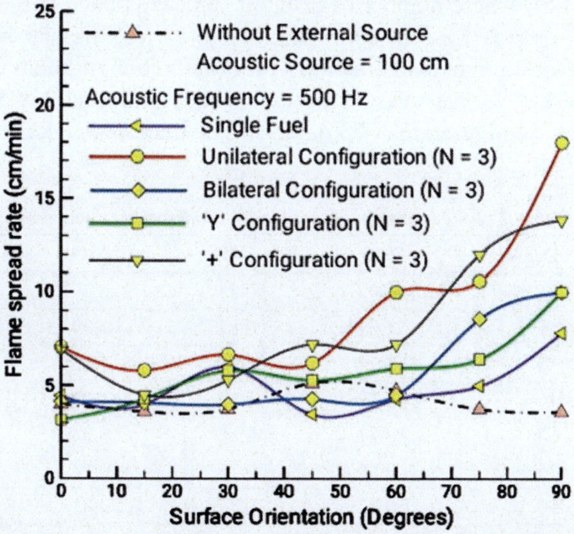

Fig. 10 For 500 Hz Frequency at N = 3

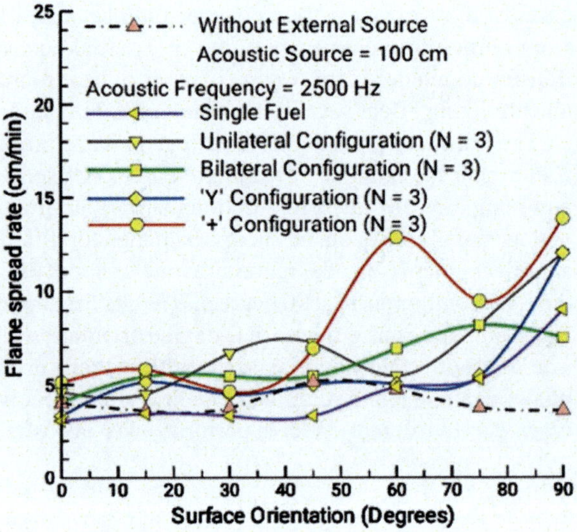

Fig. 11 For 2500 Hz Frequency at N = 3

no neutralizing effect was observed in the above scenario. Figure 11 represents the thermoacoustic energy interaction at 2500 Hz. The significant rise of **284.72% (FS > 1)** was seen at 90 degrees for + configuration were as a maximum drop of **16.75%** (FS < 1) at 0-degree orientation for Y configuration. At 0 and 15 degrees, respectively, there were few instances of blow-off for unilateral and Y configurations. Because of buoyancy, which carries heat from burned to unburned fuel due to external acoustic impact, cumulative heat transfer dominates in this region in the cases of Heat source impact. At 75°, the + configuration had the neutralizing effect (Table 3).

The following plots depicts the effect on flame propagation in presence of acoustic bass where number of external sources is fixed at 5 and coupled effect is seen and compared for all configurations at various frequencies. The variance of thermoacoustic energy interaction with various configurations at 500 Hz is depicted in Fig. 12. The massive spike in regression rate of around **269.33% (FS > 1)** can be seen for + configuration at 75 degrees. The maximum decrease was about **40.6% (FS < 1)**

Table 3 FS Number variation for 'Y' configuration when N = 4

Degree	500	1000	1500	2000	2500
0	0.9	0.71	0.78	0.78	0.71
15	1.02	0.84	0.89	0.89	0.89
30	1.24	1.08	1.13	1.62	1.68
45	1.19	1.19	1.03	0.97	0.89
60	0.71	0.86	0.96	0.66	1.09
75	0.86	0.75	0.78	1.06	0.55
90	0.87	0.72	0.81	0.81	1.08

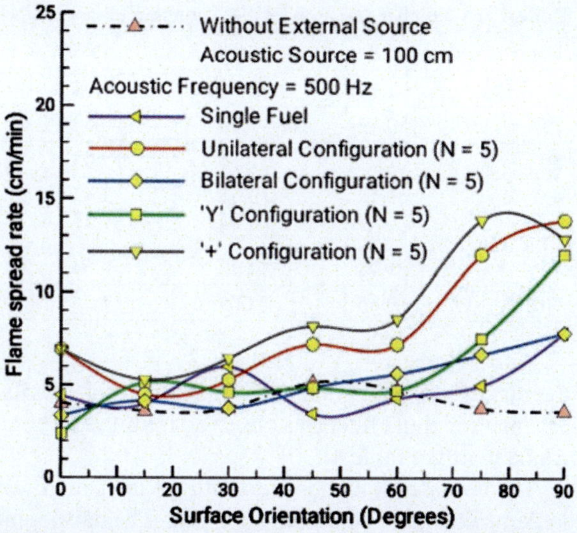

Fig. 12 For 500 Hz frequency at N = 5

for the Y configuration at 0-degree orientation, where the fire was extinguished. The blow-off condition was seen only for 0° and 15° orientation for unilateral, bilateral, and + configurations. The neutralizing effect was seen for bilateral configuration at 75° orientation. Figure 13 represents the thermoacoustic energy interaction at 2500 Hz. For regression rate at 90 degrees for + configuration, a significant increase of about **354.44% (FS > 1)** was observed, with a maximum decrease of **25% (FS < 1)** for Y configuration at 0-degree orientation. For the Y configuration at 0 and 15 degrees, there were just a few cases of blow-off. The heat sink effect was observed in

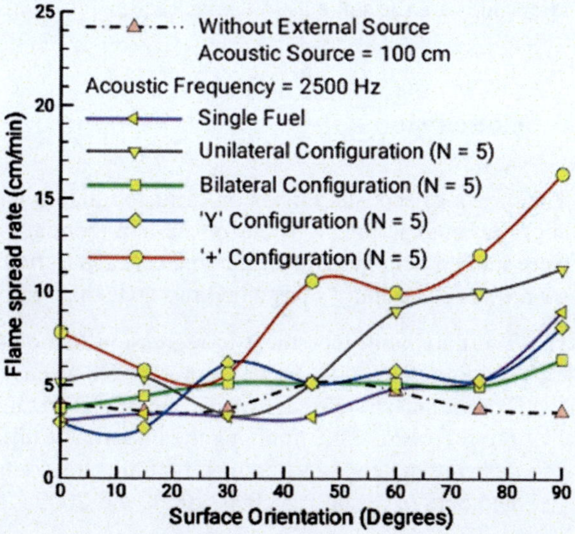

Fig. 13 For 2500 Hz frequency at N = 5

Table 4 FS Number variation for ' + ' configuration when N = 5

Degree	500	1000	1500	2000	2500
0	1.15	1.31	0.91	0.97	1.31
15	0.79	0.93	0.82	1.04	0.87
30	1.07	0.86	1.04	0.88	0.94
45	1.12	1.18	1.45	1.03	1.45
60	1.05	1.37	1.46	0.95	1.22
75	1.08	1.17	0.64	1.08	0.93
90	0.86	1.09	1	0.86	1.09

the majority of cases due to thermal effect superiority in terms of acoustic capacity. The neutralizing effect was observed for bilateral configuration at 30 degrees, which is noteworthy (Table 4).

The compression and rarefaction of the sound wave could explain the variation in flame strength and propagation rate. The development of distributed pressure and velocity fields around the pilot gasoline is assumed to influence forward heat transfer when a sound wave is present. Acoustic energy takes precedence over thermal energy under the influence of sound, resulting in a decrease in localized temperature and pressure, which results in a decrease in regression rate and is referred to as the heat sink effect. The regression rate increases as thermal energy dominates acoustic energy due to the influence of additional external sources. It alludes to the influence of a heat source. In each of the preceding examples, the flame flickers as a result of the interaction of thermal and sound energy, which has a destabilizing effect. It is instructive to note that in some situations, a neutralizing effect is seen where all acoustic and thermal energy interactions are predicted to be the same at the macroscopic scale. The flame should not flutter in this instance, and the regression rate values are the same as in the base case.

4 Conclusion

The merits of acoustic effects on diffusion flames were investigated using incense sticks in an experiment. The acoustic and thermal effects of different frequencies were studied with monitoring parameters such as fuel surface orientation and sound source size. The study's basic findings state that:

1. The maximum increment in regression rate of **400%** was seen for unilateral configuration at 90 degrees and 1500 Hz when 5 external sources are attached. This demonstrates that thermal energy outperforms acoustic energy in terms of fire regression rate, implying the importance of forward heat transfer.
2. The most noteworthy finding is for unilateral configurations with external source numbers of 1 and 2 at 500, 1000, and 2000 Hz for 15^0 and 0^0 orientations,

respectively. Although the fire was completely extinguished before reaching the 1 cm mark, the flame propagation rate decreased by **100%**, meaning that no flame was observed. The center of compression and odd fraction are often located in the center of sound as a wave. Based on the results of the tests, it can be concluded that sound has an extraneous effect and that acoustic energy exceeds thermal energy in close vicinity.

3. As the orientation was shifted from 0 to 15 degrees, several cases of blow-off and reappearance were observed for unilateral, bilateral, and 'Y' configurations. It shows that acoustics is directly accustomed to the presence of acoustics.
4. This advancement's governing physics includes—(a) reaction zone change (b) the creation of an energy interaction area (c) the relationship between Thermal and sound energy.

Applications of work: Physical instinct is used to frame modern viewpoints within fire protection protocols on the traditional and additional geographical levels. The work includes combustion and propulsion, defense system validation, research, and upgrades, as well as rocket engines, industrial power generation systems, operational, operating, and scientific applications. Acoustics can further reduce the number of casualties caused by explosions.

4.1 Suggested Applications Based on Results

1. For the case of Industrial defines systems and home usage installation of large network of small automated permeant devices which can detect and control the fire using sound waves. Which will result in a faster response to fires and a lower rate of propagation. The instrument will be calibrated to a certain frequency depending on the orientation of impingement of sound. The devices can move and change their location which will result in higher efficiency.
2. Sound waves could be very effective in fire suppression in environments that are impossible for humans to access. Long-distance sound waves may be aimed toward the origin of the explosion preventing it from spreading.
3. Acoustic waves may be used to suppress fires in wider areas, such as forests, with help of a group of UAVs. Drones may be equipped with sound generators, enabling them to reach a wide area in a short period while remaining at a safe distance from explosions.

Acknowledgements Without the significant efforts of former global researchers and academics, our forensic work may not have been feasible. Furthermore, we would like to dedicate our efforts to the firefighters in recognition of their commitment and unending support over the years. Thousands of lives have been saved as a result of their quick response time in forest fires, building and compartment fires, warehouse and household fires.

References

1. Rayleigh, L.: The theory of sound. Macmillan, London, U.K. (1896)
2. Oran, E.S., Gardner, J.H.: Chemical-acoustic interactions in combustion systems. Prog. Energy Combust. Sci. **11**(4), 253–276 (1985)
3. Johannesen, N.: *The Dynamics of Real Gases*, by J.F. Clarke and M. McChesney. Butterworths, 419pp. 90s. J. Fluid Mech. **21**(4), 761–762 (1965).
4. Clarke, J.F.: Amplification at a disturbance wave-head in a homogenous explosion. Acta Astronaut. **5**(7–8), 543–556 (1978)
5. Toong, T.Y.: Chemical effects on sound propagation. Combust. Flame **18**(2), 207–216 (1972)
6. DARPA: Instant Flame Suppression Phase II: Final Report. Defense Advanced Research Projects Agency (DARPA).
7. Nair, V., Sujith, R.: Multifractality in combustion noise: Predicting an impending combustion instability. J. Fluid Mech. **747**, 635–655 (2014)
8. Nair, V., Thampi, G., Sujith, R.: Intermittency route to thermoacoustic instability in turbulent combustors. J. Fluid Mech. **756**, 470–487 (2014)
9. Bauerheim, M., Nicoud, F., Poinsot, T.: Theoretical analysis of the mass balance equation through a flame at zero and non-zero Mach numbers. Combust. Flame **162**(1), 60–67 (2015)
10. Bennewitz, J.W., Valentini, D., Plascencia, M.A., Vargas, A., Sim, H.S., Lopez, B., Smith, O.I., Karagozian, A.R.: Perodic partial extinction in acoustically couple fuel droplet combustion. Combustion Flame **189**, 46–61 (2018)
11. Pratik, T., Vikram, R., Vinayak, M.: An experimental insight into thermoacoustic smoldering. j. Space Exploration. **6**(1), 119 (2017)

The Contraction of Froude's Number Due to Inclined Weir on the Downstream of Cut Throat Flume

S. M. Shravankumar, Urlaganti Krishna Gopika, Nikhil Sharma, and T. Sirichandana

Abstract The weirs and notches are the devices used to measure the discharge in the field of Irrigation engineering, Chemical engineering, and Hydraulic engineering also they are used for controlling the flow in conduits. The presence of obstruction after the weirs affects the coefficient of discharge by distributing the pressure through the formation of a Hydraulic Jump in between the weir and other obstructions along the flume. This paper analyses the impact of the hump on the flow characteristics of the flow and also on the flow dynamics over the cutthroat flume used to measure the discharge. The flow axis and the weir axis orientation are important toward reducing the erosion and other disadvantages downstream of the cutthroat venturi flume. The experiments were conducted and found that the inclined hump placed at an angle of 15°–30° on the downstream reduced the Froude's number at the downstream of the cut throat flume. The CFD analysis was done and was matching the experiments conducted in the laboratory, the 30° angle hump reduced the erosion and also narrowed the Froude's number at the crest of the cutthroat flume, due to which the height of hydraulic jump formed reduced.

Keywords Inclined barrier · Open channel flow · Subcritical flow · Varying Froude's number · Varying specific energy · Cut throat

1 Introduction

Canals are built worldwide for the transportation of water for irrigation, municipal water supply, and power generation [1]. Supercritical flow in canals leads to high velocity and shallow depth which causes erosion of canal lining, if a passenger falls into the canal, he/she gets washed away due to the high velocity of water, and maintenance operations can't be carried out easily if the flow is supercritical [2]. So, canals are built in such a way that flow is always subcritical.

S. M. Shravankumar (✉) · U. K. Gopika · N. Sharma · T. Sirichandana
Sreenidhi Institute of Science and Technology, Hyderabad, Hyderabad 501301, India
e-mail: shravankumarsm@sreenidhi.edu.in

R. P. Bharti and K. M. Gangawane (eds.), *Recent Trends in Fluid Dynamics Research*, Lecture Notes in Mechanical Engineering,
https://doi.org/10.1007/978-981-16-6928-6_17

The advantages of flumes include minimal head loss, low construction cost, adaptability to a variety of channel types, and the ability to measure wide ranges of flows with a custom-designed structure [3]. Cutthroat flumes are advantageous over all other measuring flumes because of their simplicity and accuracy in discharge measurement. Cutthroat flumes were first developed in India (Punjab) [4]. Experimental work carried out by the researcher recommended a maximum allowed convergence of 3:1 for flume inlet section for developing a flat-bottomed flume and the divergence should not exceed 6:1, to prevent flow separation [5]. It was found that in cutthroat flume free flow regime transits from the subcritical in inlet section, to critical in the throat section, and then to supercritical flow in the outlet section [6].

A weir is a barrier across the width of the river that alters the flow characteristics of water and usually results in a change in the height of the river level [7]. Weirs have also been used to control the flow of water for outlets of lakes, ponds, and reservoirs. There are many weir designs, but commonly water flows freely over the top of the crest before cascading down to a lower level [8].

Canals are not always constructed straight, there might be some turns and obstructions. Due to turns and obstructions, the Froude's number and specific energy varies. In such cases, it is difficult to maintain subcritical flow in canals. So, an inclined barrier/weir is placed downstream of the cutthroat to study flumes with varying Froude's number and specific energy, and the effect of the inclined barrier on flow profile is also studied. The inclined barriers of angles 15°and 30° are placed to study the effect of the angle of inclined barrier on the flow profile [9].

From Figs. 1 and 2, it is observed that the Froude's number and specific energy are varying with distance when a 15°inclined barrier is placed.

From Figs. 3 and 4, it is observed that the Froude's number and specific energy are varying with distance when a 30°inclined barrier is placed.

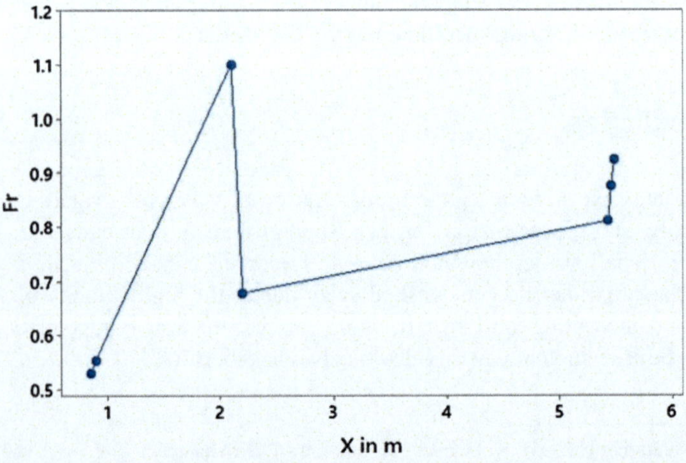

Fig. 1 Variation of Froude's number with distance for flume 1 based on experimental data

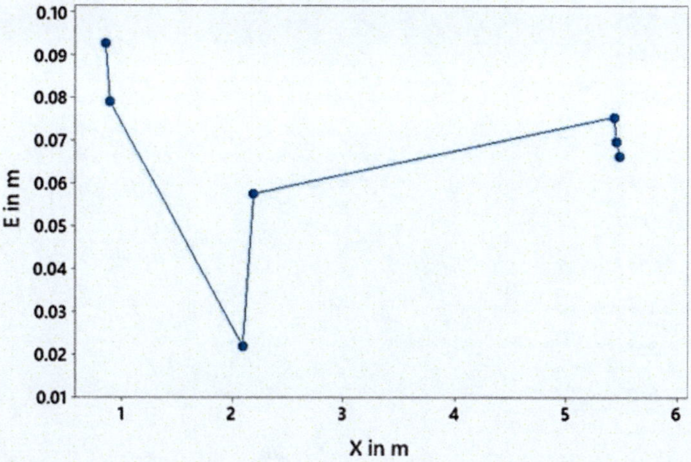

Fig. 2 Variation of specific energy with distance for flume 1 based on experimental data

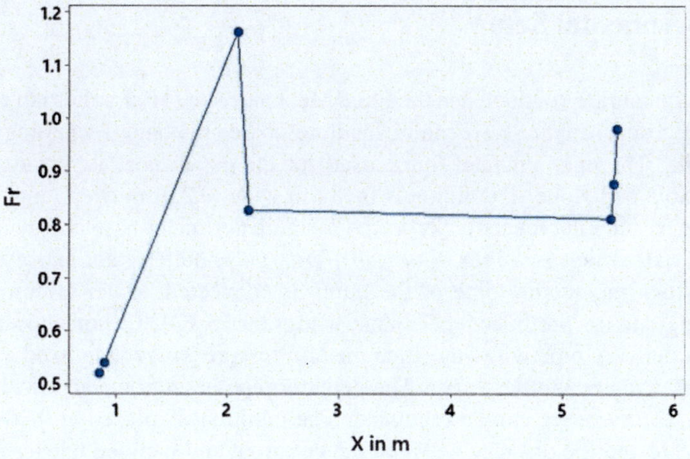

Fig. 3 Variation of Froude's number with distance for flume 2 based on experimental data

At the downstream of the cutthroat, the flow is subcritical, after that, an undular to weak hydraulic jump is formed after that flow is subcritical at the upstream of the inclined barrier and an undular to weak inclined hydraulic jump is formed at the downstream of the inclined barrier. The angle of the hydraulic jump depends on the angle of the inclined.

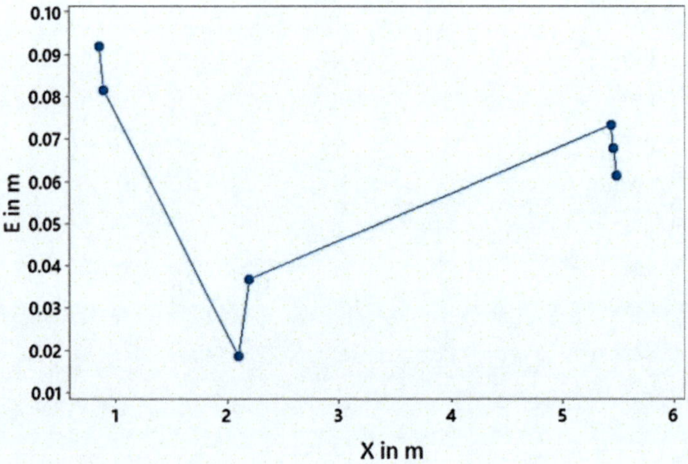

Fig. 4 Variation of Specific Energy with distance for flume 2 based on experimental data

2 Experimental Setup

The experiments are conducted in the Fluid Mechanics and Hydraulics lab of the civil engineering department at Sreenidhi Institute of Science and Technology, Hyderabad, India. The open channel flume used for the experiment is rectangular with an adjustable bed slope. The flume is 0.184 m wide, 0.220 m deep, and 6 m long. The flume has an inlet feeder tank, braced metallic perimeter base and walls (except the front wall which is a Plexiglas wall), inlet gate, outlet gate, and a collecting tank to which the suction pipe of the pump is connected. Water is pumped from the storage tank to the inlet feeder tank using a 1.13 KW (2 horsepower) pump, through a delivery pipe with an orifice meter. Pressure gauges are fixed at the inlet of the orifice meter and the throat. The delivery pipe has a flow valve with the help of which the discharge can be regulated. The cutthroat is placed at 0.709 m from the inlet gate and the distance between the cutthroat and inclined barrier is fixed to 3 m. The dimensions of the cutthroat are mentioned in Fig. 5 and the dimensions of inclined barriers are mentioned in Table 1. The depth of water is measured using a mobile venturi point gauge with 1 mm accuracy, the depth of the water is measured at 7 locations as shown in Fig. 5.

3 Experimental Procedure

The following steps are performed to carry out the experiment.

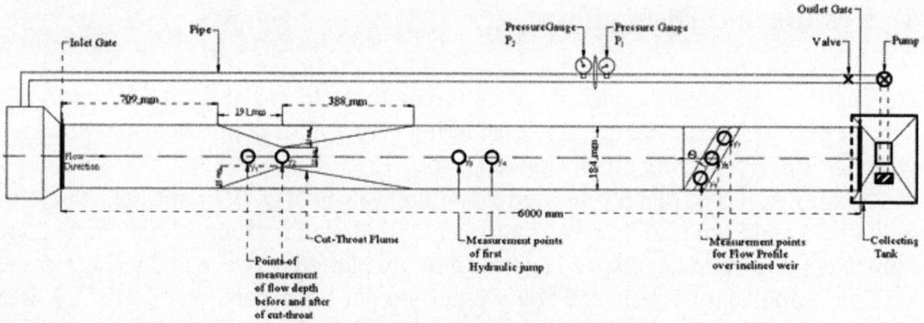

Fig. 5 Top view of experimental setup with dimensions of the cutthroat

Table 1 Dimensions of inclined barrier

Flume No	Angle of inclined barrier	Height of inclined barrier in mm
1	15°	25
2	30	25

1. The channel bed slope is maintained (since the rectangular flume has an adjustable bed) as required, in the case of 15° barrier and 30° barrier the bed slope is 0.0084 and 0.005.
2. A cutthroat venturi is placed at 0.709 m from the inlet gate and the inclined barrier is placed at a distance of 3 m from the cutthroat.
3. The pump is turned on and water is allowed to the channel (priming is done if required).
4. The flow is regulated using the flow valve. When a 15° barrier is used, the slope is 0.0084, the flow rate is maintained at full capacity to obtain experimental values (depth of water at 7 different locations).
5. Similarly, for the 30° barrier (0.005 slope) experimental values are obtained by following the above step.
6. Finally, two sets of observations are obtained.
7. When everything is at its place, first thing is to take note of pressure values from the pressure gauges P1 and P2, and then
8. The depth of flow is measured at seven different locations as shown in Fig. 5. Depth of water is measured with the help of Vernier depth gauge, which is used to calculate flow cross-section and wetted perimeter which will be used to calculate the velocity at different locations using Manning's equation.
9. Steps number 7 and 8 are repeated in order to get two sets of observations for two inclined barriers.

4 Results and Discussion

The inclined barrier downstream of cutthroat leads to varying Froude's number, Specific energy, and discharge. By considering b, Fr, y, E as parameters the following equation is derived using dimension analysis.

By using observed depth y, the corresponding values of Q, E, Fr and are calculated for flume 1 and flume 2. Observed values of Q, E, Fr and for flume 1 and flume 2 are shown in Tables 1 and 2, respectively. Graphs are plotted between and y (Figs. 6 and 7). Regression analysis between and y is carried out for flume 1 and flume 2. Based on regression analysis, the following relationship is obtained.

$$\varphi = \frac{Q}{b \times Fr \times E \times \sqrt{gy}} \tag{1}$$

$$Q_{15} = (0.137 + 7.73 \times y)\sqrt{gy} \times E \times b \times Fr \tag{2}$$

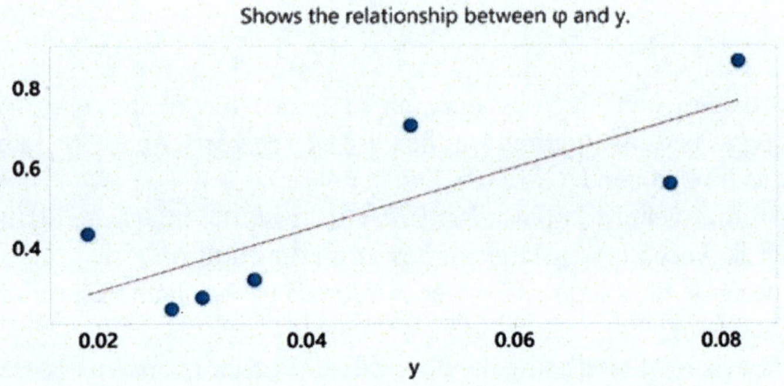

Fig. 6 Variation of φ with y for flume 1 based on experimental data

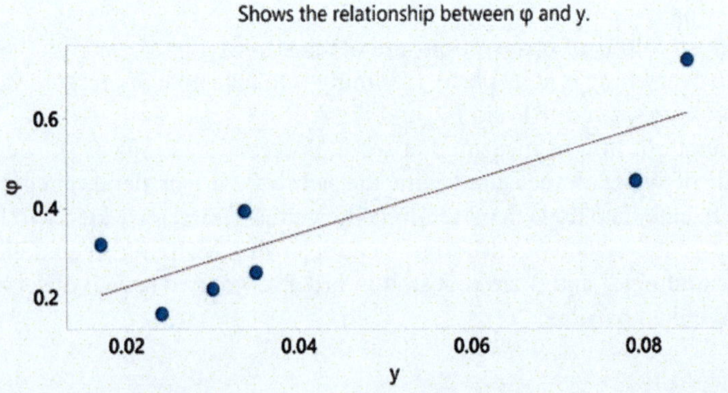

Fig. 7 Variation of φ with y for flume 2 based on experimental data

Equation 2 is the discharge equation of flume 1 with 15°inclined barrier and the coefficient of regression (R^2) = 0.6490.

$$Q_{30} = (0.1034 + 5.98 \times y)\sqrt{gy} \times E \times b \times Fr \tag{3}$$

Equation (3) is the discharge equation of flume 1 with 30°inclined barrier and the coefficient of regression (R^2) = 0.7170.

With the help of manning's formula velocity (v) can be calculated using observed depth values.

$$v = \frac{1}{n} \times R^{2/3} S^{1/2} \tag{4}$$

Corresponding velocity values calculated from formula 4 are substituted into formula 5 and formula 6 and the values of E and Fr are calculated.

$$E = y + \frac{v^2}{2g} + z \tag{5}$$

$$Fr = \frac{v}{\sqrt{gy}} \tag{6}$$

5 Conclusion

The flow region from downstream of the cutthroat and upstream of the barrier for flume 1 and flume 2 is studied.

1. The height of hydraulic jump in flume 1 is 0.05 m and the height of hydraulic jump for flume 2 is 0.0336 m. So, there is a reduction in the height of the hydraulic jump by 48.8% with increase in angle of inclined barrier from 15°to 30°.
2. Figure 8 represents a graph plotted between velocity 'v' and discharge 'Q' and it is observed that there is a significant reduction in velocity in flume 2 which proves that increase in angle reduces the velocity of water and creates a subcritical flow (Table 3).

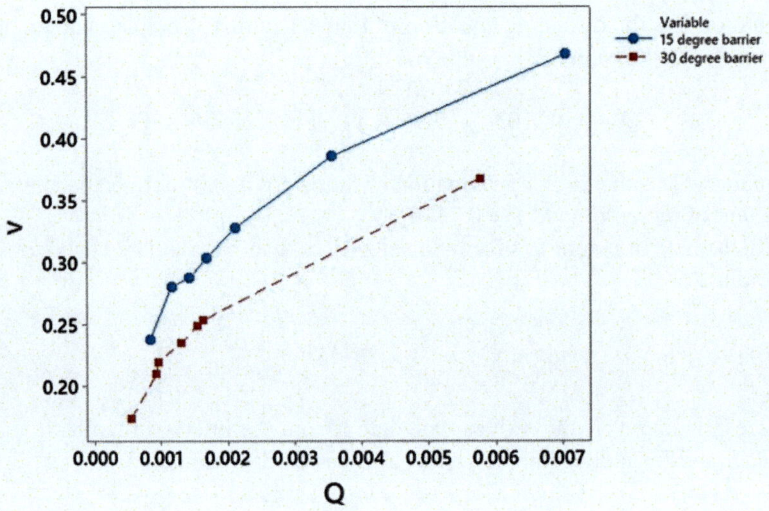

Fig. 8 Variation of velocity v with discharge Q

3. The formula for discharge for 15°barrier is

$$Q_{15} = (0.137 + 7.73 \times y)\sqrt{gy} \times E \times b \times Fr \tag{3}$$

4. The formula for discharge for 30°barrier is

$$Q_{30} = (0.1034 + 5.98 \times y)\sqrt{gy} \times E \times b \times Fr \tag{4}$$

6 Appendix

Table 2 The observed values of depth of water y and corresponding calculated values of Q, E, Fr, Cd, φ for flume 1 with channel bed slope S = 0.0084

Y	Q	E	Fr	φ	Cd
0.0816	0.007027	0.092763	0.530907	0.866685	0.985252
0.075	0.001159	0.079025	0.553775	0.561445	0.592863
0.019	0.000832	0.021887	1.100239	0.434955	0.501047
0.05	0.003551	0.057594	0.678233	0.705474	0.812623
0.035	0.002112	0.075483	0.810643	0.320178	0.690519
0.03	0.001678	0.06971	0.875595	0.275423	0.639993
0.027	0.001431	0.066228	0.922958	0.247183	0.606309

Table 3 The observed values of depth of water y and corresponding calculated values of Q, E, Fr, Cd, φ for flume 2 with channel bed slope S = 0.005

Y	Q	E	Fr	φ	Cd
0.085	0.005756	0.091902	0.520181	0.731317	1.030485
0.079	0.000956	0.081467	0.539873	0.460949	0.616050
0.017	0.000544	0.018543	1.16316	0.317661	0.487240
0.03365	0.001542	0.03681	0.826745	0.393121	0.697257
0.035	0.001636	0.073288	0.810643	0.255369	0.711259
0.03	0.001297	0.067815	0.875595	0.218860	0.658054
0.024	0.000956	0.081467	0.978945	0.163330	0.588048

References

1. Shravankumar, S.M., Grace, R.A., Venkatesh, K.S., Gujjula, R., Gujjari, A.: The fluctuation of pressure due to bends in venturimeter. In: Recent Trends in Civil Engineering, pp. 833–846. Springer, Singapore (2021)
2. Shravankumar, S. M., Sandeep, M., Raj, M.A.: Performance of cut throat flume of 2.16 throat-width. Int. J. Multidisciplin. Stud. Res. **1**(1), (2020)
3. Berner, E.K., Berner, R.A.: The Global Water Cycle: Geochemistry and Environment, p. 1990. Prentice-Hall, Englewood Cliffs (1987)
4. Masalvad, S.K.S., Sakare, P.K.: Application of photo Fenton process for treatment of textile Congo-red dye solution. Mater. Today: Proc. **46**, 5291–5297 (2020)
5. Franc, J.M., Michel, J.M.: Cavitation in Chutes and Spillways. USBR Engineering Monograph, No. 42. USBR, Denver (2004)
6. IAHS: Fundamentals of Cavitation. Springer Series Fluid Mechanics, and Its Applications, Vol. 76, 306p. Springer, New York (1984)

7. Knapp, R.T., Daily, J.W., Hammitt, F.G.: World catalog of maximum observed floods. IAHS Publication No. 143, 354p. International Association of Hydrological Sciences (1970)
8. Lempérière, F.: Cavitation. McGraw-Hill Book Company, New York (1993)
9. Lempérière, F., Vigny, J.P., Deroo, L.: Dams that have failed by Flooding: an Analysis of 70 Failures. Int. Water Power Dam Constr. **45**(9/10), 19–24 (2012)
10. Lesleighter, E.: New methods and criteria for designing spillways could reduce risks and costs significantly. Hydropower Dam Constr. **3**, 120–128 (1983)
11. Novak, P., Moffat, A.I.B., Nalluri, C., Narayanan, R.: Cavitation in high-head gated outlets – prototype measurements and model simulation. In: Proceedings of 20th IAHR Biennial Congress, Moscow, Vol. 3, pp. 495–503 (2001)
12. Peterka, A.J.: The Effect of Entrained Air on Cavitation Pitting. IAHR/ASCE Joint Meeting Paper, Minneapolis, Minnesota, 507–518 (1953). Peterka, A.J.: Hydraulic Structures. Spon Press, London, 3rd edn, 666pp (1958)
13. Russell, S.O., and Sheehan, G.J.: Hydraulic Design of Stilling Basins and Energy Dissipators. USBR Engineering Monograph, No. 25, 239pp. USBR, Denver (1974)
14. Smith, N.: Effect of entrained air on cavitation damage. Can. J. Civil Eng. **1**, 97–107 (1971)
15. USBR: A History of Dams. The Chaucer Press, Peter Davies, London, UK (1965)
16. Design of Small Dams. US Department of the Interior, Bureau of Reclamation, Denver CO, USA, 1st edition, 3rd printing. WMO (1994)
17. World Meteorological Organization: Guides to Hydrological Practices. WMO No. 168, 5th edn., 735pp

Modeling of Enhanced Oil Recovery Using Polyaniline

Lomas Rishi and Monisha Mridha Mandal

Abstract It is essential to adopt newer and advanced technologies such as enhanced oil recovery (EOR) to improve the oil recovery process, encouraging sustainable development and growth in the energy sector. It has been found that polymers augment the EOR process because it improves mobility ratio, reduces interfacial tension of the fluid and changes the wettability. Present work was carried out using the IMEX Computer Modeling Group simulator (CMG) to evaluate the usage of polyaniline (PANI) on the EOR process because of its favorable properties. Computational simulations were carried out to understand and evaluate recovery efficiency using PANI in water. Present computational work was validated with existing literature for given process conditions. Contours for polymer adsorption and oil saturation within the given period in an oil field were developed and analyzed. The present study confirmed enhanced oil recovery using polyaniline nanofluid.

Keywords Crude oil · Enhanced oil recovery · Modeling · Polyaniline

1 Introduction

Non-conventional energy resources cannot fulfill the demand for global energy. Petroleum crude oil is being the largest source of energy for the present time. The current global demand of energy is about 500 EJ. This is expected to increase by more than 50% in next 10 years. This shows that the global demand of energy will be nearly 700 EJ by the year 2030. However, hydrocarbons sources are depleting. Fulfillment of energy demand for present and future scenarios could be done by finding new wells. In the oil and gas industry, exploration of any new reservoir is the most expensive process of the whole oil production process. The detection of reservoirs is being done by different seismic surveys. Therefore, the cost of exploring and development of new wells increases significantly. On the other hand, for the use of technologies such as enhanced oil recovery (EOR), the cost of modification of process may be quite low

L. Rishi · M. M. Mandal (✉)
University School of Chemical Technology, GGSIPU, Dwarka, New Delhi, India
e-mail: monishamridha1@ipu.ac.in

R. P. Bharti and K. M. Gangawane (eds.), *Recent Trends in Fluid Dynamics Research*, Lecture Notes in Mechanical Engineering,
https://doi.org/10.1007/978-981-16-6928-6_18

and may help in fulfilling exponential demand of the crude oil in market. Enhanced oil recovery is one such process that recovers crude oil hold by rocks or by capillaries in the earth. Chemicals are injected into the core by injection to reduce the interfacial tension between the oil and the rock and affect rock wettability. Nanotechnology is being currently explored in the petroleum industry. EOR is adopting nanotechnology to capture oil from the oil wells. Nanoparticles are the particles that have diameter less than 100 nm. They show different behaviors in comparison to particles with larger size because of their higher surface area. Most of the oil is trapped in capillaries or reservoir's pores. High capillary force and governing capillary pressure are influenced by interfacial tension and wettability. Performance of EOR chemicals is evaluated based on flow behavior, interfacial tension, rock of reservoirs' wettability conditions and mobility ratio [1–5]. Mobility ratio can be defined as the displacing phase (water) mobility to the displaced phase (oil) mobility. Mobility ratio < 1 can be obtained using polymers that are soluble in water.

It has been reported in literature that polymers may be used to enhance fluid's viscosity. It is injected into the reservoir to reduce media's permeability. In the case of water flooding, water forms finger-like structure in the reservoir [1]. Due to this, a large area of the reservoir remains unused and mitigates oil recovery performance. However, when polymer is added along with the water, the viscosity of fluid is increased. The increase in viscosity of injected fluid reduces mobility of the fluid. The problem of oil recovery from unused area during the water flooding may be reduced and higher oil recovery may be obtained. Therefore, effective use of injection of polymer is to control the mobility of fluid by reduction in mobility ratio between oil and the water. The sweeping area may be increased which results in increase of the oil recovery. Moreover, the pulling effect comes into picture and pulls residual oil from the capillaries and push for oil recovery. This pushing overcomes the capillary force and makes the oil mobilize [1]. There are two leading types of polymers which are being used for oil recovery operations, i.e., synthetic polymers and biopolymers. Chief examples of synthetic polymers which are being used are polyacrylamides and their derivatives such as partially hydrolyzed polyacrylamide (HPAM), and copolymers of acrylamide [1, 3, 4]. On the other hand, examples of biopolymers which are being used may include xanthan gum [5]. These commonly used polymers have their own limitations. HPAM is very sensitive to external factors such as pH, temperature, salinity, shear forces and hardness [6, 7]. Acrylamide molecules are trapped by divalent ions present in rocks. This process causes reduction in viscosity of polymer solution and the permeability of reservoir. Furthermore, xanthan gum polymers cost more, are subject to microbial degradation, and have a greater potential for wellbore plugging [1]. Therefore, there is need to study more effective polymers for EOR. It has been cited in literatures that polyaniline (PANI) is conducting polymers which lack oxygen single bonds (-O-) in the polymer backbone (carbon chain), providing thermal stability. It has been reported that PANI nanofibers are amphiphilic in nature and has excellent capability of creating emulsions at a wide range of pH. PANI nanofibers can be easily wetted by water and oil and are easy to synthesis. To the best of knowledge, no work has been carried out to investigate the performance of PANI based in water nanofluid on EOR process. Therefore, the objective of the present study was to investigate enhanced oil recovery using PANI nanofluid.

It is difficult to conduct many experiments to find the optimum operating conditions for better oil recovery. The core flooding experiments are laborious, expensive and time-consuming, so to overcome these problems, computational methods are being used. The vital purpose of conducting polymer flooding simulation is to explore the potential of polymer to recover the oil from oil field and compare it with other polymers. Three common simulators cited in literature were Eclipse by Schlumberger, IMEX designed by CMG and UTCHEM, which have been used for research applications [8]. It has been reported by many researchers that UTCHEM simulator can model polymer behavior of shear thickening and is specifically designed to model the polymer and chemical behavior. However, the disadvantage of this simulator is that industries outside the academic world do not generally use it. Eclipse is one of the most useable simulation tool software in petroleum industry. It is used by different companies in industry for field application, and it could be proved to be an ideal simulator for experiment work simulation. But at the time of investigation, the ability of Eclipse simulator was limited to polymer solution with shear-thinning nature. So, it was not used for the present work. The simulation software that was taken into consideration and selected to perform the modeling was the IMEX by CMG. Many companies in petroleum industry use this simulator. It is because of its facility to model the lab-scale experiments as well as the field scale and the handling of complex chemical behaviors. Several researchers have used this simulator to perform their work [8, 9].

In the present work, flooding simulations have been carried out to investigate the contours of polymer adsorption and oil saturation with polymers such as hydrolyzed polyacrylamide (HPAM) and PANI using CMG-IMEX simulation software. Effect of concentrations of polymers [2 to 4 Conc. (lb/bbl)] and temperature (127 to 147 °F) on oil recovery were investigated for HPAM and PANI.

2 Methodology

In the present work, modeling was carried out using CMG simulator in which IMEX mode of simulation, units and starting date of project was selected. 3D structured grid was created and is shown in Fig. 1. Length of each grid was considered to be 41ft in i and j directions. Properties of the oil considered for present work were taken from literatures [3, 10]. The literature reported the size of PANI was less than 100 nm. The methodologies of preparation of polymers are also mentioned in the above literature. The viscosity of polymers for different concentrations and temperature is shown in Table 1. Two types of wells are present. The first one was injection well and second was producer well. The injection well was the well which was used to inject nanofluid in earth so that it can pressurize the oil for extraction. Producer well was the well which was used to extract the oil from the reservoir. Reservoir properties and process conditions were same as cited in literature [7].

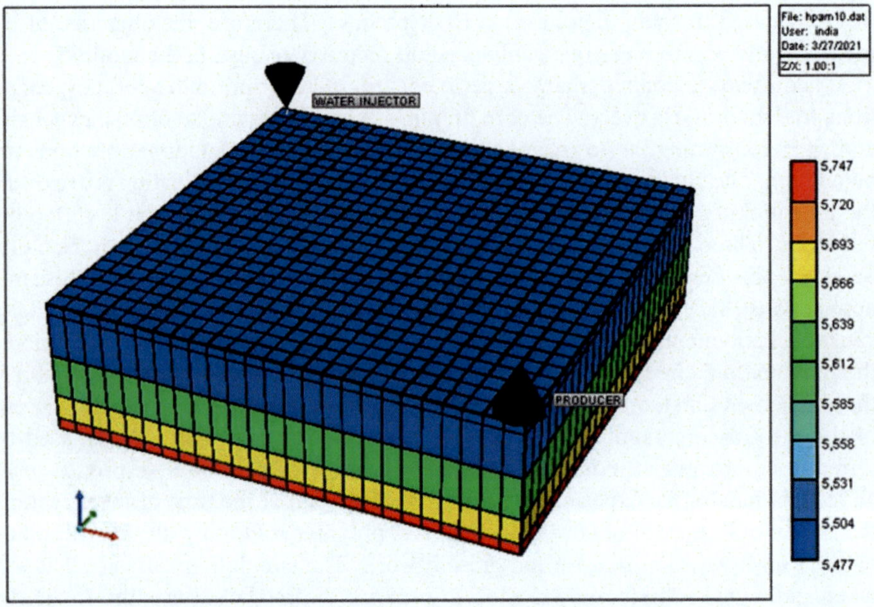

Fig. 1. 3D structured grid of reservoir

Table 1 Viscosity of polymers at different concentration and temperature

Polymer	Viscosity (cP)			
	Conc. (lb/bbl)	Temperature (°F)		
HPAM	–	127	137	147
	2	1.76	1.408	1.056
	3	2.6	2.21	1.6575
	4	3.4	2.72	2.04
PANI	–	127	137	147
	2	1.52	1.3224	1.091
	3	2.2	1.914	1.58
	4	2.95	2.57	2.09

Polymer model was considered for solving the conservation of mass equations and Darcy's law for each component. [8, 13] In this model, fluid satisfies Darcy's law, which establishes a relationship among flow rate, reservoir properties, fluid properties and phase pressure differences, and it is written as:

$$Q = -\frac{\kappa A \Delta p}{\mu L} \tag{1}$$

where κ is permeability of rock (or reservoir), mD; A is cross-section area, ft^2; Δp is pressure difference, psi; μ is viscosity of the fluid, cP and L is length of a porous media in the direction, ft.

3 Results and Discussion

Grid independence test was carried out with different grid sizes as shown in Table 2 to find the optimal grid condition with which there is no difference in the numerical results. In the grid test, different grids with different number of nodes were created and verified the pressure drop data obtained from present computational work with data of HPAM cited in literature [7]. Maximum error of 2.8% was obtained. It was found that 25×25×5 numbers of grids were sufficient to have results in agreement with the value cited in literature. Further increase in grids did not significantly affect the numerical results. Hence, this grid was considered for further computational work.

Further, computations were carried out to validate the present computations for other different concentrations of HPAM and compared with existing literature [7]. It was found that present computed results were in good agreement with existing literature as shown in Fig. 2.

3.1 *Polymer Adsorption*

When the polymers are flooded in the reservoir for EOR process, an amount of polymer may get trapped in reservoir due to the polymer adsorption, mechanical entrapment and hydrodynamic retention [1]. An attempt was made to investigate the region of polymers being adsorbed using polymers of concentration 2 lb/bbl at 127 °F. Figure 3 shows the polymer adsorption in lb/bbl at the end of test period of ten years. Contours of polymers adsorbed in reservoir are shown for HPAM as well as PANI. Red color denotes 0.044 lb/bbl of polymer adsorbed, whereas the minimum value is represented by blue color. It can be seen from Fig. 3a and b, red region in HPAM was more than PANI near the injection well. This showed that HPAM which

Table 2 Grid test at 194 °F in presence of HPAM

Grid	Pressure (literature) in psi	Pressure (present study) in psi	Error %
10×10×5	4014.009	3898.935	2.87
15×15×5	4053.93	3939.338	2.83
20×20×5	4079.449	3964.641	2.81
25×25×5	4096.792	3982.31	2.80
30×30×5	4109.168	3995.601	2.76

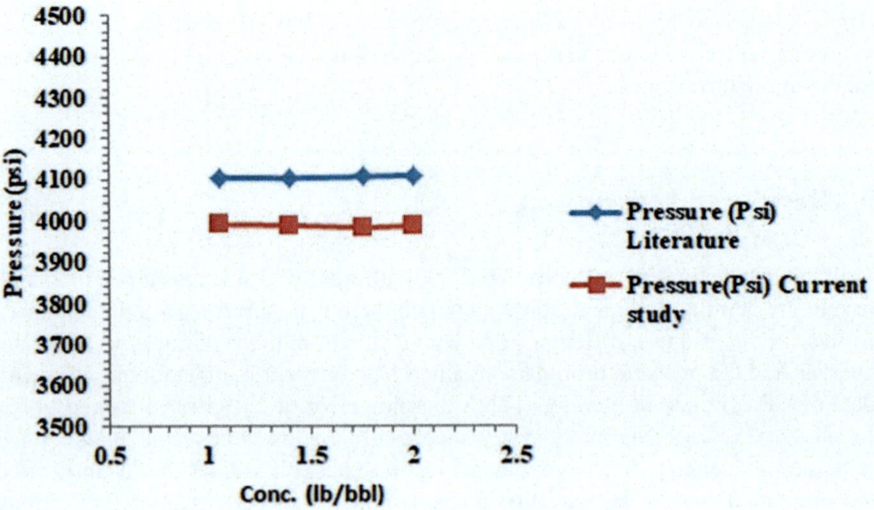

Fig. 2 Validation of present numerical work with literature

had more viscosity got more adsorbed and may be trapped near the injection well as compared to PANI. However, it has been cited in literature that adsorption of polymer can severe the flooding process as it is irreversible process [11]. If adsorbed polymer occupies in fine pores and capillaries, it may be difficult to remove the polymers. This may affect the permeability of HPAM polymer to sweep further in the reservoir. In case of PANI, the sweeping area was found to be more as compared to HPAM.

Figure 4 shows the effect of concentrations of polymer as well as temperature of reservoir on polymer adsorption. It was observed that amount of HPAM adsorbed increased with increase in temperature. This may be due to the reason that as the temperature increased in reservoir, the viscosity of polymers decreased which might help in increasing permeability of polymer in reservoir. It was further noticed that with increase in polymer concentration, the amount of polymer adsorbed decreased. It may be due to the reason that as the concentration was increased, the mobility of the fluid reduced which may lead to reduced chances of the fluid flowing in the non-desired regions and hydrodynamic retention. So, due to this reason, the usage or adsorption of polymer was may be less [12].

3.2 *Oil Saturation and Production*

Contours of oil saturation of both polymers, i.e., HPAM and PANI, were investigated and are shown in Fig. 5a and b, respectively. When polymers with water were injected in the reservoir, oil-saturated space became water-saturated space. Figure 5 shows the contours of oil saturation in the reservoir after the completion of ten years. Green

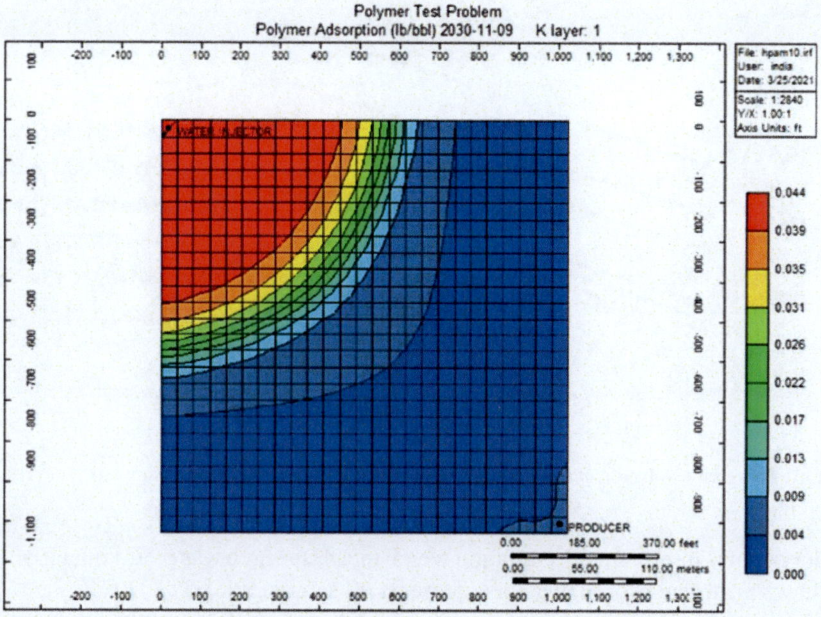

(a)

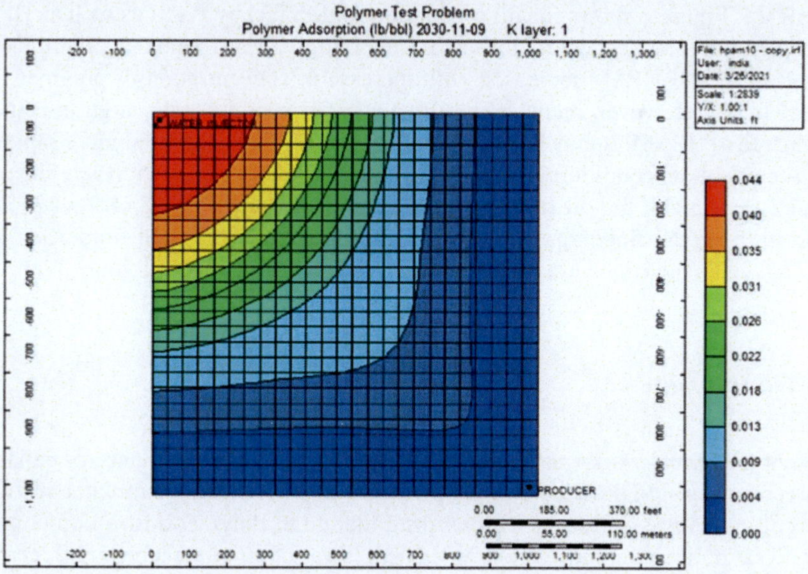

(b)

Fig. 3 Contours of adsorption of 2 lb/bbl. **a** HPAM. **b** PANI

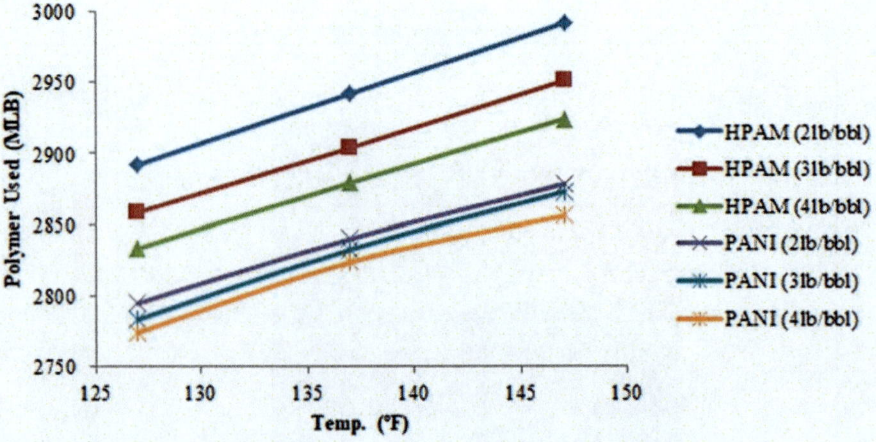

Fig. 4 Polymer adsorption at different polymer concentrations and temperature

color denotes minimum oil saturation which means the area was predominantly filled with water, whereas orange color represents oil saturation of 0.8.

It was observed that PANI was much able to saturate the oil near the producer well area, which may enhance the oil production. This may be due to reason that PANI was able to sweep more area of reservoir as it was able to permeate more as compared to HPAM. Figure 6 shows the oil production (in MSTB) by PANI as well as HPAM flooding at different concentrations and temperature. It was found that oil production using PANI and HPAM increased with increase in concentration of polymer from 2 to 4 lb/bbl. However, increase in oil recovery was significantly high in PANI as compared to HPAM. It may be because HPAM viscosity is degraded more rapidly at high-temperature conditions. In the HPAM flooding, the concentration was increased from 2 to 4 lb/bbl, the oil recovery percentage increased from 28.68% to 31.22%. Also in the PANI flooding, the oil recovery percentage increased from 30.50% to 32.77% when the concentration was increased from 2 lb/bbl to 4 lb/bbl.

4 Conclusion

Present numerical work was carried out to investigate the performance of enhanced oil recovery using HPAM and PANI polymers in CMG-IMEX simulator software. Present model was validated with existing literature. Polymer adsorption contours and oil saturation contours were investigated. It was observed that better oil recovery was found in the case of PANI than HPAM. 30.05% of oil recovery in PANI 2 lb/bbl flooding was obtained, whereas 28.60% for HPAM. This study showed that PANI can be efficiently used for EOR process as it was more stable at higher temperature and had improved sweeping area of reservoir. Future works may be carried out to investigate the EOR performance of polymer flooding along with

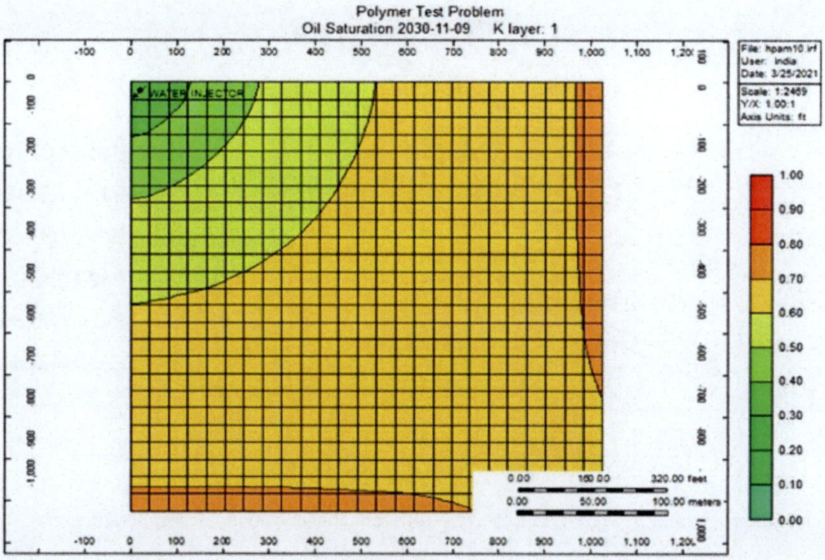

(a)

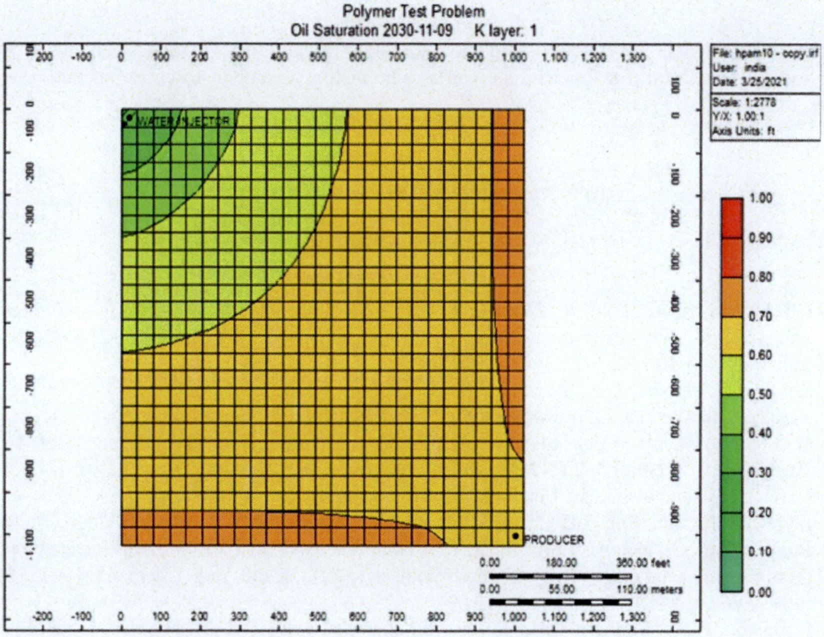

(b)

Fig. 5 Contour of oil saturation by 2 lb/bbl of (**a**) HPAM and (**b**) PANI

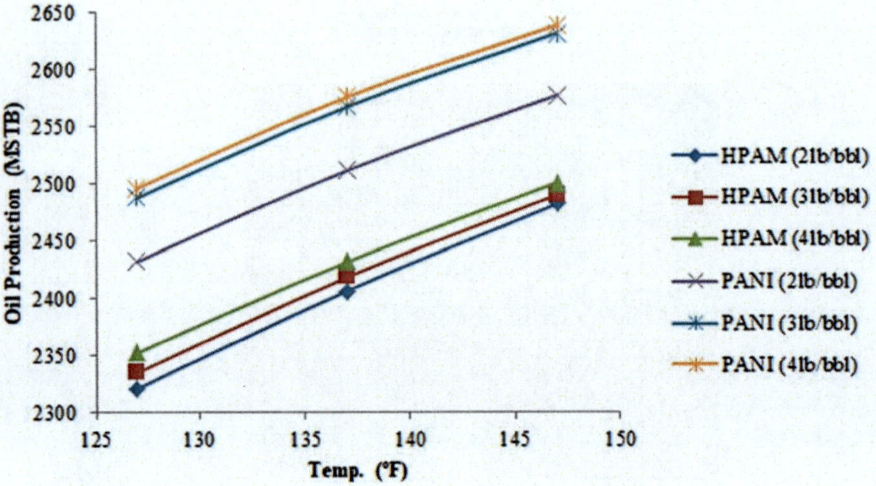

Fig. 6 Oil production by HPAM and PANI at different concentration and temperature

the surfactant system (polymer–surfactant flooding) and alkali–surfactant system (alkali–surfactant–polymer flooding).

Acknowledgements The authors would like to acknowledge Guru Gobind Singh Indraprastha University, New Delhi, for financial as well as administrative support to carry out this research work.

References

1. Gbadamosi, A.O., Junin, R., Manan, M.A., Agi, A., Yusuff, A.S.: An overview of chemical enhanced oil recovery: recent advances and prospects. Int. Nano Lett. **9**, 171–202 (2019). https://doi.org/10.1007/s40089-019-0272-8
2. Sheng, J.J., Leonhardt, B., Azri, N.: Status of polymer-flooding technology. J Can Pet Technol **54**(02), 116–126 (2015). https://doi.org/10.2118/174541-PA
3. Gao, C.: Viscosity of partially hydrolyzed *polyacrylamide* under shearing and heat. J. Pet. Explor. Prod. Technol **3**, 203–206 (2013). https://doi.org/10.1007/s13202-013-0051-4
4. Raffa, P., Broekhuis, A.A., Picchioni, F.: Polymeric surfactants for enhanced oil recovery: a review. J. Pet. Sci. Eng. **145**, 723–733 (2016). https://doi.org/10.1016/j.petrol.2016.07.007
5. Pu, W., Shen, C., Wei, B., Yang, Y., Li, Y.: A comprehensive review of *polysaccharide* biopolymers for enhanced oil recovery (EOR) from flask to field. J. Ind. Eng. Chem. **61**, 1–11 (2018). https://doi.org/10.1016/j.jiec.2017.12.034
6. Firozjaiia, A.M., Saghafi, H.R.: Review on chemical enhanced oil recovery using polymer flooding: Fundamentals, experimental and numerical simulation. Petroleum **6**, 115–122 (2020). https://doi.org/10.1016/j.petlm.2019.09.003
7. Zarepakzad, N.: Development of a screening model for polymer flooding in multi-layer reservoirs. Master's Thesis, Middle East Technical University, Ankara (2018)

8. Rai, S.K., Bera, A., Mandal, A.: Modeling of surfactant and surfactant–polymer flooding for enhanced oil recovery using STARS (CMG) software. J Petrol Explor Prod Technol **5**, 1–11 (2015). https://doi.org/10.1007/s13202-014-0112-3
9. Chaipornkaew, M., Wongrattapitak, K., Chantarataneewat, W., Boontaeng, T., Opdal, S.T., Maneeintr, K.: Preliminary study of in situ combustion in heavy oil field in the North of Thailand. Proc Earth Planet Sci **6**, 326–334 (2013). https://doi.org/10.1016/j.proeps.2013.01.043
10. Bhanvase, B.A., Kamath, S.D., Patil, U.P., Patil, H.A., Pandit, A.B., Sonawane, S.H.: Intensification of heat transfer using PANI nanoparticles and PANI-CuO nanocomposite based nanofluids. Chem. Engg. Proc. **104**, 172–180 (2016). https://doi.org/10.1016/j.cep.2016.03.004
11. Hajri, S.A., Mahmood, S.M., Abdulelah, H., Akbari. S.: Overview on polymer retention in porous media. Energies 11, 2751 (2018).
12. Chauveteau, G., Kohler, N.: Polymer flooding: The essential elements for laboratory evaluation. In: Proceedings of the SPE Improved Oil Recovery Symposium. Society of Petroleum Engineers, Texas (1974).
13. Mahmud, H.B., Tan, B.C., Giwelli, A., Al-Rubaye, A.F., Shafiq, M.U.: Numerical analysis of SiO_2-SDS surfactant effect on oil recovery in sandstone reservoirs. Energy Geosci. **2**(4), 238–245 (2021). https://doi.org/10.1016/j.engeos.2021.06.003

Effect of Different Shock Generator Configurations on Ethylene-Fuelled Transverse Injection-Based Scramjet Combustor

Pabbala Monish Yadav, Gautam Choubey, and Sumit Tripathi

Abstract Even though hydrogen is considered and promoted as a prospective fuel for combustion in a scramjet engine, ethylene can operate better under low Mach number speeds and can produce greater thrust at such speeds than hydrogen. This study is focussed on determining the parametric variations in the flow-field, arising due to the adaptation of single and dual shock generators, and the subsequent effects on the combustor. This is a computational study in which ANSYS Fluent software has been used to perform the simulations. Moreover, a simulation of a previously completed experiment is run to verify the solver's validity, and the findings are remarkably close to the experimental results. Four geometries with shock generators located at different positions, two for each single and dual shock generator modules, are employed for the study. A two-dimensional design with fine-structured mesh is constructed for the study. To register the flow-field dynamics of the simulation, 2D coupled implicit Reynolds's-averaged Navier–Stokes equations and typical $k - \varepsilon$ turbulent model are utilized to evaluate the interaction between the supersonic airflow and the injected ethylene.

Keywords Scramjet combustor · Ethylene fuel · Shock generator · Transverse injection

1 Introduction

To withstand the high temperatures and pressure ratios, there was a prominent devoir to fabricate a novel type of engine that does not have any rapid swivelling apparatus. Although scramjets have their own intricacies, their design is very lucid and is a solution to the above-mentioned hindrance [1]. They are designed for high velocities and thrust generation [2]. As the scramjets work at extreme conditions, the conditions are not favourable for slow mixing and combustion; hence, they are to be constructed in a way to ensure very fast mixing and combustion of fuel likely in a few milliseconds

P. M. Yadav · G. Choubey (✉) · S. Tripathi
Department of Mechanical and Aerospace Engineering, Institute of Infrastructure Technology Research and Management (IITRAM), Ahmedabad 380026, India

R. P. Bharti and K. M. Gangawane (eds.), *Recent Trends in Fluid Dynamics Research*, Lecture Notes in Mechanical Engineering,
https://doi.org/10.1007/978-981-16-6928-6_19

[3, 4]. Therefore, for fabricating a robust scramjet engine, it is imperative to cogitate the injection and mixing of propellant in supersonic air currents. For congruous mixing, flaming and oxidation, a lot of experimental and computational investigations have been done employing various injection techniques [5–9]. To bring the best out of a scramjet combustor, it is vital to choose a felicitous injection technique [10]. The next step after researching the injection techniques was devising designs and instruments for enriched diffusion and their evaluation [7].

The parallel injection and transverse injection both are revered as prominent injection methods for scramjet engines, but the transverse injection out wins the parallel injection as it has better fuel penetration than the other [5, 7, 11–16]. The transverse injection also surpasses the parallel injection in faster near-field mixing, higher flame retention capability, drag force reduction and requires low cooling than the latter one. However, it comes with some inadequacies like higher pressure loss, very strong shock wave generation and perplexing flow domain. Transverse injection performed with an oblique injection angle is a transcended version of transverse injection (i.e. at only 90° with the wall) as it has superior near- and far-field mixing with low thrust loss and also outdoes the hurdles occurring in parallel injection [17–22].

Though hydrogen fuel [23–29] is a renowned fuel for its lightness and net calorific value, ethylene produces greater thrusts at low Mach number flights when compared to hydrogen and many other hydrocarbon fuels [30]. It can be said after analysing the preceding studies that there is a significant opportunity to explore the division of combustors benefited by shock generators. Thus, the aforementioned study targets research on using a unique and new shock generator which is like a reverse-shaped wedge in a scramjet combustor and the resulting variations are caused by them in the flow dynamics. A simulation using 2D coupled implicit RANS equations and the standard k–ε model [31] solver performed on ANSYS Fluent software is implemented for the current work. Two geometric models with a single shock generator located at different positions and other two geometric models with dual shock generators at various locations are generated and used for the simulation.

2 Geometry and Mesh Generation

In the present study, the 2D model is a rectangular channel of length 25.4 cm and of height 7.62 cm with a 0.0559 cm width slot on the bottom wall at a distance of 17.8 cm from the entrance of the channel as shown in Fig. 1a. The model is taken from the experimental investigation performed by Weidner and Drummond [32]. A structured mesh, with cells densely positioned at walls and injection slot for capturing the boundary layer and its interaction with the shock wave, is constructed using ANSYS Software. A grid independence test is also performed for the present case. The mesh used for the simulations is shown in Fig. 1b.

For determining the effects of various shock generator configurations on ethylene-fuelled scramjet engines, four schematic models are used: model 1, model 2, model 3 and model 4. The shock generators in models 1 and 2 are independent, while dual

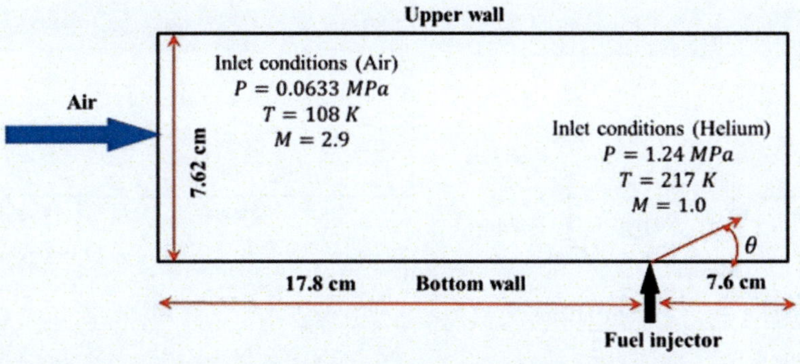

(a)

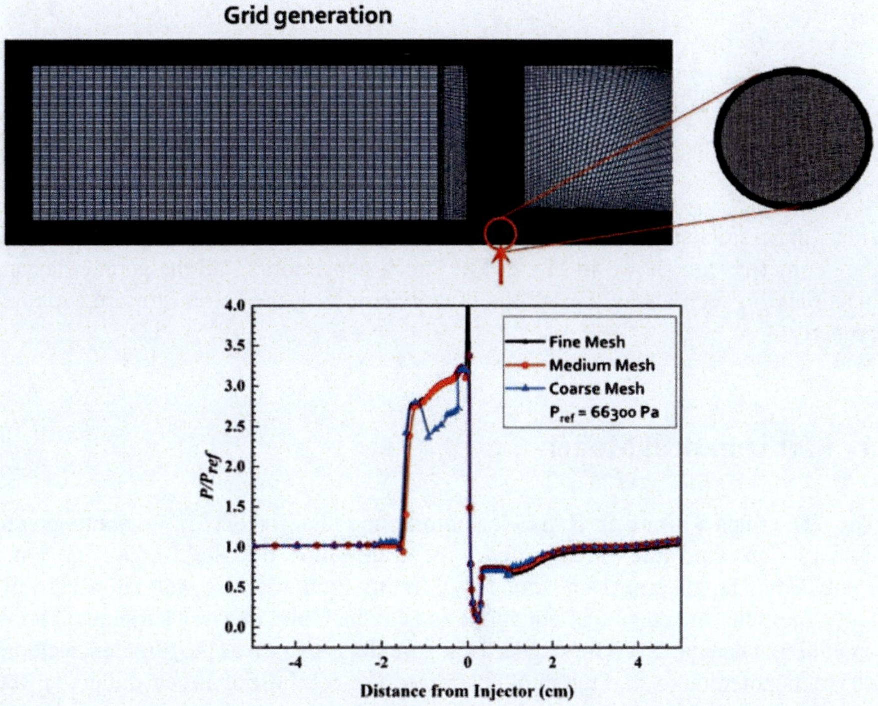

(b)

Fig. 1 a Schematic of the model employed for validation [32]. (All units are in mm). b Illustration of the grid used along with grid independence test

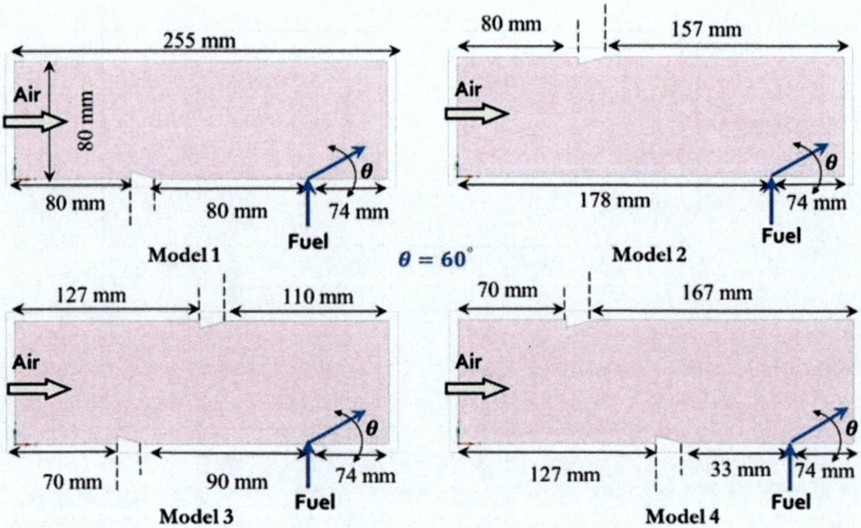

Fig. 2 Geometric models for the parametric study (all dimensions are in mm)

shock generators are introduced in the rest of the geometries. The schematics of all the geometries are shown in Fig. 2. The shock generators in all the geometries are positioned in such a way that all the flow dynamics in the combustor are captured properly.

3 Mathematical Model

The 2D coupled implicit RANS equations and the standard $k - \varepsilon$ model are used to study and numerically model the combustion flow-field of a supersonic combustor. The governing equations used in this simulation are well known (continuity, momentum and energy equation) and can be found in every subsequent book on fluid mechanics [33]. The standard $k - \varepsilon$ model is chosen as the turbulence model due to its potential to start the initial iteration. The combustion model adopted is the combination of species transport + finite rate chemistry/eddy dissipation reaction model.

4 Validation of the Model

The validation uses the data from the experimental work provided by Weidner and Drummond [32]. From Fig. 3a, it can be observed that there are certain discrepancies such as underestimation of static pressure surge vicinal to the upstream region

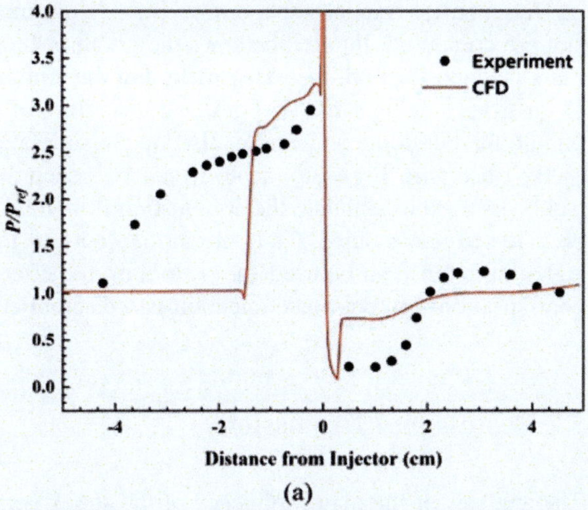

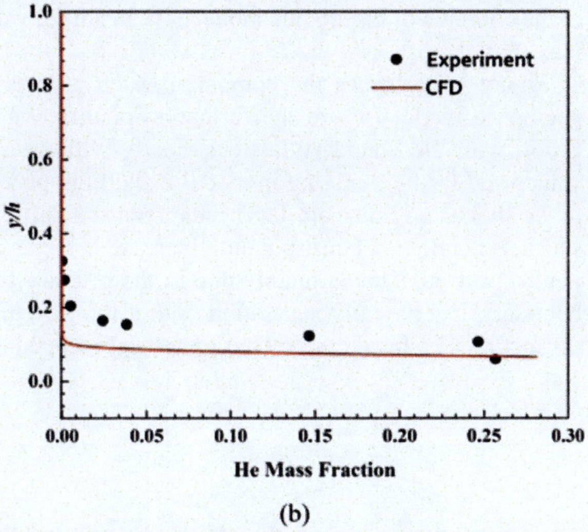

Fig. 3 Comparison between the experimental data of Weidner and Drumond [32] (**a**) numerically simulated pressure variation along the wall (**b**) helium mass fraction at a distance of 3.81 cm downstream of the injector

of the injection slot by $k - \varepsilon$ turbulent model, due to its shortcomings, constant flow inference at fuel exit and the complexity of flow [10]. Putting aside these few disparities, there is a major concurrence between the experimental results and the values obtained from the simulation as shown in Fig. 3a. One can observe the formation of a strong shock wave due to the crossing of the bow shock wave, boundary layer and bottom wall causing high pressures in both the separation and reattachment zones near the injection slot. It can also be seen that the pressure gets reduced downstream the injector and the boundary layer and shock wave intermesh to create separation regions on either side of the injector slot. At a distance of 3.81 cm downstream the fuel slot, it is observed that the shock wave dies in the combustor.

Helium mass fractions on a vertical line at a distance of 3.81 cm from the injection slot are computed. Figure 3b shows the graph of the normalized height vs. helium mass fraction for both the experiment and the simulation work. We see the nature of the graphs being same for both, and the values of both being close to each other though there is some underestimation by the simulation work.

By observing Figs. 3a and b, it can be concluded that computational models can be utilized to simulate the combustion flow-field, and shock waves formed can be detected and studied. Thus, it can be deduced that computational and numerical simulations can be used for examining the effect of different shock generator configurations on transverse injection-based scramjet combustor.

5 Results and Discussion

The analysis of the major influence of the shock generator on the effective mixing inside the flow-field requires a clear understanding of the flow pattern. The images of the contours of the studied model have localized values (i.e. localized minima and maxima).

Figure 4 illustrates the computed Mach number contour profile for different geometric models where the formation of different shock waves and their interaction with the boundary layer is clearly witnessed. We can observe high Mach values and far-field mixing in model 2 and high penetration of fuel in model 3 due to the shock and boundary layer interactions. Additionally, with the addition of two shock generators, a corresponding increase in the expansion of injecting jet plume is also noticed. This is mainly due to the primary flow's lack of strength and the boundary layer's early separation within the flow-field. The development of two distinct shocks inside the model may be seen in Mach number contour, and these

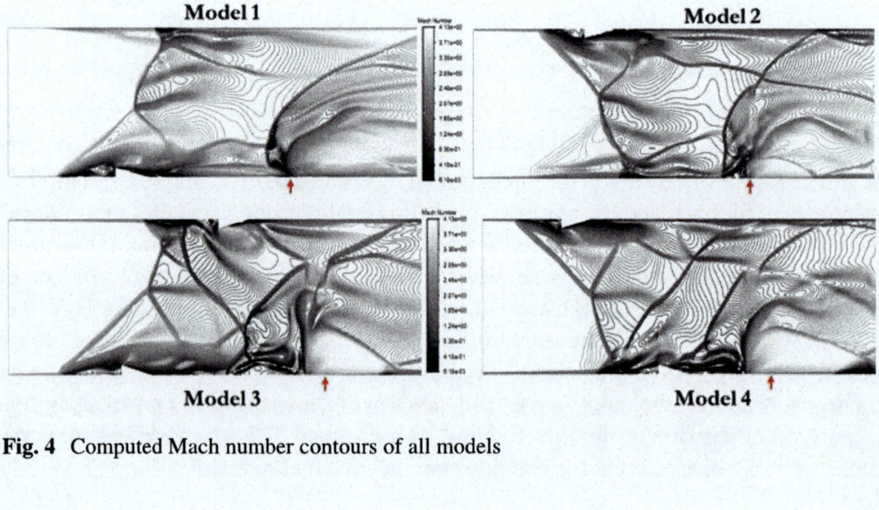

Fig. 4 Computed Mach number contours of all models

are known as oblique shock and separation shock, respectively. The addition of two shock generators substantially accelerates the intensity of these shocks, as shown in Fig. 4.

The velocity streamlines of all the geometries are captured for exploring the flow-field dynamics of the combustor. The resulting streamlines are shown in Fig. 5. From model 1, it can be seen that a small recirculation zone is formed just upstream of the injection slot at a distance of 1.9 cm from the bottom wall, which can help in the mixing of fuel with airstream and small insignificant recirculation zone at the upper wall slightly opposite to the shock generator where the strong shock touches the upper wall.

Model 2 also has a small recirculation zone formed just in front of the shock generator and another one on the bottom wall slightly opposite to the shock generator on the upper wall just like model 1. The formation of a coupled vortex at the injection slot is also observed which helps in separation of the air stream and induces more penetration of fuel.

A significant recirculation zone is witnessed ahead of the injection slot in model 3 which is primarily achieved because of the greater availability of space between the bottom wall shock generator and the injector location. This recirculation zone is bigger in size than the ones formed in models 1 and 2 because of the increased amount of interaction between the free-stream and the shock waves in model 3 which is obvious as it has two shock generators rather than models 1 and 2 where only one shock generator is introduced. The distance between the bottom wall shock generator and the injector is larger in model 3 as compared to model 4, and that is why the scope for increased interaction is more in model 3 than that of model 4. This recirculation

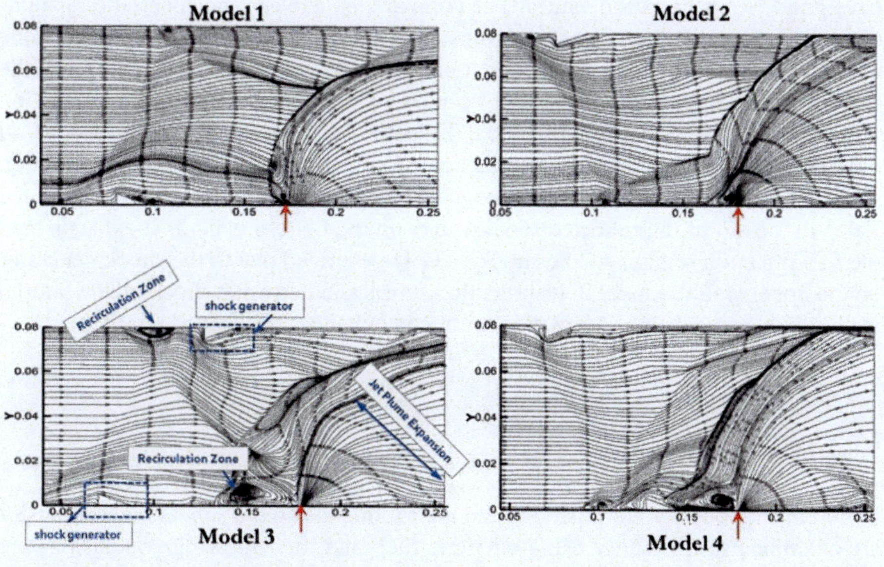

Fig. 5 Velocity streamlines of all models

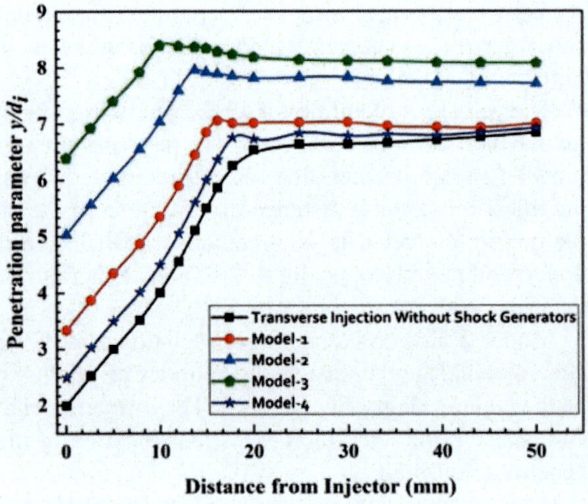

Fig. 6 Penetration parameter plot for different models

zone has a greater flame holding capacity and also helps in enhanced penetration and mixing. The jet plume expansion is also witnessed in model 3. Additionally, one more recirculation zone is observed ahead of the shock generator on the upper wall, which itself produces a shock and reduces the shock created by the upper wall shock generator. Thus, the interactions between the different shock waves generated by the upper and lower walls create vortices and recirculation zone, which is beneficial for effective mixing and flame holding.

To decrease the temperatures on the bottom wall, it is required for the combustor to have good fuel penetration values. The comparison between the penetration parameters of all the models and a normal transverse injection-based model has been done and is shown in Fig. 6. From Fig. 6, it can be easily noticed that model 3 has the highest penetration parameter value because of the vortices and the recirculation zone formed in it. Additionally, model 4 has the least penetration parameter of all the four models studied due to the effect of the strong shock wave and its interaction with fuel jet and boundary layers. Model 2 has higher penetration values than model 1 due to the aid of the reflected shock wave, on the bottom wall, as it expands near the fuel injection region. As the shock wave is generated due to the shock generator on the upper wall in model 2, its reflection forms another weak shock on the bottom wall, which expands at the injection point and enhances the fuel penetration.

6 Conclusions

Different shock generator positions and their influence on the supersonic combustor are examined in this study using ethylene fuel, and the following conclusions are drawn based on numerical analysis.

- From the velocity streamlines plot, it can be concluded that the best flame holding capacity is seen in the model 3 due to the formation of strong recirculation zone in it than any other model investigated in this study.
- Penetration parameter graph shows that the penetration of the fuel is highest in model 3, placing the model 2 after it, and least being the model 4. It also shows that the penetration is better in every model when compared to the conventional transverse injecting model.

Hence, from the above observations, it can be concluded that the most suitable model for adaptation is the model 3 and the next setup being the model 2 in terms of Mach number generation, penetration parameters and the flame holding capacity for efficient combustion for ethylene-based fuel injection technique.

References

1. Curran, E.T., Murthy, S.N.B. (eds.): Scramjet propulsion, 1st edn. American Institute of Aeronautics and Astronautics, Reston (2001)
2. Segal, C.: The Scramjet Engine: Processes and Characteristics, 1st edn. Cambridge University Press, Cambridge (2009)
3. Hyungseok, S.: A study of the mixing characteristics for cavity sizes in scramjet engine combustor. J. Korean Phys. Soc. **55**(5), 2180–2186 (2009)
4. Huang, W., et al.: Investigation on the flameholding mechanisms in supersonic flows: backward-facing step and cavity flameholder. J. Visualization **14**(1), 63–74 (2011)
5. Bogdanoff, D.W.: Advanced injection and mixing techniques for scramjet combustors. J. Propul. Power **10**(2), 183–190 (1994)
6. Curran, E.T.: Scramjet engines: the first forty years. J. Propul. Power **17**(6), 1138–1148 (2001)
7. Seiner, J.M., et al.: Historical survey on enhanced mixing in scramjet engines. J. Propul. Power **17**(6), 1273–1286 (2001)
8. Huang, W., et al.: Flame propagation and stabilization in dual-mode scramjet combustors: a survey. Prog. Aerosp. Sci. **101**, 13–30 (2018)
9. Huang, W., et al.: Supersonic mixing in airbreathing propulsion systems for hypersonic flights. Prog. Aerosp. Sci. **109**, 100545 (2019)
10. Manna, P., Chakraborty, D.: Numerical investigation of transverse sonic injection in a non-reacting supersonic combustor. Proc. Inst. Mech. Eng., Part G: J. Aerosp. Eng. **219**(3), 205–215 (2005)
11. Cecere, D., et al.: Hydrogen/air supersonic combustion for future hypersonic vehicles. Int. J. Hydrogen Energy **36**(18), 11969–11984 (2011)
12. Choubey, G., Pandey, K.M., Sharma, D., Debbarma, A.: Computational simulation of multi-strut central lobed injection of hydrogen in a scramjet combustor. Perspect Sci. **8**, 222–224 (2016)
13. Choubey, G., Suneetha, L., Pandey, K.M.: Composite materials used in scramjet—A review. Mater Today Proc. **5**, 1321–1326 (2018)
14. Dong, M., Liao, J., Choubey, G., Huang, W.: Influence of the secondary flow control on the transverse gaseous injection flow field properties in a supersonic flow. Acta Astronaut. **165**, 150–157 (2019)
15. Choubey, G., Yuvarajan, D., Huang, W., Shafee, A., Pandey, K.M.: Recent research progress on transverse injection technique for scramjet applications-a brief review. Int J Hydrogen Energy. **45**(51), 27806–27827 (2020)
16. Choubey, G., Pandey, K.M.: Numerical studies on the performance of scramjet combustor with alternating wedge-shaped strut injector. Int J Turbo & Jet engines. **34**(1), 11–22 (2017)

17. Aso, S., et al.: A study on supersonic mixing by circular nozzle with various injection angles for air breathing engine. Acta Astronautica **65**(5–6), 687–695 (2009)
18. Abdel-Salam, T., et al.: Investigation of mixing and flow characteristics in a dual-mode combustor. In: 41st Aerospace Sciences Meeting and Exhibit, 372 (2003).
19. Abdelhafez, A., et al.: Evaluation of oblique and traverse fuel injection in a supersonic combustor. In: Proceedings of 43rd AIAA/ASME/SAE/ASEE Joint Propulsion Conference & Exhibit, pp. 8–11. American Institute of Aeronautics and Astronautics, Cincinnati (2007).
20. Choubey, G., Pandey, K.M.: Effect of different wall injection schemes on the flow-field of hydrogen fuelled strut-based scramjet combustor. Acta Astronaut. **145**, 93–104 (2018)
21. Choubey, G., Pandey, K.M.: Effect of variation of angle of attack on the performance of two-strut scramjet combustor. Int J Hydrogen Energy. **41**(26), 11455–11470 (2016)
22. Choubey, G., Pandey, K.M.: Effect of parametric variation of strut layout and position on the performance of a typical two-strut based scramjet combustor. Int J Hydrogen Energy. **42**(15), 10485–10500 (2017)
23. Choubey, G., Devarajan, Y., Huang, W., Yan, L., Babazadeh, H., Pandey, K.M.: Hydrogen fuel in scramjet engines-A brief review. Int J Hydrogen Energy. **45**, 16799–16815 (2020)
24. Choubey, G., Devarajan, Y., Huang, W., Mehar, K., Tiwari, M., Pandey, K.M.: Recent advances in cavity-based scramjet engine-a brief review. Int J Hydrogen Energy. **44**(26), 13895–13909 (2019)
25. Choubey, G., Pandey, K.M.: Investigation on the effects of operating variables on the performance of two-strut scramjet combustor. Int J Hydrogen Energy. **41**(45), 20753–20770 (2016)
26. Li, Z., Moradi, R., Marashi, S.M., Babazadeh, H., Choubey, G.: Influence of backward-facing step on the mixing efficiency of multi microjets at supersonic flow. Acta Astronaut. **175**, 37–44 (2020)
27. Du, Z., Shen, C., Huang, W., Yan, L., Liang, J., Choubey, G.: Parametric study on mixing augmentation mechanism induced by cantilevered ramp injectors in a shock-induced combustion ramjet engine. Aerosp Sci Technol. **108**, 106413 (2020)
28. Huang, W., Wu, H., Du, Z., Yan, L., Choubey, G.: Design exploration on the mixing augmentation induced by the oblique shock wave and a novel step in a supersonic flow. Acta Astronaut. **180**, 622–629 (2021)
29. Liu, X., Moradi, R., Manh, T.D., Choubey, G., Li, Z., Bach, Q.V.: Computational study of the multi hydrogen jets in presence of the upstream step in a Ma= 4 supersonic flow. Int J Hydrogen Energy. **45**(55), 31118–31129 (2020)
30. Zhi-hua, W., et al.: Influence of different fuels on scramjet engine performance. In: 2019 IEEE 10th International Conference on Mechanical and Aerospace Engineering (ICMAE), pp. 434–438. IEEE, Brussels (2019).
31. Huang, W., Wang, Z.G.: Numerical study of attack angle characteristics for integrated hypersonic vehicle. Appl. Math. Mech. **30**(6), 779–786 (2009)
32. Weidner, E.H., Drummond, J.P.: A parametric study of staged fuel injector configurations for scramjet applications. In: Proceedings of the AAS/AIAA Astrodynamics Conference, pp. 3–5. American Institute of Aeronautics and Astronautics (AIAA), Nevada (1981)
33. Ferziger, J.H., Peri, M., Street, R.L.: Computational methods for fluid dynamics. vol. 3, pp. 196–200. Springer, Berlin (2002)
34. Huang, W., et al.: Effect of cavity flame holder configuration on combustion flow field performance of integrated hypersonic vehicle. Sci. China Technol. Sci. **53**(10), 2725–2733 (2010)
35. Kim, K.M., et al.: Numerical study on supersonic combustion with cavity-based fuel injection. Int. J. Heat Mass Transf. **47**(2), 271–286 (2004)
36. Cheng, T.S., Yang, W.J.: Numerical simulation of three-dimensional turbulent separated and reattaching flows using a modified turbulence model. Comput. Fluids **37**(3), 194–206 (2008)

CFD Simulation of EOR Technique, by Gas Injection of CO_2-LPG Along with the Nanoparticles by Using the Eulerian–Eulerian Approach

Shrutika Kashetti, Greeshma K. Anand, and Priya C. Sande

Abstract CO_2 injection into shale reservoirs has become an area of interest in recent studies in the field of enhanced oil recovery techniques. The use of CO_2 reduces the viscosity of oil, making it easier to flow through the pores of the rocks. To increase the sweep efficiency, addition of nanoparticles into the gas stream is a novel approach. The small size of nanoparticles helps in better penetration into pores where traditionally injected fluids cannot reach. Injection of LPG along with the CO_2 stream is said to enhance the EOR technique as it leads to minimum miscible pressure. Computational fluid dynamics (CFD) simulation study is used for the first time to model the nanoparticles by combining them with the injection mixture of LPG and CO_2. Geometry for the shale porous media is created using ANSYS SpaceClaim software, to resemble the actual rock, with a tortuous flow path. Eulerian–Eulerian multiphase approach is used to simulate the flow of each of the three phases (oil, gas mixture, and nanoparticles). The main parameter studied is the velocity of gas, as this has a major effect on the sweep efficiency. CFD results show that a gas velocity of just greater than 0.5 m/s is optimal for an efficient gas sweep.

Keywords CO_2 and LPG injection · Nanoparticles · Enhanced oil recovery (EOR) · Discrete phase model

Nomenclature

ε_g =	volume fraction of gas phase (–)
ε_s =	volume fraction of solid phase (–)
ρ_s =	density of solid phase (kg/m^{-3})
$\overrightarrow{v}_s$ =	velocity of solid phase (m/s)
p =	gas pressure (Pa)

S. Kashetti · G. K. Anand · P. C. Sande (✉)
Department of Chemical Engineering, Birla Institute of Technology and Science, Pilani, Pilani-Campus RJ 333031, India
e-mail: priya@pilani.bits-pilani.ac.in

R. P. Bharti and K. M. Gangawane (eds.), *Recent Trends in Fluid Dynamics Research*, Lecture Notes in Mechanical Engineering,
https://doi.org/10.1007/978-981-16-6928-6_20

$p_s =$ solids pressure (Pa)
$\vec{\tau}_s =$ solid phase stress tensor (N/m^2)
$g =$ acceleration due to gravity (m/s^2)

1 Introduction

Exploration of new techniques to extract oil to its maximum extent is an ultimate goal for many researches. Oil recovery operations involve three stages, which are primary, secondary, and tertiary [1]. While primary and secondary oil recovery produces oil that is easily extractable using the pressure difference between surface and oil reservoir, tertiary techniques employ oil recovery processes at the level of the porous pathways in the strata of the oil containing rock. This is also called the enhanced oil recovery (EOR). Studies show that primary and secondary methods extract about 20–40% of the oil in the reservoir. This is maximized to 30–60% by the EOR technique [1]. EOR utilizes gases, chemicals, or thermal energy to displace the unrecovered oil by modifying the chemical composition of the oil. There are three types of EOR—gas injection, thermal injection, and chemical injection.

1.1 Gas Injection

Gas injection is one of the most commonly used EOR methods [1]. It is associated with the introduction of gases like carbon dioxide, nitrogen, or natural gas to retrieve oil. Gas injection focuses on the light and medium crude oils by reducing the interfacial tension between the injected gas and oil. The carbon dioxide-enhanced oil recovery technique is a green and most reassuring way to recover oil as it also reduces the CO_2 content in the atmosphere [2]. The use of CO_2 for enhanced oil recovery (EOR) reduces the viscosity of oil, making it easier to flow through the pores. The oil recovery is intensified by improving the sweep efficiency [3]. The injected gas helps in oil swelling and reduction of its viscosity consequently making the oil more mobile and easier to retrieve. This technique is considered economical in the perspective that less amount gas will be required to be injected into the reservoir [4].

Ronghao Cui et al. (2019) illustrated oil–water two-phase flow in shale in an integrated pore network model framework, and CFD simulation was employed with different geometric pores (circular, square, and equilateral triangle [5]). Di Hea et al. (2019) illustrated a better understanding of the flow of two-phase supercritical CO_2–water through porous media under geological conditions. CFD simulation was done by incorporating the VOF model (non-steady-state laminar flow model) to obtain the permeability saturation curve [6]. M. Krzaczek et al. (2019) designed a 2D model of hydrofracking in rocks at mesoscale. High-pressure injection of water was modeled realistically, and DEM was used to represent the mechanical behavior of rocks in mesoscale [7].

1.2 Thermal Method

The thermal method works on the principle of increasing the temperature of the reservoir to diminish the viscosity, specific gravity, and interfacial tension of oil for recovery. It is considered to be one of the most advanced EOR processes. Besides the physical changes, the chemical properties of the oil are altered sometimes via cracking and dehydrogenation processes. A few noted thermal techniques are cyclic steam stimulation, hot water injection, steam flooding, electric heating, and electromagnetic heating [8].

A 2D pore geometry was considered for the simulation of the process of hot water displacement in a single pore. VOF multiphase model was utilized by Lv and Wang (2015) to solve the continuity equation, the momentum equation, the energy equation, and the volume fraction equation [9].

1.3 Polymer Injection

Introduction of polymers into the reservoir is a form of the chemical EOR method. Chemical techniques are the least utilized compared to the former methods. It is categorized by the usage of three chemicals—alkaline, surfactant, and polymer [10]. Polymer flooding has been a prominent method, and it is said to improve the sweep efficiency by regulating the mobility ratio of injected fluid, escalating the recovery factor by 30% [11].

A micromodel of pore geometry as designed by Clemens et al. (2013) in the scanning electron microscope (SEM) and CFD analysis was performed considering the flow of non-Newtonian polymer fluid in the pore. The volume of fluid (VOF) model was utilized to study the governing equations and compared with the experimental results of polymer injection [12]. Jackson et al. (2011) designed new models for pressure in horizontal wells, in which polymers are introduced for oil recovery. CFD modeling was done coupling the well and the reservoir flow, studying the Navier–Stokes equations plus Darcy's law for Newtonian and non-Newtonian flows. The influence of non-uniform pressure on sweep efficiency and hydrocarbon recovery was analyzed [13].

1.4 Injection of Nanoparticles

Nanoparticle's properties are employed in conjunction with the conventional EOR methods to eliminate drawbacks and enhance the sweep efficiency of the displacing fluid. The inert nature of nanoparticles is beneficial in this aspect. Addition of nanoparticles to the injected fluids further optimizes the properties of displaced fluids (mainly oil) such as density, viscosity, oil–water interfacial tension, thermal conductivity, and specific heat, thereby reducing the fingering effect. It is a condition in which the interface of the fluids in the reservoir flows in an uneven/fingered profile. Due to their ultrasmall size, they have the ability to penetrate into minor pores, where conventionally injected fluids cannot reach. Nanoparticles also prevent the injected gas from getting trapped in minor pores.

Geometry of a porous medium was designed in Gambit software by Dezfully et al. (2015). The finite volume method with the Eulerian approach (granular) was used to solve the continuity, momentum, and volume fraction equations in the micromodel pore. The impact of nanoparticle addition (supercritical CO_2) in the gas injection method was investigated [14]. Gharibshahi et al. (2015) conducted a study on the CFD technique in which nanosilica/supercritical CO_2 was considered as the injected fluid. The finite volume method was employed to elucidate the continuity, momentum, and volume fraction equations. The granular model in the Eulerian model was used for the same as it is accurate and also enables the direct application of nanoparticles diameter [15]. Gharibshahi et al. (2019) designed a 2D micromodel of a randomly distributed porous medium to analyze the enhanced extraction of heavy oil using CFD simulation. The flow pattern of fluid and oil was investigated. The effect of Al_2O_3, CuO, and Fe_3O_4 nanoparticles on steam injection was studied [16]. VOF Model was incorporated in a 2D porous geometry by Zhao and Wen (2017) to illustrate the oil–water two-phase system. Pore-scale flooding was examined for oil-saturated and oil-unsaturated pores for enhanced oil recovery [17].

Computational fluid dynamics (CFD) has gained enormous popularity in this field due to its power of prediction and provides an alternative to optimizing parameters via experimental EOR study, which is expensive. This CFD simulation study, for the first time, models the nanoparticles by combining them with the injection mixture of LPG and CO_2. Geometry for the shale porous media is created using ANSYS SpaceClaim software, to resemble the actual rock, with a tortuous flow path. The Eulerian–Eulerian multiphase approach to simulate the flow of the three phases (oil, gas mixture, and nanoparticles) is used. The parametric study for this proposed model includes the velocity of gas as this parameter has a huge effect on the sweep efficiency. In addition, the inlet ratio of CO_2 and LPG will play a vital role in enhancement, as miscible condition reduces gas saturation. This in turn will cause a decrease in gas relative permeability. This study will add value to the literature due to the novelty of modeling the EOR technique with simultaneous injection of CO_2–LPG along with the nanoparticles.

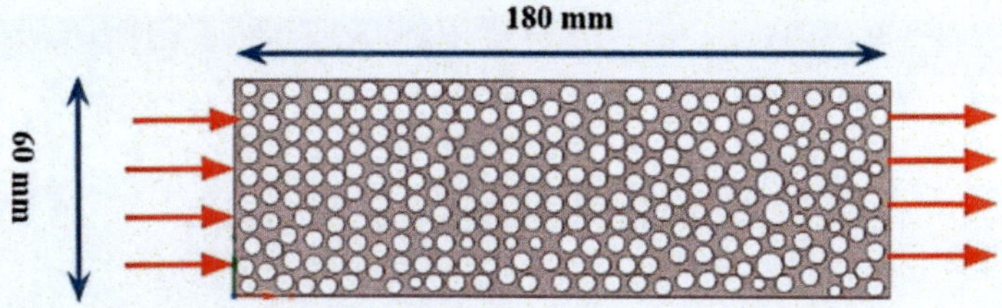

Fig. 1 Geometry of domain representing shale rock with a tortuous internal flow path

2 Numerical Implementation (Method)

2.1 *CFD Model and Geometry*

The Eulerian–Lagrangian techniques offer far too detailed flow information and thus may require averaging to extract useful results. This is because particles in the granular phase are tracked explicitly. In the Eulerian–Eulerian approach, the equations formulated are averaged for each of the phases which amounts to considering the phases as interpenetrating continua, without having to track each particle. The law of conservation of momentum, energy, and mass is satisfied for each phase individually in this approach. Thus, the dispersion of nanoparticles (with a particular diameter) in the primary phase can be modeled adequately without requiring huge CPU power.

In this study, the shale rock layer was modeled in ANSYS SpaceClaim. A 2D geometry was created to mimic dimensions of 6 cm × 18 cm. A porous medium was created using random circles of slightly varying size inside a rectangular geometry as shown in Fig. 1. The circles themselves are impervious, while the remaining area allows for flow. The diameter of these circles was randomly chosen between 0.25 cm and 0.5 cm.

The resulting flow domain represents a tortuous path for the flow of oil and gases through it, similar to the actual sedimentary rock from which oil is recovered. It adequately resembles porous media which is permeable to oil and gas, as previously studied [18]. The domain has a velocity inlet (left edge) and pressure outlet (right edge). The other two remaining and parallel edges are impervious walls (wall boundary condition).

2.2 *Meshing*

The type and size of mesh have an impact on CFD results. Therefore, a mesh of suitable size and shape needs to be chosen. The domain studied was meshed using tri/pave mesh with 3,76,407 cells and 2,35,638 nodes. The orthogonal quality of the mesh was acceptable at 0.82. The average size of one side of a single tri mesh was approximately 2 mm which is approximately two orders of magnitude smaller than

Fig. 2 Triangular mesh on section of domain

domain length. The mesh dependency considered in previous studies [15] with size of the order of few mm was found to be adequate. Figure 2 represents a section of the domain to show details of the mesh.

2.3 Governing Equations

The governing equations of mass, momentum, and granular energy was solved by finite volume method. The conservation equations were integrated using the accurate second-order upwind differencing in space and second-order implicit formulation in time. The set of these algebraic equations was solved by the tridiagonal matrix algorithm (TDMA), and for the volume fraction, point integration method was used. The Euler–Euler model is the most accurate multi-phase model. It uses a single pressure field for all the phases and solves momentum, enthalpy, continuity, and turbulence equations for each of the phases separately. Three phases were considered here for high accuracy and resolution of flow, with gas being the primary phase and oil and nanoparticles being the two secondary phases. The Euler–Euler model also tracks the volume fractions of the respective phases [1]. Unsteady-state formulation was given with ambient temperature and pressure boundary conditions. The following set of equations were solved, where the symbols have their standard meanings. The volume fraction of the phase q is defined as:

$$V_q = \int_v \alpha_q d\mathbf{V} \tag{1}$$

where, $\sum_{q=1}^{n} \alpha_q = 1$

Continuity equation for phase q is:

$$\frac{\partial}{\partial x}(\rho_q\alpha_q) + \nabla(\rho_q\alpha_q\overrightarrow{v_q}) = \sum\nolimits_{p=1}^{n}(\dot{m}_{pq} - \dot{m}_{pq}) + S_q \quad (2)$$

Momentum equations for gas and solid phase are given as:

$$\begin{aligned}\frac{\partial}{\partial t}(\alpha_q\rho_q\vec{v}_q) + \nabla.(\alpha_q\rho_q\vec{v}_q\vec{v}_q) &= -\alpha_q\nabla P + \nabla.\underline{\tau}_{-q} + \alpha_q\rho_q\vec{g} \\ &+ \sum\nolimits_{p=1}^{n}(\vec{R}_{pq} + \dot{m}_{pq}\vec{v}_{pq} - \dot{m}_{pq}\vec{v}_{pq}) \\ &+ (\vec{F}_q + \vec{F}_{lift,q} + \vec{F}_{vm,q})\end{aligned} \quad (3)$$

where $\underline{\tau}_{-q}$ - stress strain tensor of the *qth* phase.

$\vec{R}_{pq}$—interphase force.
$\vec{F}_{lift,q}$—lift force on secondary phase.

Conservation of granular energy is:

$$\frac{3}{2}\left(\frac{\partial}{\partial t}(\rho_s\varepsilon_s\theta_s) + \nabla.(\rho_s\varepsilon_s\theta_s\vec{v}_s)\right) = \left(\left(-p_s\vec{I}\right) + (\vec{\tau}_s)\right) : \nabla\vec{v}_s + \nabla.(k_{\theta_s}\nabla\theta_s) - \gamma_{\theta_s} + \phi_{gs} \quad (4)$$

$\left(\left(-p_s\vec{I}\right) + (\vec{\tau}_s)\right) : \nabla\vec{v}_s =$ Generation of energy by solid stress tensor.
$k_{\theta_s}\nabla\theta_s =$ Energy diffusion (k_{θ_s} is diffusion coefficient of the particles).
$\gamma_{\theta_s} =$ Collision dissipation of energy.
$\phi_{gs} =$ Energy exchange between gas and solid phase.

For steady state conditions, the convective and diffusive terms are usually neglected to give:

$$\left(\left(-p_s\vec{I}\right) + (\vec{\tau}_s)\right) : \nabla\vec{v}_s = -\gamma_{\theta_s} + \phi_{gs} \quad (5)$$

Inlet boundary condition used was the Dirichlet's boundary condition:

$$y(\text{atagivenboundary}) = C_1 \quad (6)$$

Here, C_1 is a constant velocity magnitude. Further, the impervious walls were used with no-slip boundary condition, and the outlet was given only a constant pressure boundary condition. All the equations were solved with a minimum residual accuracy of 10e-3 during unsteady state simulation, and the steady state results have a residual accuracy of approximately 10e-5.

3 Results and Discussion

The gas was modeled as a mixture material, comprising propane, butane, CO_2, and nitrogen whose properties are mentioned in Table 1. The gas mixture was defined as the primary phase while oil and nanoparticles were considered to be the two secondary phases.

Simulations for the gas velocities of 0.2 m/s, 0.4 m/s, and 0.5 m/s were studied. Addition of 5 vol. % of nanoparticles was done to improving the oil recovery. These estimates were taken from literature [14]. The designed tortuous paths were patched with 95 vol. % of oil and 5 vol. % of nanoparticles to model the flow of oil and nanoparticle phases as fluid phases in the same flow domain.

Figure 3a–c shows the oil volume fraction swept by gas velocities of 0.2, 0.4, and 0.5 m/s respectively. Each column of simulations in Fig. 3 represents flow behavior for one gas velocity. The topmost simulation in the column is early in time, which evolves to the approximate steady state (bottom simulation in the column). These results show how increase in gas velocity sweeps away the oil present in the porous

Table 1 Properties of fluid and nanoparticles

Components	Density (kg/m^3)	Viscosity (kg/ms)	Diameter (nm)	Mole fraction
CO_2	1.7878	1.37 × 10–5	–	0.78
Propane	1.91	7.95×10^{-6}	–	0.104
Butane	2.46	7×10^{-6}	–	0.088
Nitrogen	1.138	1.663×10^{-5}	–	0.028
Oil	970	0.007	–	Single phase
Nanoparticle	3880	–	14	Single phase

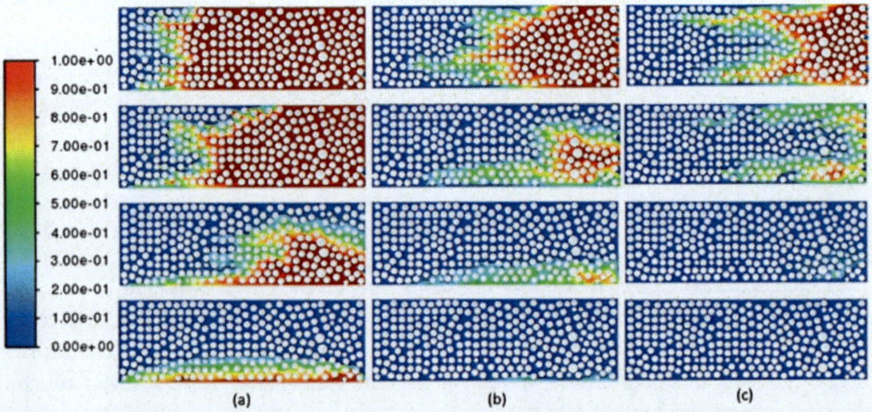

Fig. 3 Oil volume fraction for gas velocities. **a** 0.2 m/s. **b** 0.4 m/s. **c** 0.5 m/s showing evolution with time till steady state, where the top simulation of the column is early in the sweep, evolving into the bottom most simulation in the column which is approximately steady state

rocks more quickly. The final simulations (bottom simulation in the column) for each gas velocity, are steady-state estimates, and clearly show a significant oil retention for 0.2 m/s as seen toward the base of the rock in Fig. 3a. This represents unrecovered oil still retained in the shale or sedimentary rock. This oil is called 'unrecoverable oil' representing a fraction of the oil which cannot be removed even by EOR. Again, for 0.4 m/s gas velocity, there is a small amount of upswept or retained oil visible in the steady-state simulation of Fig. 3b. However, for 0.5 m/s, all the oil is removed as seen from the steady-state simulation of Fig. 3c. Clearly, 0.5 m/s gas velocity is a threshold value for a complete and successful gas sweep as shown by the CFD simulations. It means that if large times are given for the gas to sweep at 0.5 m/s, then a major fraction of the oil is removed. Since pumping costs increase with gas velocity, the velocities higher than the threshold are not recommended, and may add unnecessarily to costs without much benefit. However, if the time for the sweep is not long enough to approach steady state, then higher velocities are recommended.

The findings of Fig. 3 are confirmed by Fig. 4. In Fig. 4, the contours of gas volume fraction are illustrated for 0.5 m/s and evolving in time from (a) through (c). In Fig. 4a, which is early in the sweep, contours show gas just entering the rock layer from the left domain edge, which is the inlet. With time, the gas advances deeper into the rock layer as seen in Fig. 4b. Here, the gas front has a partial core annular profile. The gas is clearly seen advancing through the tortuous pathways of the rock layer. By the time depicted in contour Fig. 4c, the gas fills a major fraction of the rock layer. Figure 4d is the steady-state situation. Here, the rock layer is filled with the gas indicating that the major oil fraction is driven out from the exit. Again, this confirms that 0.5 m/s gas velocity is the minimum gas velocity to apply for an efficient sweep, which leaves only traces of oil, assuming that sufficient time is given for the sweep.

Figure 5 shows the contours of nanoparticle volume fraction for the inlet gas velocity of 0.5 m/s. The contours evolve in time from the top simulation in Fig. 5 to a steady state as shown in the bottom simulation in Fig. 5. The contours of the nanoparticles closely follow those of the oil, since the nanoparticles stay in contact

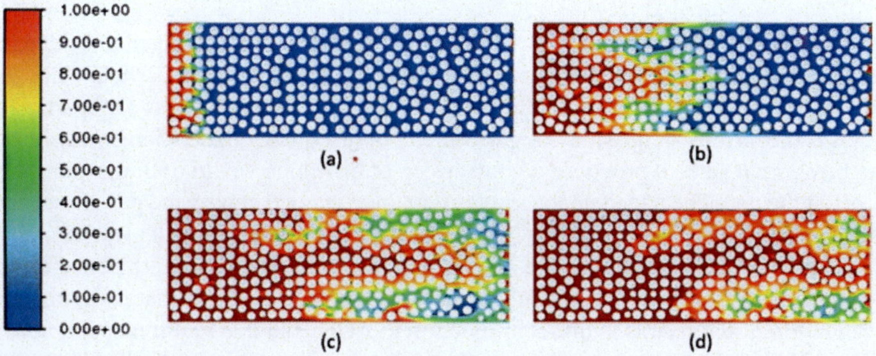

Fig. 4 The tortuous flow of gas through the layer of rock geometry (volume fraction) at 0.5 m/s velocity, showing evolution with increasing time from (**a**) up to (**d**)

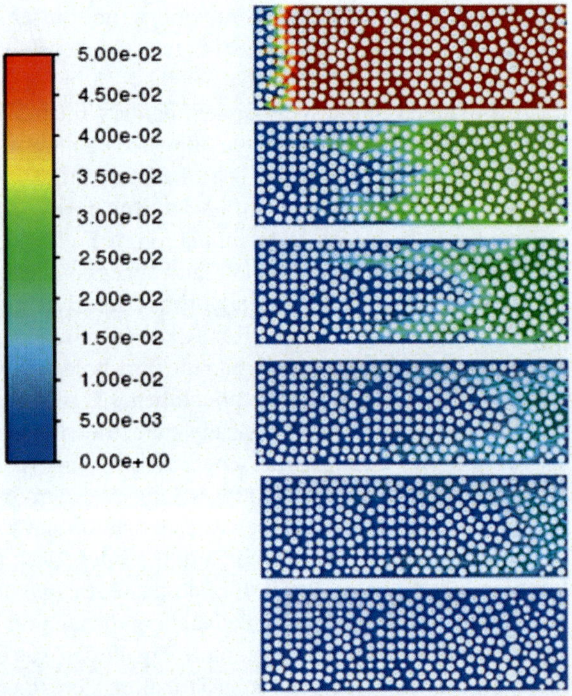

Fig. 5 Nanoparticle sweep (volume fraction) for 0.5 m/s gas velocity

with the oil and facilitate smoother flow. This seems to confirm that by adding nanoparticles to the fluid phase, the average properties of the nanoparticles and fluid alter to improve oil recovery.

4 Conclusion

A CFD model was built in ANSYS 19 to study EOR of oil in rock layer and aided by nanoparticles. Eulerian approach for all the three phases of oil, nanoparticles, and gas allows the definition of the granular phase diameter for the particles as nanoparticles. This is the novelty of the study. A gas velocity of just greater than 0.5 m/s was shown to be optimal since it produced a good sweep of oil which would optimize pumping cost of the gas. The added nanoparticles remained in contact with the gas phase, and this suggested that they aided the flow of the oil. No turbulence model was used, but the tortuous path modeled in the domain was able to capture the turbulence effect of the gas and oil flow. The main drawback of the CFD study is that experimental validation of the results is difficult to obtain in EOR, and this requires much future work. Variation of gas composition and nanoparticle size were not studied here, and thus offers scope for future simulation work.

References

1. Sun, X., Zhang, Y., Chen, G., Gai, Z.: Application of nanoparticles in enhanced oil recovery: A critical review of recent progress. Energies **10**(3), 345 (2017). https://doi.org/10.3390/en10030345
2. Zhu, G., Yao, J., Li, A., Sun, H., Zhang, L.: Pore-Scale Investigation of Carbon Dioxide-Enhanced Oil Recovery. Energy Fuels **31**(5), 5324–5332 (2017). https://doi.org/10.1021/acs.energyfuels.7b00058
3. Heidari, P., Kharrat, R., Alizadeh, N., Ghazanfari, M.H.: "A comparison of WAG and SWAG processes: Laboratory and simulation studies", *Energy Sources*. Part A Recover. Util. Environ. Eff. **35**(23), 2225–2232 (2013). https://doi.org/10.1080/15567036.2010.532189
4. Afzali, S., Rezaei, N., Zendehboudi, S.: A comprehensive review on Enhanced Oil Recovery by Water Alternating Gas (WAG) injection. Fuel **227**(February), 218–246 (2018). https://doi.org/10.1016/j.fuel.2018.04.015
5. Cui, R., Feng, Q., Chen, H., Zhang, W., Wang, S.: Multiscale random pore network modeling of oil-water two-phase slip flow in shale matrix. J. Pet. Sci. Eng. **175**(2018), 46–59 (2019). https://doi.org/10.1016/j.petrol.2018.12.026
6. He, D., Jiang, P., Lun, Z., Liu, X., Xu, R.: "Pore scale CFD simulation of supercritical carbon dioxide drainage process in porous media saturated with water", *Energy Sources*. Part A Recover. Util. Environ. Eff. **41**(15), 1791–1799 (2019). https://doi.org/10.1080/15567036.2018.1549155
7. Krzaczek, M., Nitka, M., Kozicki, J., Tejchman, J.: Simulations of hydro-fracking in rock mass at meso-scale using fully coupled DEM/CFD approach. Acta Geotech. **15**(2), 297–324 (2020). https://doi.org/10.1007/s11440-019-00799-6
8. Mokheimer, E.M.A., Hamdy, M., Abubakar, Z., Shakeel, M.R., Habib, M.A., Mahmoud, M.: A comprehensive review of thermal enhanced oil recovery: Techniques evaluation. J. Energy Resour. Technol. Trans. ASME **141**(3), 030801 (2019). https://doi.org/10.1115/1.4041096
9. Lv, M., Wang, S.: Pore-scale modeling of a water/oil two-phase flow in hot water flooding for enhanced oil recovery. RSC Adv. **5**(104), 85373–85382 (2015). https://doi.org/10.1039/c5ra12136a
10. Abidin, A.Z., Puspasari, T., Nugroho, W.A.: Polymers for Enhanced Oil Recovery Technology. Procedia Chem. **4**, 11–16 (2012). https://doi.org/10.1016/j.proche.2012.06.002
11. Jafari, A., Hasani, M., Hosseini, M., Gharibshahi, R.: Application of CFD technique to simulate enhanced oil recovery processes: current status and future opportunities. Pet. Sci. **17**(2), 434–456 (2020). https://doi.org/10.1007/s12182-019-00363-7
12. Clemens, T., Tsikouris, K., Buchgraber, M., Castanier, L., Kovscek, A.: Pore-scale evaluation of polymers displacing viscous oil-computational- fluid-dynamics simulation of micromodel experiments. SPE Reserv. Eval. Eng. **16**(2), 144–154 (2013). https://doi.org/10.2118/154169-PA
13. Jackson, G.T., Balhoff, M.T., Huh, C., Delshad, M.: CFD-based representation of non-Newtonian polymer injectivity for a horizontal well with coupled formation-wellbore hydraulics. J. Pet. Sci. Eng. **78**(1), 86–95 (2011). https://doi.org/10.1016/j.petrol.2011.05.007
14. Ghanad Dezfully, M., Jafari, A., Gharibshahi, R.: CFD simulation of enhanced oil recovery using nanosilica/supercritical CO_2. Adv. Mater. Res. **1104**, 81–86 (2015)
15. Gharibshahi, R., Jafari, A., Haghtalab, A., Karambeigi, M.S.: Application of CFD to evaluate the pore morphology effect on nanofluid flooding for enhanced oil recovery. RSC Adv. **5**(37), 28938–28949 (2015). https://doi.org/10.1039/c4ra15452e
16. Gharibshahi, R., Jafari, A., Ahmadi, H.: CFD investigation of enhanced extra-heavy oil recovery using metallic nanoparticles/steam injection in a micromodel with random pore distribution. J. Pet. Sci. Eng. **174**(2017), 374–383 (2019). https://doi.org/10.1016/j.petrol.2018.10.051
17. Zhao, J., Wen, D.: Pore-scale simulation of wettability and interfacial tension effects on flooding process for enhanced oil recovery. RSC Adv. **7**(66), 41391–41398 (2017). https://doi.org/10.1039/c7ra07325a

18. Maghzi, A., Mohammadi, S., Ghazanfari, M.H., Kharrat, R., Masihi, M.: Monitoring wettability alteration by silica nanoparticles during water flooding to heavy oils in five-spot systems: A pore-level investigation. Exp. Therm. Fluid Sci. **40**, 168–176 (2012). https://doi.org/10.1016/j.expthermflusci.2012.03.004

Numerical Instability Assessment of Natural Circulation Loop Subjected to Different Heating Conditions

Srivatsa Thimmaiah, Tabish Wahidi, Ajay Kumar Yadav, and M Arun

Abstract Natural circulation loop (NCL) is a passive system in which the driving action of the buoyancy force establishes fluid circulation by overcoming the frictional force without the help of any external power source. NCLs are prone to several kinds of instabilities due to the nonlinearity of the natural convection process. In fact, it is an inability of NCLs to sustain themselves against small perturbations to which any physical system is subjected. This instability in fluid flow creates flow oscillation, chaotic non-linear dynamic behaviour and flow reversal. In this article, three-dimensional computational fluid dynamics (CFD) numerical simulations have been carried out for a range of supercritical pressures (80 bar to100 bar) and heat inputs (250 W to 2500 W) to do the comparative investigation of instability phenomenon in supercritical CO_2-based regular natural circulation loop configured with two different types of heat sources, i.e. heater and isothermal wall at the source with a cold heat exchanger (CHX) at sink. Results show higher instabilities for heater-exchanger loop (Heater-CHX) than an isothermal heater with heat-exchanger loop (ISO-CHX). With an increase in heat input, loops attain stability at a faster rate for a given operating pressure. At a lower heat input, both the loops show bidirectional fluctuation, whereas it is unidirectional at high heat input. Nusselt number shows that the Heater-CHX loop's heat transfer capability is more compared to ISO-CHX loops. Obtained results are validated with the existing correlations, and a good agreement is obtained.

Keywords Natural circulation loop · Supercritical CO_2 · Instability · CFD

Nomenclature

Q	Heat generation per unit mass, (W kg^{-1})
A	Area, (m^2)
Cp	Specific heat at constant pressure, J/(kg. K)

S. Thimmaiah (✉) · T. Wahidi · A. K. Yadav · M. Arun
Department of Mechanical Engineering, National Institute of Technology Karnataka, Surathkal, Mangalore 575 025, India

R. P. Bharti and K. M. Gangawane (eds.), *Recent Trends in Fluid Dynamics Research*, Lecture Notes in Mechanical Engineering,
https://doi.org/10.1007/978-981-16-6928-6_21

d	diameter, (m)
g	acceleration due to gravity, (ms^{-2})
h	Enthalpy, $(J\ kg^{-1})$
m	mass ow rate, $(kg\ s^{-1})$
P	Pressure, (Pa)
Pr	Prandtl number, $Pr = \rho\ C_p/\ \lambda$
Q	Heat input, (W)
Re	Reynolds number, $Re = \rho vd/\mu$
sCO_2	supercritical carbon dioxide
sCO_2	NCL supercritical carbon dioxide based NCL
T	Temperature, (K)
Teff	Temperature reference = 305 K, (K)
u, v, w, V	Velocity, (ms^{-1})

Greek symbols

β	Volume expansion coefficient, (1/K)
κ	Turbulent Kinetic energy, (m^2s^{-2})
λ	Thermal conductivity, $(W\ m^{-1}K^{-1})$
μ_T	Turbulent viscosity, (Pa-s)
μ	Viscosity, (Pa-s)
ρ_r	local density along the radial direction.
ρ	Density, (kgm^{-3})
ε	Turbulent Kinetic energy dissipation rate, (m^2s^{-3})
φ	Viscous dissipation function, $(W\ m^{-3})$

1 Introduction

The fundamental objective of a natural circulation loop (NCL) is to transport heat energy from a heat source to a heat sink which are connected by adiabatic vertical raiser and down comer to form a closed circuit. Natural circulation loops are very simple geometrical structures driven by the differential density across the source and sink, which is generated thermally and have numerous applications in engineering energy equipment.

The main attraction of NCLs is its geometrical simplicity and in-built passive safety feature. NCLs are very promising options in many engineering applications due to their high reliability in terms of safety as pumps, compressors and their associated mechanical devices are absence in the system. Further, this feature makes NCLs economically more attractive over a forced circulation system in terms of operating and maintenance costs. NCLs are more suitable and viable for the applications where operational safety is a major concern. Nuclear reactor core cooling system [1] is one of the safest applications of NCLs and electronic components cooling systems [2] is one of its evolving applications. The most common and wide spread applications of NCLs are solar water heaters [3], natural circulation boilers, internal combustion engines cooling, transformers core cooling, geothermal processes [4], refrigeration and air conditioning systems [5], etc. In spite of all these benefits, NCLs have certain disadvantages such as low Reynolds number flow [6], which reduces its thermal transport capabilities and also its low driving force increases the instabilities in the system. The basic principles and the performance evaluation of NCLs have been explored by numerous researchers. By determining the variation of the friction and heat-transfer coefficients in laminar flow regime, Mertolet et al. [7] obtained the distribution of pressure, temperature and velocity in NCLs. Lifshitz et al. [8] found that the initial conditions govern the characteristics of fluid flow in the NCLs. Acosta et al. [9] study on NCLs tilt angles show that maximum fluid velocity in single-phase NCL can be achieved when loop is kept in vertical position.

The loop fluid, carbon dioxide, employed in the present study is gaining more popularity as a primary and secondary working fluid in both natural as well as forced circulation systems as it is non-flammable and non-toxic compared to other widespread conventional working fluids. Moreover, CO_2 possess excellent environmental friendly properties such as zero-ozone depletion potential and negligible global warming potential, and hence it can safely replace CFCs, HCFCs, hydrocarbons and even ammonia in many applications. Due to some favourable properties of carbon dioxide, it is being extensively used as a secondary loop fluid [10] in many applications like refrigeration and air-conditioning, solar heaters, etc. Supercritical CO_2 as working fluid will have advantages both in improved heat transport capacities [11], better heat-to-power ratios [12] and highly compact system design [13]. Yamaguchi [14] experimental study shows that even at lowest level of solar radiation condition, heat recovery efficiencies of 65.0% could be achieved with sCO_2. Yoshikawa et al. [15] simulation on measuring flow velocities of sCO_2 shows that with a temperature differences of 3 to 8 °C could induce an average flow velocities of 4 m/min in the loop. Chen et al. [16] show that as the diameter of the pipe increases, flow get stabilizes. 3D computational study of Sarkar et al. [17] show that sCO_2-based NCL can be a superior choice over water.

Many theoretical and experimental works revel that oscillatory instability is inherent in NCLs under some conditions. Welander [18] concluded that the system is always unstable with small-amplitude oscillations and also developed stability maps for the same. Creveling et al. [19] also confirmed the oscillatory behaviour of flow experimentally. Bau et al. [20] investigated the stability in NCLs and concluded that loop is unstable. Zvirin et al. [21] shown that unstable prevails in the rest state only. Prediction of Keller et al. [22] on NCLs shows unidirectional oscillations of flow without any reversals in the direction. Vijayan et al. [23] noticed repetitive flow reversals along with the unstable oscillatory behaviour. Nishihara et al. [24] observed only oscillatory behaviour in the loop without flow reversals. Chen et al. [25] numerical study shows that at certain temperature of source, loop fluid flow transition starts, and further, if the source temperature increases, the flow reaches to stable unidirectional flow. Instability behaviours of the NCL are still not clearly understood because of its inherent complexities. Efforts have been put by various researcher to understand these behaviours both theoretically as well as experimentally. 3D CFD model developed by Yadav et al. [26] on sCO_2-based NCLs show that change in orientation of the loop can tackle the flow instability issues. The experimental results of Yadav et al. [27] show that the system reaches to steady state at a faster rate with the increase in operating pressure of sCO_2 in the loop. Chen et al. [28] show that circulation rate and heat transfer is better when the inclination angle is larger. Sharma et al. [29] conducted experiments on sCO_2-based NCL and observed instability only in horizontally oriented heater and cooler.

Most of the works on the natural circulation loops study have been limited to specific geometry of the loop and attention is paid only to study the heat transfer capability, instability behaviours, stability maps etc., in that particular loop. Comparison of oscillatory instability of different geometry and boundary conditions is not available in the literature. The author had studied various NCLs configured with different boundary conditions and presented comparison between two NCLs in this paper. The analysis is carried out on this rectangular loop to investigate the basic stability characteristics of velocity, temperature and natural circulation mass flow at same loop pressure and quantum of heat input source in both loops. In this research, attention is focussed on transient 3D numerical analysis of sCO_2-based NCLs configured with Heater-CHX isothermal wall-CHX.

2 Model and Simulation

Two different sCO_2-based NCLs subjected two different boundary conditions have been compared by imposing same quantity of heat at source and keeping the same boundary condition at sink in both NCLs. One NCL imposed with constant heat fluxes at source (Fig. 1a) and other NCL with isothermal wall (Fig. 1b) and cold heat exchanger with constant temperature of 305 K at inlet are considered in the sinks of both NCLs. Both right and left legs are assumed to be adiabatic, and the loops are kept in a vertical plane to maximize the mass flow rate of working fluid. Metallurgy of loop piping considered is stainless steel with 10 mm inner diameter and 3 mm thick, height and width of both the NCLs is 500 mm and length of source and sink is 400 mm. Mass flow rate, temperature and velocity of the working fluid in these are studied and compared in these loops at different heat input at source. Further, change in mass flow rate with Fig. 2 respect to different operating pressure is also analysed and compared. Heat flux is varied from 250 to 2500 W for heater, and temperature at isothermal wall is varied from 306 K to 315.25 K for all the operating pressure of the loop fluid, which is varied from 80 to 100 bar in a step of 10 bar.

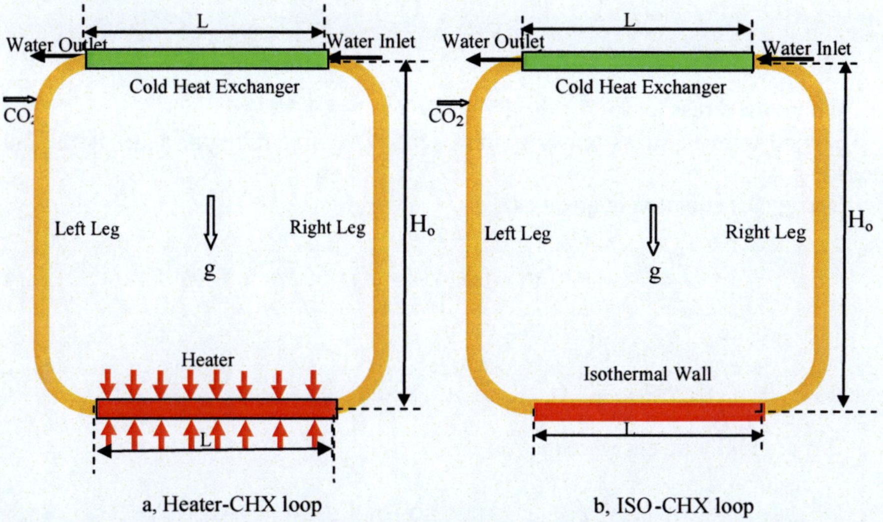

Fig. 1 **a** Heater-CHX loop, **b** ISO-CHX loop

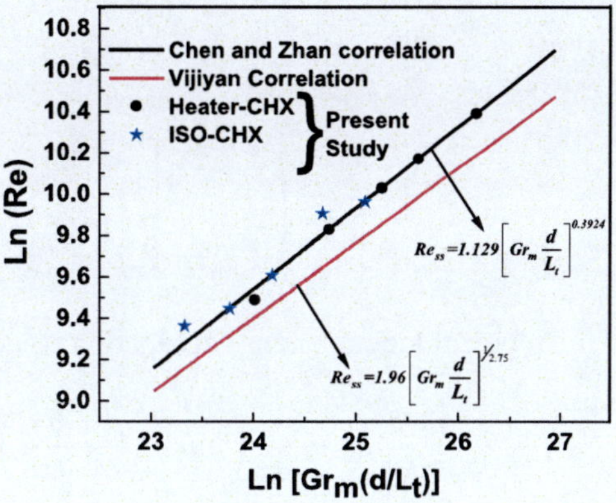

Fig. 2 Validation of the obtained result with correlations

3 Governing Equations

The fundamental conservation equations (mass, momentum and energy) solved for the simulation are given below. The commercial software ANSYS (FLUENT) V-19.0 is employed to solve these equations with the associated boundary specifications.

The mass conservation equation can be given as:

$$\frac{\partial \rho}{\partial t} + \nabla.(\rho \mathrm{V}) = 0 \tag{1}$$

Momentum conservation:

$$\frac{\partial(\rho \mathrm{V})}{\partial t} + \nabla.(\rho \mathrm{VV}) = -\nabla p + \nabla.(\bar{\bar{\tau}}) + \rho g \tag{2}$$

Where the stress tensor $\bar{\bar{\tau}} = \mu\left[\left(\nabla \mathrm{V} + \nabla \mathrm{V}^{\mathrm{T}}\right) - \frac{2}{3}\nabla.\mathrm{VI}\right]$

Second term in stress tensor is the effect of volume dilation, and I is the unit tensor.

Energy (E) equation is given as

$$\frac{\partial(\rho E)}{\partial t} + \nabla.\left[\mathrm{V}(\rho \mathrm{E} + \mathrm{p})\right] = \nabla.\left(\lambda_{eff}\nabla T + \bar{\bar{\tau}}.\mathrm{V}\right) \tag{3}$$

where $E = \int_{T_{eff}}^{T} C_p dT + \frac{\mathrm{V}^2}{2}$

Turbulent kinetic energy equation

$$\begin{aligned} &\frac{\partial}{\partial t}\rho k + \frac{\partial}{\partial x}\left(\rho u k - \frac{\mu_T}{\sigma_\kappa}\frac{\partial \kappa}{\partial x}\right) + \frac{\partial}{\partial y}\left(\rho \upsilon \kappa - \frac{\mu_T}{\sigma_\kappa}\frac{\partial \kappa}{\partial y}\right) \\ &+ \frac{\partial}{\partial z}\left(\rho w \kappa - \frac{\mu_T}{\sigma_\kappa}\frac{\partial \kappa}{\partial z}\right) = G - \rho\varepsilon \end{aligned} \tag{4}$$

where

$$G = \mu_T\left(\begin{array}{c} 2\left[\left(\frac{\partial u}{\partial x}\right)^2 + \left(\frac{\partial v}{\partial y}\right)^2 + \left(\frac{\partial w}{\partial z}\right)^2\right] + \\ \left(\frac{\partial u}{\partial y} + \frac{\partial v}{\partial x}\right)^2 + \left(\frac{\partial u}{\partial z} + \frac{\partial w}{\partial x}\right)^2 + \left(\frac{\partial v}{\partial z} + \frac{\partial w}{\partial y}\right)^2 \end{array}\right)$$

Turbulent kinetic energy dissipation equation

$$\begin{aligned} &\frac{\partial}{\partial t}\rho\varepsilon + \frac{\partial}{\partial x}\left(\rho u\varepsilon - \frac{\mu_T}{\sigma_\varepsilon}\frac{\partial}{\partial x}\varepsilon\right) + \frac{\partial}{\partial y}\left(\rho \upsilon\varepsilon - \frac{\mu_T}{\sigma_\varepsilon}\frac{\partial}{\partial y}\varepsilon\right) \\ &+ \frac{\partial}{\partial z}\left(\rho w\varepsilon - \frac{\mu_T}{\sigma_\varepsilon}\frac{\partial}{\partial z}\varepsilon\right) = c_1\frac{\varepsilon}{\kappa}G - c_2\rho\frac{\varepsilon^2}{\kappa} - R \end{aligned} \tag{5}$$

where $R = \frac{c_\mu \eta^3 \rho(1-\eta/\eta_0)\varepsilon^2}{\kappa(1+\beta\eta^3)}$,

$$\eta = \frac{S\kappa}{\varepsilon}, and, S = \frac{1}{\sqrt{2}}\left(\frac{\partial u}{\partial y} + \frac{\partial u}{\partial z} + \frac{\partial v}{\partial x} + \frac{\partial v}{\partial z} + \frac{\partial w}{\partial x} + \frac{\partial w}{\partial y}\right), \mu_T = c_\mu\rho\frac{\kappa^2}{\varepsilon}$$

where the values of constants are as follows:

$$\eta_0 = 4.8, \beta = 0.012, c_\mu = 0.0845, \sigma_\varepsilon = \sigma_\kappa = 0.7178, c_1 = 1.42 and c_2 = 1.68,$$

Modified Grashof number $Gr_m = \frac{g\beta d^3 \rho^2 Q H_0}{A\mu^3 C_p}$.

4 Simulation Detail

ANSYS fluent version 19.0 is used to carry out the three-dimensional transient simulation. The geometries shown in Fig. 1 is used to perform the simulations. Adiabatic condition is considered for both right and left legs of the loop by perfectly insulating both the legs. In simulation activity, 3 mm wall thickness at the source and sink is considered. Pressure–velocity terms coupled in the Navier–Stokes equation is solved by using SIMPLE (Semi-Implicit Method for Pressure-Linked Equations) algorithm. A second-order upwind scheme method is utilized to iterate the velocity and temperature parameters in the momentum and energy governing equations. Similarly same technique of second-order upwind scheme is used to iterate the turbulence parameters (k, ε, etc.). The momentum equation is discretized to obtain the pressure term by using PRESTO (Pressure staggering option). No-slip boundary condition is considered for pipe walls. The turbulence in the working fluid is taken into account by using general RNG (Renormalization group), i.e. k-ε model. Further, the least effective conditions such as axial conductivity and viscous dissipation in working fluid are also considered in the simulation. Convergence conditions for the governing equations are considered to be achieved when residuals iteration are below 10^{-6}. Supercritical condition of sCO_2 in the loop is ensured by keeping the operating temperature at 305 K and pressure at 80, 90 and 100 bar. A NIST REFPROP property table [39] is utilized to obtain all the standard properties of CO_2 for every interval of 2 K temperature and piecewise-linear interpolation is used to obtain property between two data points.

5 Validation of the Results

Correlation between Reynolds number and modified Grashof number, i.e. Re–Gr_m developed by Chen and Zhang [32] and P.K. Vijayan [37] as depicted in Fig. 4a has been used for validating the results obtained from the 3Dimentional CFD analysis. The CFD simulation results obtained for 90 bar is validated using these correlations. Good agreement was found between the generated results and existing correlations as shown in Fig. 4a. The maximum discrepancies are found to be less than 2% with the correlation. Obtained results are closer to Chen and Zhang [32] correlation than the Vijayan correlation as sCO_2 is the loop fluid in both Chen's study and the present study. The value of Reynolds number evaluated from Vijayan [37] correlation is lesser compared to the results obtained from the present study as this correlation developed for water. Further, the high Reynolds number is a very common phenomenon for supercritical flow conditions [31, 38].

Chen and Zhang [32] Simulation correlation

$$Re = 1.129\left(\frac{Gr_m d}{L_t}\right)^{0.3924} \tag{6}$$

P.K. Vijayan [37] experimental correlation,

$$Re = 1.96(Grmd/Lt)^{1/2.75} \tag{7}$$

6 Results and Discussion

Stability behaviour of two different sCO_2-based natural circulation loop subjected to two different kinds of boundary conditions at source has been analysed by using three-dimensional CFD simulation techniques. Figure 1a and b depict these two natural circulation loops, wherein first loop is source is subjected to a constant heat flux and the other one is subjected to an isothermal temperature. Simulation has been conducted by imposing a heat flux in the range of 250 W to 2500 W in the *Heater-CHX* loop at various pressures, i.e. 80, 90 and 100 bar. Similarly, for the *ISO-CHX* loop, equal quantity of heat is injected by maintaining the constant temperature corresponding to these fluxes at various pressures. The corresponding constant temperature values for the *ISO-CHX* loop have been derived from the steady-state condition of *Heater-CHX* loop. The transient oscillations of loop fluid mass flow rate, temperature and velocity at various pressures and heat fluxes along with the effect of pressure on mass flow rate have been recorded and discussed in the following sections.

6.1 Transient Variation of Temperature

Figure 3 shows the transient variation of loop fluid temperature in both *Heater-CHX* loop and *ISO-CHX* loop at various heat input at 80 bar pressure. In the heater section, constant heat flux of 250 W, 750 W, 1200 W and 2500 W is imposed and the corresponding equal quantity of heat at constant temperature of 306 K, 308.75 K. 310.25 K and 315.25 K is imposed on the ISO thermal section of *ISO-CHX* loop. At lower differential temperature, the amplitude of temperature fluctuation is very consistent throughout the operation in both the loops, and it never reaches to a steady state. However, at higher differential temperature, the fluctuation of temperature in both the loop is very mild in nature and also reaches to a steady state over a period of time. Further, in case of *ISO-CHX* loop, the temperature reaches to a steady state at a faster rate compared to *Heater-CHX* loop. At intermittent heat input, the fluctuation of temperature is very high in *Heater-CHX* loop compared to *ISO-CHX* loop. In case of *ISO-CHX* loop as the heat input at source increases, the amplitude of temperature fluctuation keeps on reducing. However, in case of *Heater-CHX* loop, the temperature fluctuation increases at lower as well as at intermediate level of heat

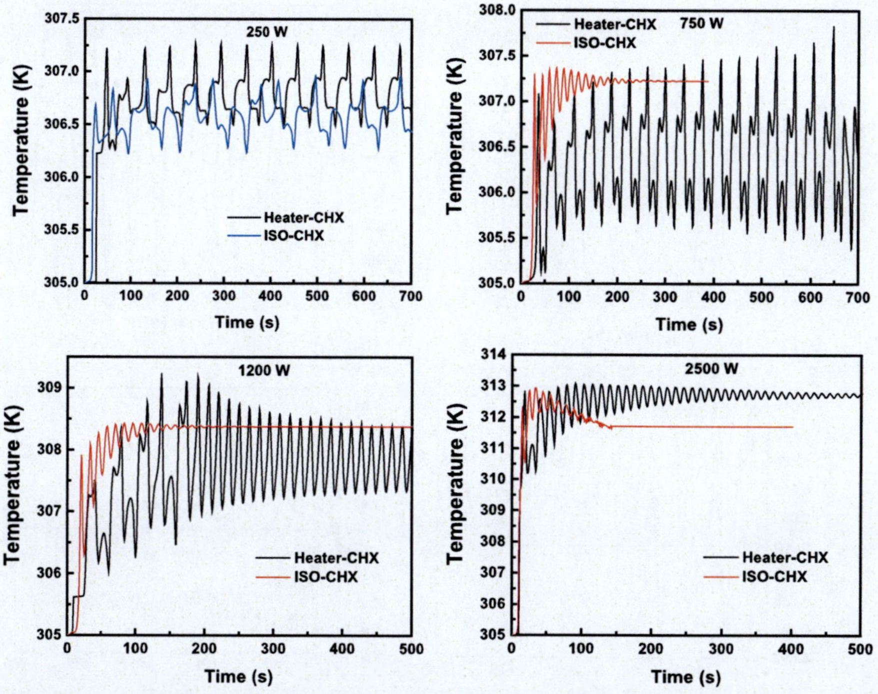

Fig. 3 The transient variation of loop fluid temperature

input, but it decreases and eventually reaches to a steady state at higher level of heat input.

6.2 Transient Variation of Mass Flow Rate

Figure 4 shows the transient variation of mass flow with respect to time, in both NCLs at lower, intermittent and higher levels of heat input at source. The NCLs considered in the current study are bilaterally symmetrical geometrical structures, and also the loop receives uniform heat energy throughout the heating section. This geometrical as well as thermal symmetry of the loop makes the flow direction in NCL more chaotic. Further, the flow direction reversal is found to be very high at lower heat input due to very weak driving force in the loop. The mass flow reversal phenomenon is very frequent in both the loops at lower heat input; however, at higher heat input, both loops take unidirectional path due to stronger driving force in the loop. At higher heat input, it is interesting to note that the direction of flow is opposite to each other. *ISO-CHX* loop follows unidirectional path at all levels of heat input except at lowest level, whereas *Heater-CHX* loop follows bidirectional path at all levels of heat input except at highest level. Amplitude of mass flow oscillation in

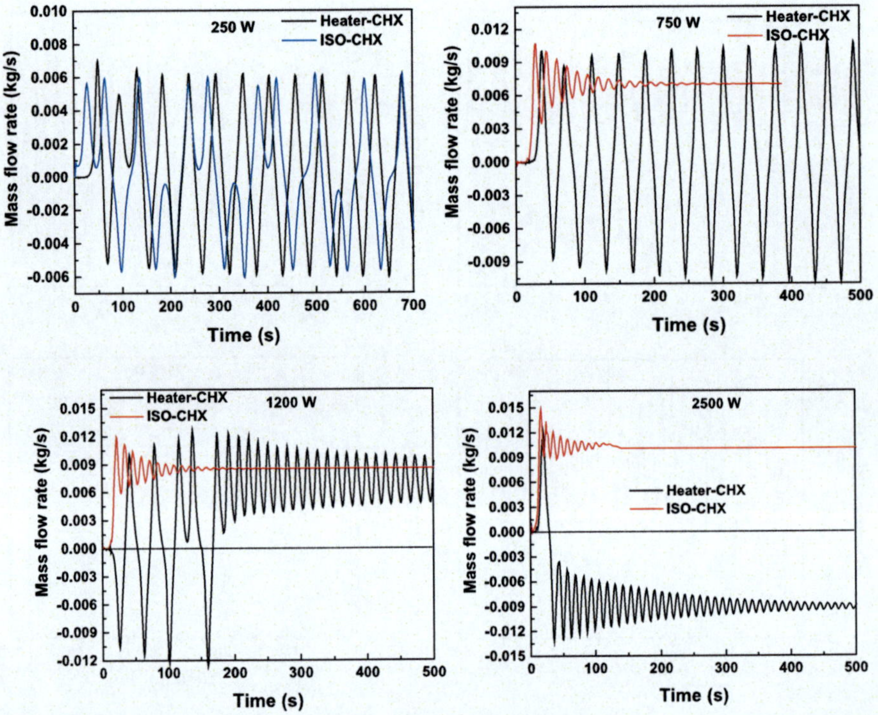

Fig. 4 The transient variation of loop fluid mass flow rate

Heater-CHX loop is very high compared to *ISO-CHX* loop, and also it takes very long duration to reach steady state at all levels of heat input.

6.3 Transient Variation of Velocity

Figure 5 shows the transient variation of velocity with time in *Heater-CHX* and *ISO-CHX* loops at various levels of heat input at source. In NCLs, the fluid velocity is a function of potential difference, i.e. the density difference between two locations as external driving force is absent. Higher the temperature difference across the loop, higher will be the density difference and hence higher velocity, which is very much visible in the graphs. Similar to mass flow and temperature fluctuations, the velocity oscillation is also very high in both the loops at lower heat input. At higher heat input, the velocity oscillation is very high in the initial stage of energy input, but within a short period of time, it gradually reaches to a steady state. At intermediate level of heat input, the *ISO-CHX* loop reaches to a steady state even though there exists oscillations in the initial stage, whereas *Heater-CHX* loop continue to oscillate at

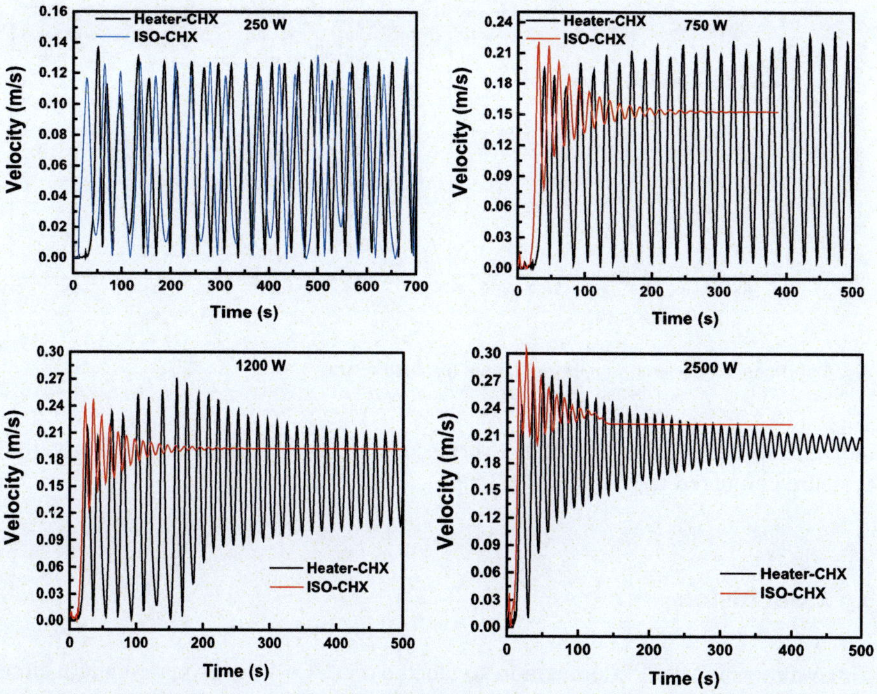

Fig. 5 The transient variation of loop fluid mass flow rate

very high amplitude throughout the operation. The velocity keep on increases with increase in heat input at source for given operating pressure.

6.4 *Effect of Operating Pressure on Mass Flow Rate and Flow Instability*

Influence of operating pressure on the mass flow rate of working fluid is checked at 2000 W of heat input at source and the same is plotted in Fig. 6. As the loop fluid pressure increases, the mass flow rate increases, and it is very much obvious from the results obtained from the simulations. Further, dampening of mass flow oscillation as well as time required to reach a steady state reduces at a faster rate as the loop fluid pressure increases. The mass flow rate is very high in *ISO-CHX* loop at all levels of loop fluid pressure compared to the *Heater-CHX* loop. In all level of loop fluid pressure, the direction of flow is clockwise in case of *Heater-CHX* loop, whereas in case of *ISO-CHX* loop, at lower and intermediate pressure, the fluid follows in clockwise direction, and it changes to counter-clockwise direction at higher pressure. The mass flow rate in *ISO-CHX* loop reaches to steady state at a faster rate at all level of loop fluid pressure compared to *Heater-CHX* loop. The simulation results

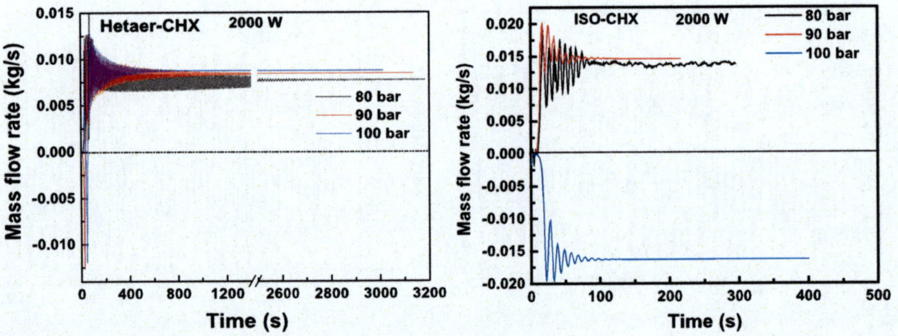

Fig. 6 Influence of operating pressure on the mass flow rate

indicates that the *ISO-CHX* loop is very stable at all level heat input and loop fluid pressure compared to *Heater-CHX* loop.

7 Conclusion

Three-dimensional CFD simulation conducted on sCO_2-based square-natural circulation loops to study the influence of change in boundary conditions, at source, on transient and instability behaviour reveals the following,

1. The mass flow rate of loop fluid increases with increase in fluid pressure and the velocity keeps on increasing with increase in heat input at source.
2. Dampening of mass flow oscillation as well as time required to reach a steady state reduces at a faster rate as the loop fluid pressure increases.
3. Simulation results indicates that the *ISO-CHX* loop is very stable at all levels of heat input and loop fluid pressure compared to *Heater-CHX* loop.
4. At a lower heat input both the loops show bidirectional fluctuation, whereas it is unidirectional at high heat input.
5. The geometrical as well as thermal symmetry of the loop makes the flow direction more chaotic and it will be very high at low heat input due very weak driving force in the loop.
6. In *ISO-CHX* loop as the heat input increases at source the amplitude of temperature fluctuation keeps on reducing and time required to reach steady state is also keeps on reducing in comparison to *Heater-CHX* loop.

References

1. Basu, D.N., Bhattacharyya, S., Das, P.K.: A review of modern advances in analyses and applications of single-phase natural circulation loop in nuclear thermal hydraulics. Nucl. Eng. Des.

280, 326–348 (2014)
2. Usman, H., et al.: An experimental study of PCM based finned and un-finned heat sinks for passive cooling of electronics. Heat Mass Transf. **54**(12), 3587–3598 (2018)
3. Chen, L., Zhang, X.R.: Experimental analysis on a novel solar collector system achieved by supercritical CO_2 natural convection. Energ. Convers. Mgmt. **77**, 173–182 (2014)
4. Kreitlow, D.B., Reistad, G.M., Miles, C.R., Culver, G.G. 1978 Thermosyphon models for downhole heat exchanger applications in shallow geothermal systems.
5. Kupireddi, K.K. 2010. Carbon dioxide based natural circulation loops: Application to refrigeration and air conditioning systems (Doctoral dissertation, IIT Kharagpur).
6. Yadav, A.K., Bhattacharyya, S., Maddali, R.G.: On the suitability of carbon dioxide in forced circulation-type secondary loops. International Journal of Low-Carbon Technologies **9**(1), 85 (2014)
7. Mertol, A., Greif, R., Zvirin, Y.: Two dimensional analysis of transient flow and heat transfer in a natural circulation loop. Wärme-und Stoffübertragung **18**(2), 89–98 (1984)
8. Lifshitz, S., Zvirin, Y.: Transient heat and mass transfer in a natural circulation loop. Wärme-und Stoffübertragung **28**(7), 371–380 (1993)
9. Acosta, R., Sen, M., Ramos, E.: Single-phase natural circulation in a tilted square loop. Wärme-und Stoffübertragung **21**(5), 269–275 (1987)
10. Kumar, K.K., Gopal, M.R.: Carbon dioxide as a secondary fluid in natural circulation loops. Proc Inst MechEng Part E J Process Mech Eng. **223**(3), 189 (2009)
11. Yadav, A.K., Gopal, M.R., Bhattacharyya, S.: CO_2based natural circulation loops: new correlations or friction and heat transfer. Int. J. Heat Mass Transf. **55**(17–18), 4621–4630 (2012)
12. Ahn, Y., Bae, S.J., Kim, M., Cho, S.K., Baik, S., Lee, J.I., Cha, J.E.: Review of supercritical CO_2 power cycle technology and current status of research and development. Nucl. Eng. Technol. **47**(6), 647–661 (2015)
13. Kumar, K.K., Gopal, M.R.: Steady-state analysis of CO_2 based natural circulation loops with end heat exchangers. Appl. Therm. Eng. **29**, 1893–1903 (2009)
14. Yamaguchi, H., Sawada, N., Suzuki, H., Ueda, H., Zhang, X.R.: Preliminary study on a solar water heater using supercritical carbon dioxide as working fluid. ASME J. Solar Energ. Eng. **132**, 101–106 (2010)
15. Yoshikawa, S., Smith Jr., R.L., Inomata, H., Matsumura, Y., Arai, K. 2005. Performance of a natural convection circulation system for supercritical fluids.
16. Chen, L., Deng, B.L., Jiang, B., Zhang, X.R.: Thermal and hydrodynamic characteristics of supercritical CO_2 natural circulation in closed loops. Nucl. Eng. Des. **257**, 21–30 (2013)
17. Sarkar, M.K.S., Basu, D.N.: Numerical comparison of thermal hydraulic aspects of supercritical carbon dioxide and subcritical water-based natural circulation loop. Nucl. Eng. Tech. **49**, 103–112 (2017)
18. Kumar Yadav, A., Ram Gopal, M., Bhattacharyya, S. E_ect of tilt angle on subcritical/supercritical carbon dioxide-based natural circulation loop with isothermal source and sink, J. Therm. Sci. Eng. Appl. 8(1) (2016)
19. Yadav, A.K., Ramgopal, M., Bhattacharyya, S.: Transient analysis of subcritical/supercritical carbon dioxide based natural circulation loop with end heat exchangers:experimental study. Heat Mass Transf. **53**(9), 2951 (2017)
20. Welander, P.: On the oscillatory instability of a differentially heated fluid loop. J. Fluid Mech. **29**(1), 17 (1967)
21. Creveling, H., De Paz, J., Baladi, J., Schoenhals, R.: Stability characteristics of a single-phase free convection loop. J. Fluid Mech. **67**(1), 65 (1975)
22. Bau, H.H., Torrance, K.E.: Transient and steady behavior of an open, symmetrically-heated, free convection loop. Int. J. Heat Mass Transf. **24**(4), 597–609 (1981)
23. Zvirin, Y.: The instability associated with the onset of motion in a thermosyphon. Int. J. Heat Mass Transf. **28**(11), 2105–2111 (1985)
24. Keller, B.: Periodic oscillations in a model of thermal convection. J. Fluid Mech. **26**(3), 599 (1966)

25. Vijayan, P.K.: Experimental observations on the general trends of the steady state and stability behavior of single phase natural circulation loops. Nucl. Eng. Des. **215**, 139–152 (2002)
26. Nishihara, T.: Study on the stability of a single-phase natural circulation flow in a closed loop Demonstrative experiments on the higher-mode density wave oscillation. Denryoku Chuo Kenkyusho Hokoku **U96034**, 1–29 (1997)
27. Chen, L., Zhang, X.R., Yamaguchi, H., Liu (Simon), Z.S.,: Effect of heat transfer on the instabilities and transitions of supercritical CO_2 flow in a natural circulation loop. Int. J. Heat Mass Trans. **53**, 4101–4111 (2010)
28. Chen, L., Zhang, X.R., Deng, B.L., Jiang, B.: Effects of inclination angle and operation parameters on supercritical CO_2 natural circulation loop. Nucl. Eng. Des. **265**, 895–908 (2013)
29. Sharmaa, M., Vijayana, P.K., Pilkhwala, D.S.: Yutaka Asako, Steady state and stability characteristics of natural circulation loops operating with carbon dioxide at supercritical pressures for open and closed loop boundary conditions. Nucl. Eng. Des. **265**, 737–754 (2013)
30. Hahne, E.W. Natural convection heat transfer through an enclosed horizontal layer of supercritical carbon dioxide, W¨arme-und Stoff ¨ubertragung 1(3), 190 (1968)
31. Zhang, X.R., Chen, L., Yamaguchi, H.: Natural convective flow and heat transfer of supercritical CO_2 in a rectangular circulation loop. Int. J. Heat Mass Transf. **53**(19–20), 4112 (2010)
32. Chen, L., Zhang, X.R.: Simulation of heat transfer and system behavior in a supercritical CO_2 based thermosyphon: effect of pipe diameter. J. Heat Transfer **133**(12), 122 (2011)
33. Vijayan, P., Sharma, M., Pilkhwal, D., Saha, D., Sinha, R. A comparative study of singlephase, two-phase, and supercritical natural circulation in a rectangular loop, J. Eng Gas Turb. Power 132(10) (2010)
34. Jiang, Y., Shoji, M.: Spatial and temporal stabilities of flow in a natural circulation loop: influences of thermal boundary condition. J. Heat Transfer **125**(4), 612 (2003)
35. Vijayan, P., Sharma, M., Saha, D.: Steady state and stability characteristics of single-phase natural circulation in a rectangular loop with different heater and cooler orientations. Exp. Thermal Fluid Sci. **31**(8), 925 (2007)
36. Chen, L., Zhang, X.R., Jiang, B. Effects of heater orientations on the natural circulation and heat transfer in a supercritical CO_2 rectangular loop, J. Heat Transf. 136(5), 052501 (2014)
37. Roache, P.J., Ghia, K.N., White, F.M. Editorial policy statement on the control of numerical accuracy (1986)
38. Lisboa, P.F., Fernandes, J., Sim˜oes, P.C., Mota, J.P., Saatdjian, E. Computational-fluiddynamics study of a kenics static mixer as a heat exchanger for supercritical carbon dioxide, The Journal of Supercritical Fluids 55(1), 107 (2010)
39. NIST, Nist standard reference database 23: Refprop version 9.0,, Reference Fluid Thermodynamic and Transport Properties, Standard Reference Data Program, National Institute of Standards and Technology, Gaithersburg, MD (2010)

Comparative Numerical Appraisal of Subcritical and Supercritical CO_2-Based Natural Circulation Loop

Tabish Wahidi and Ajay Kumar Yadav

Abstract A natural circulation loop (NCL) is a passive heat transfer system in which circulation occurs solely due to density differences caused by thermal imbalance and the elevational difference between the source and sink of the loop. Carbon dioxide (CO_2)-based NCL is highly sensitive to operating conditions and vulnerable to unstable behaviour, mainly due to intense changes in the thermo-physical properties of CO_2. Therefore, NCLs always require precise design assessment that focusses on the interaction of all the transient responses of buoyancy and friction forces, ensuring a stable zone of operation. In this article, a three-dimensional computational fluid dynamics study has been carried out for over a range of pressures (30 to 100 bar) and heat inputs (500 to 1500 W) to do the comparative investigation of fluid flow and heat transfer phenomenon of subcritical/supercritical CO_2-based NCLs with water-based NCL. The simulations quantify the degree of instability and heat transfer rate for subcritical/supercritical CO_2 and water. A possible mechanism for continuous flow oscillation and measurement of instability with different pressure in unstable loops is also proposed in this study. Obtained results are validated with the correlations available in the literature; it shows an amicable agreement.

Keywords Supercritical CO_2 · Subcritical CO_2 · Instability · Natural convection Loop

T. Wahidi (✉) · A. K. Yadav
Department of Mechanical Engineering, National Institute of Technology Karnataka, Surathkal, Mangalore 575 025, India

R. P. Bharti and K. M. Gangawane (eds.), *Recent Trends in Fluid Dynamics Research*, Lecture Notes in Mechanical Engineering,
https://doi.org/10.1007/978-981-16-6928-6_22

Nomenclature

Acronyms

A	Area, (m^2)
C_p	Specific heat at constant pressure, J/(kg. K)
d	diameter, (m)
g	acceleration due to gravity, (ms^{-2})
h	Enthalpy, (J kg^{-1})
P	Pressure, (Pa)
Q	Heat input, (W)
Re	Reynolds number, Re= $\rho vd/\mu$
T	Temperature, (K)
V	Velocity, (ms^{-1})
t	time, (s)

Greek symbols

β	Volume expansion coefficient, (K^{-1})
κ	Turbulent Kinetic energy, (m^2s^{-2})
λ	Thermal conductivity, (W $m^{-1}K^{-1}$)
μ_T	Turbulent viscosity, (Pa−s)
μ	Viscosity, (Pa−s)
ρ	Density, (kgm^{-3})
ε	Turbulent Kinetic energy dissipation rate, (m^2s^{-3})
φ	Viscous dissipation function, (W m^{-3})
G	Rate of generation of turbulent kinetic energy, (kg m^{-1} s^{-3})

Subscripts

r	radial direction
ss	steady state

1 Introduction

NCL offers a very efficient option to transport heat energy within a closed system without employing mechanical drives like a pump/compressor. NCL simple geometry and passive operation improve fundamental safety and offer an inherent benefit in price, loop reliability and maintenance. NCLs have many traditional industrial and engineering applications in the geothermal process, cooling turbine blades, cooling of electronic components, nuclear reactor cooling systems, cryogenic refrigeration, solar heater etc.

When an NCL is subjected to disturbance due to thermal imbalance, then forces acted on the loop fluid senses the destabilization, then irregularities developed due to destabilization grow exponentially with time and amplitude. The inability of NCL to sustain themselves against these irregularities created due to small perturbations is known as instability. This instability in fluid flow creates flow oscillation, chaotic non-linear dynamic behaviour and flow reversal. Flow instability in NCL is highly unacceptable as it can induct potent mechanical vibration in the system components, and eventually, it may lead to catastrophic incidences due to fatigue development in the components. Further, if oscillations get augmented, it can affect the heat transfer characteristics, which detriments the system efficiency. Keller [1] analytically studied the instability and found that NCL behaves as self-excited oscillator when subjected to certain operating conditions. Later, Welander [2] explained that unstable motions are related with thermal anomalies and the anomalies amplify through the correlated variations in flow rate in the fluid. Chen [3] defined a single dimensionless parameter, i.e. modified Grashof number; when it is less than a critical value, flow is stable; above a critical value, oscillatory instability exists. Vijayan [4] developed a correlation between steady-state Reynolds number and modified Grashof number by considering the geometric ratio. Vijayan [5] divides the stability of the loop in three major zones, stable, unstable and neutrally stable. With advancements in computational techniques and higher-order analytical models, several codes have been developed for linear and non-linear methods. For linear stability analysis, NUFREQ [6] and DENSITY-PARA [7] codes were developed, while for non-linear analysis, SPORTS [8] and RAMONA [9] codes were designed. Application of commercial software, e.g. RELAP5, CATHARE2, ATHLET, FLUENT etc., were also used by various authors to do instability analysis.

CO_2 offers a compelling alternative secondary working fluid in NCLs amid to its environmental benign nature and thermophysical properties. A supercritical CO_2 ($scCO_2$)-based loop compared to a water-based loop yields a higher mass flow rate and higher heat transfer rate [10]. Thippeswamy and Yadav [11] experimental study concluded that the heat transfer rate of supercritical CO_2 is 900% more than water-based NCL.

For supercritical CO_2 ($scCO_2$), the underlying physics for instability is complex and not well characterized. Any substantial change in thermo-physical properties of CO_2 may lead to dramatic changes in the driving force and makes NCL susceptible to dynamic instability. Chen et al. [12] noted an intense fluctuation in mass flow rate near critical and pseudocritical regions for $scCO_2$-based NCLs. Yadav et al. [13] numerical simulation of subcritical/supercritical CO_2-based NCL with different operating pressures and temperatures show that time to reach steady-state decrease as operating pressure increases. Thimmaiah et al. [14] carried out 2D CFD analysis on two different configurations of $scCO_2$-NCLs; simulation result shows that both side heat-exchanger loop has higher instabilities than an isothermal heater with heat-exchanger loop.

Many researchers have been using different approaches to curb uncertainty in NCL. Misale and Frogheri [15] performed series of experiments to stabilize the unstable flow using orifices of different diameters located in the vertical legs. Wahidi et al. [16, 17] proposed a Tesla valve as a viable option for unidirectional flow and mitigation of instability in $scCO_2$-NCLs. Yadav et al. [18] numerical investigation proposed a solution to instability by tilting the loop, but it reduced the heat transfer rate.

As discussed above, substantial researches have been done on the different types of geometrical designs of NCLs. Still, very few studies are available in the literature on comparative assessment for subcritical and supercritical CO_2 with water-based NCL. This research focusses on loop fluid dynamic behaviour and steady-state characteristic for different pressures and boundary conditions. With the current findings, it is expected that it will help in the interpretation of NCL mechanism based on CO_2.

2 Physical Model

In this present study, NCL configured with a cold heat exchanger (CHX) as sink and heater as a source, respectively (as depicted in Fig. 1), is explored to comprehend the impact of heat input and operating pressure on the working fluid.

2.1 *Grid Independence Study*

A grid independence study has been done to certify the obtained results are reliable. Figure 2 shows the meshing of the CHX cross-section, in which for the CO_2 side, 0.2 mm minimum grid size is taken in a radial direction near the wall and 1.2 mm maximum grid size away from the wall. For the CHX fluid side (i.e. water side), in radial direction adjacent to the wall, 0.4 mm of minimum grid size is taken, increasing to the maximum grid size of 1.0 mm away from the wall. Coarse meshing is adopted in the axial direction. Mesh generated a total of 18,17,549 elements as a grid-independent element.

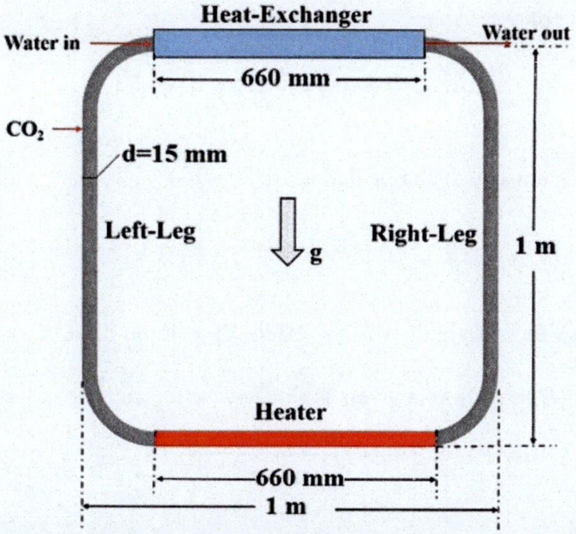

Fig. 1 Schematic of NCL analysed in the present study

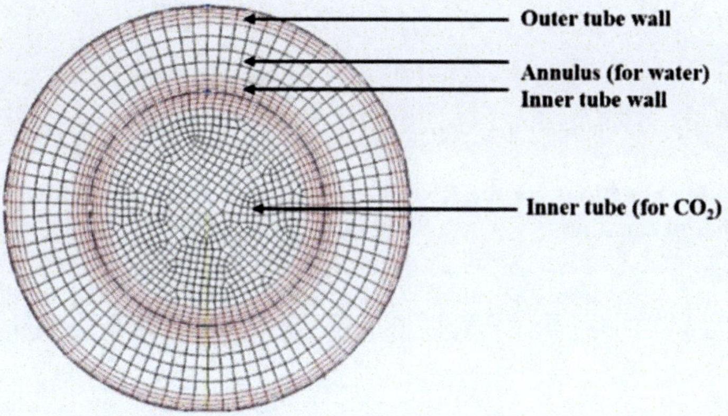

Fig. 2 The meshing of the cross-section at the heat-exchanger

3 Mathematical Formulation

The commercial software ANSYS (FLUENT) V-19.0 is employed to solve the fundamental conservation equations (mass, momentum and energy) with associated boundary specifications.

The mass conservation equation can be given as:

$$\frac{\partial \rho}{\partial t} + \nabla.(\rho \mathrm{V}) = 0 \tag{1}$$

Momentum conservation:

$$\frac{\partial(\rho \mathrm{V})}{\partial t} + \nabla.(\rho \mathrm{VV}) = -\nabla p + \nabla.(\overline{\overline{\tau}}) + \rho g \tag{2}$$

where the stress tensor (τ) can be defined as

$$\overline{\overline{\tau}} = \mu\left[\left(\nabla \mathrm{V} + \nabla \mathrm{V}^{\mathrm{T}}\right) - \frac{2}{3}\nabla.\mathrm{VI}\right] \tag{3}$$

Second term in stress tensor is the effect of volume dilation, and I is the unit tensor.

The energy (E) equation is given as

$$\frac{\partial(\rho E)}{\partial t} + \nabla.\left[\mathrm{V}(\rho \mathrm{E} + \mathrm{p})\right] = \nabla.\left(\lambda_{eff}\nabla T + \overline{\overline{\tau}}.\mathrm{V}\right) \tag{4}$$

where E can be written as

$$E = \int_{T_{eff}}^{T} C_p dT + \frac{\mathrm{V}^2}{2} \tag{5}$$

A general renormalization group (RNG) k-ε model is chosen to know the turbulent effect [12, 13, 19, 20].

Governing equations for the RNG k-ε model include two equations. Turbulent kinetic energy equation:

$$\frac{\partial}{\partial t}\rho k + \frac{\partial}{\partial x}\left(\rho u k - \frac{\mu_T}{\sigma_\kappa}\frac{\partial \kappa}{\partial x}\right) + \frac{\partial}{\partial y}\left(\rho v \kappa - \frac{\mu_T}{\sigma_\kappa}\frac{\partial \kappa}{\partial y}\right) + \frac{\partial}{\partial z}\left(\rho w \kappa - \frac{\mu_T}{\sigma_\kappa}\frac{\partial \kappa}{\partial z}\right) = G - \rho\varepsilon \tag{6}$$

where

$$G = \mu_T\begin{pmatrix} 2\left[\left(\frac{\partial u}{\partial x}\right)^2 + \left(\frac{\partial v}{\partial y}\right)^2 + \left(\frac{\partial w}{\partial z}\right)^2\right] + \left(\frac{\partial u}{\partial y} + \frac{\partial v}{\partial x}\right)^2 \\ + \left(\frac{\partial u}{\partial z} + \frac{\partial w}{\partial x}\right)^2 + \left(\frac{\partial v}{\partial z} + \frac{\partial w}{\partial y}\right)^2 \end{pmatrix} \tag{7}$$

Turbulent kinetic energy dissipation equation:

$$\frac{\partial}{\partial t}\rho\varepsilon + \frac{\partial}{\partial x}\left(\rho u\varepsilon - \frac{\mu_T}{\sigma_\varepsilon}\frac{\partial}{\partial x}\varepsilon\right) + \frac{\partial}{\partial y}\left(\rho v\varepsilon - \frac{\mu_T}{\sigma_\varepsilon}\frac{\partial}{\partial y}\varepsilon\right) + \frac{\partial}{\partial z}\left(\rho w\varepsilon - \frac{\mu_T}{\sigma_\varepsilon}\frac{\partial}{\partial z}\varepsilon\right) = c_1\frac{\varepsilon}{\kappa}G - c_2\rho\frac{\varepsilon^2}{\kappa} - R \quad (8)$$

where

$$R = \frac{c_\mu \eta^3 \rho\left(1 - \frac{\eta}{\eta_0}\right)\varepsilon^2}{k\left(1 + \beta\eta^3\right)} \quad (9)$$

$$\eta = \frac{Sk}{\varepsilon},\ S = \frac{1}{\sqrt{2}}\left(\frac{\partial u}{\partial y} + \frac{\partial v}{\partial x} + \frac{\partial v}{\partial w} + \frac{\partial w}{\partial y} + \frac{\partial u}{\partial z} + \frac{\partial w}{\partial x}\right) \quad (10)$$

where the values of constants are as follows:

$$\eta_0 = 4.8,\ \beta = 0.012,\ c_\mu = 0.0845,\ \sigma_k = \sigma_\varepsilon = 0.7178,\ c_1 = 1.42,\ c_2 = 1.68 \quad (11)$$

$$\mu_T = c_\mu \rho \frac{\kappa^2}{\varepsilon} \quad (12)$$

The following terms are defined to describe the fluid flow and heat transfer phenomena.

Mass flow rate at any cross section is defined as,

$$\dot{m} = \int_0^A \rho_r \overline{V_r}.dA' \quad (13)$$

The local flow velocity can be given as:

$$u_x = \frac{\int_0^A u_r\left|\rho_r \overline{V}.dA\right|}{\int_0^A \left|\rho_r \overline{V}.dA\right|} \quad (14)$$

Modified Grashof number [21] is defined as follows:

$$Gr_m = \frac{g\beta d^3 \rho^2 Q H_0}{A\mu^3 C_p} \quad (15)$$

4 Simulation Detail

In this study, three-dimensional geometry is prepared and meshed in Gambit 2.4.6. Transient simulations are performed using ANSYS (FLUENT) version 19.0. External walls are considered adiabatic, the wall thickness is kept 3 mm, and the material used is stainless steel. The temperature range is taken in such a way that loop fluid (CO_2) and heat exchanger fluid (water) both will be in a single phase. The coolant supply to the heat exchanger is 0.038 kg/s. The pressure implicit with the splitting of operators (PISO) algorithm is used to resolve the coupling between velocity and pressure. The momentum and energy parameters are discretized by a second-order upwind scheme, which calculates the value for a control volume using upstream values and gradients. Turbulence parameters renormalization group (RNG) k-ε are iterated with the second-order upwind scheme. To discretize the pressure term, PRESTO (Pressure staggering option) is used. The axial conduction and viscous dissipation of the flow are considered. Convergence is obtained when residuals of the all-governing equations were less than 10^{-3} for governing equation, except the energy equation which was less than 10^{-6}. Properties of CO_2 and water taken from the NIST REFPROP software [22] and implemented in Fluent database by considering piecewise-linear interpolation at 2 K temperature difference.

5 Validation

Correlations developed by (i) Vijayan [4] based on the experimental study with water-based NCL, and (ii) Yadav et al. [20] numerical correlation developed using CO_2 as loop fluid, are used to do the validation. These correlations involve non-dimensional parameters like Reynolds number (Re), modified Grashof number (Gr_{m}). Figure 3 shows a good agreement between the simulation results and existing correlations with a maximum discrepancy of less than 3% with the correlations.

Correlation suggested by Vijayan [4]

$$Re = 1.96(Grmd/Lt)1/2.75$$

Correlation suggested by Yadav et al. [20]

$$Re = 2.066(Grmd/Lt)1/2.77$$

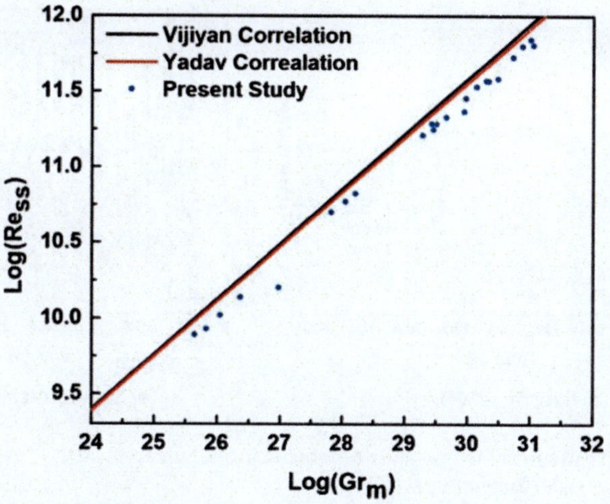

Fig. 3 Validation of the obtained result with correlations

6 Results and Discussion

Three-dimensional CFD steady and transient studies are carried out to obtain the fluid flow behaviour and performance of subcritical/supercritical CO_2 over a range of pressures (30–100 bar) and heat inputs (500–1500 W). The flow rate of coolant in the heat exchanger is 0.038 kg/s with an inlet temperature of 305 K. Operating pressure of the system is defined at the centre of the heater.

6.1 Transient Variation of Temperature

The temperature variations are shown in Figs. 4a, b against the flow time for subcritical and supercritical CO_2 for 1000 W heat input. In both cases, the temperatures oscillate vigorously after an initial overshoot and later on, it oscillates or increases with a consistent rate till neutral stability or steady state establishes in the loop. The magnitude of temperature for subcritical pressures (i.e. 30, 50 and 70 bar) are found to be more than the supercritical pressure (80, 90 and 100 bar) due to the low value of specific heat capacity (C_p) of subcritical carbon dioxide.

It is found from Fig. 4 that the rate of temperature fluctuations in supercritical is more than subcritical. Hence, supercritical CO_2 are more susceptible to instability than subcritical. The flow pattern depends on the thermo-physical properties of supercritical CO_2, mostly on its pseudo-critical point at which fluid encounters transition of specific heat capacity, volumetric expansivity, density etc. The pseudo-critical temperature of 100 bar is $\sim$ 318 K and for 90 bar is $\sim$ 313 K; at this temperature, specific heat capacity gains a high peak whereas viscosity, thermal conductivity and

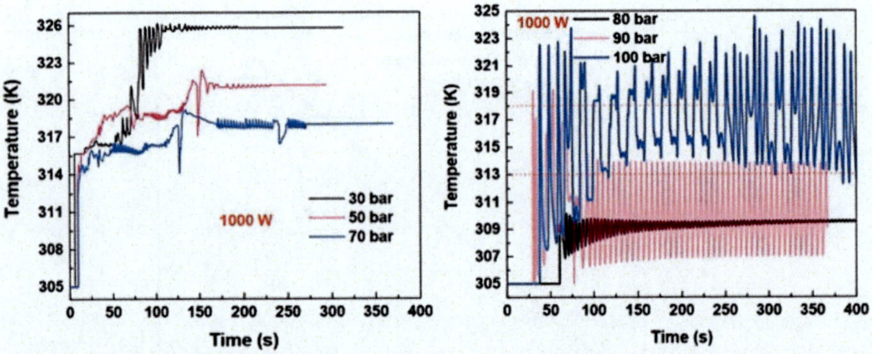

(a) Subcritical CO_2 (b) Supercritical CO_2

Fig. 4 Transient variation of temperature of subcritical and supercritical CO_2 at 1000 W heat input (**a**) Subcritical CO_2, (**b**) Supercritical CO_2

density decrease significantly. It is evident from Fig. 4b that the temperature fluctuation for the 90 and 100 bar is oscillating from its pseudo-critical value. Therefore, these operating pressure NCL systems are unstable. After the initial fluctuation, the temperature of the loop for 90 bar varies ± 3 K from a mean value of 311 K. While, for 100 bar, the temperature of the loop varies ± 4 K from its critical temperature of 318 K. Therefore, the rate of temperature oscillation is comparatively high in case of 100 bar than 90 bar.

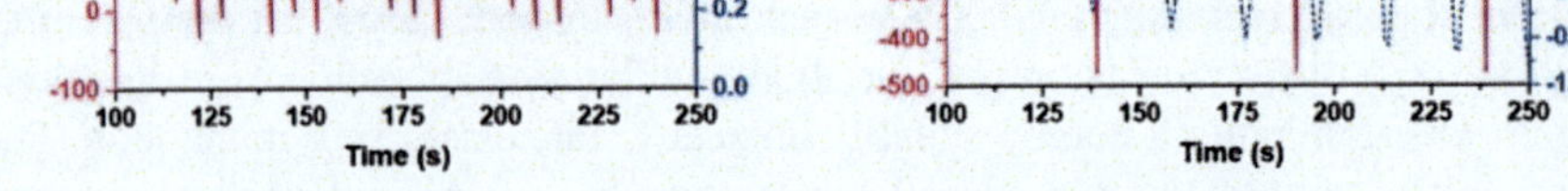

(a) 90 bar (b) 100 bar

Fig. 5 Velocity and pressure variation oscillations for heat input of 1000 W

6.2 Fluid Flow Behaviour

Transient results for the velocity and the pressure difference ($p - p_{op}$) vs time are reported in Fig. 5 for 90 bar and 100 bar operating pressure. Both the operating pressure and temperature show an unstable behaviour in their flow pattern (as shown in Fig. 4b). To immaculate the whimsical fluctuations, pressure difference and velocity graphs over time duration from 100 to 250 s drawn for both pressures. The operating pressure (p_{op}) of the loop is first kept at 90 and then at 100 bar, and hence, the values (as shown in Fig. 5) are the pressure variations considering 90 bar first (as shown Fig. 5a) and then for 100 bar (as shown Fig. 5a) as the mean value (zero at the initial condition).

The negative velocity value shows clockwise flow direction and vice-versa. Fluid fluctuates in both cases; the flow pattern changes with the pressure and attains a flow fluctuation/reversal phenomenon. This type of periodic flow variation is similar to the previous investigations [12], mainly for supercritical CO_2 having a smaller friction effect and more significant density differences. The flow pattern depends on the thermophysical properties of supercritical CO_2, mostly on its pseudo-critical point at which fluid encounters transition of specific heat capacity, volumetric expansivity, density etc. A rapid change in density generates a pressure gradient along the vertical pipe, and this pressure fluctuation induces potent natural convection flow. For both pressures (i.e. 90 and 100 bar), the temperature fluctuates near its pseudo-critical temperature (as shown in Fig. 5b.) Due to a sudden decrease in thermal conductivity at a pseudo-critical point, the fluid does not get sufficiently cooled, resulting in a frequent change in the flow direction. This may be the possible cause for the flow reversal/fluctuation associated with these operating pressures. Therefore, both pressures have whimsical flow fluctuations.

6.3 Nusselt Number

Figure 6 depicts the variation of the Nusselt number for subcritical and supercritical CO_2 with different heat inputs. A high Nusselt number is obtained for supercritical CO_2 than subcritical CO_2 and water due to the combined effect of *Re* and *Pr*. Large Nusselt numbers indicate a higher heat transfer rate and hence a more effective natural convection phenomenon.

7 Conclusions

Three-dimensional computational fluid dynamics studies are conducted to compare subcritical/supercritical CO_2 and water-based NCL. Results are obtained for different

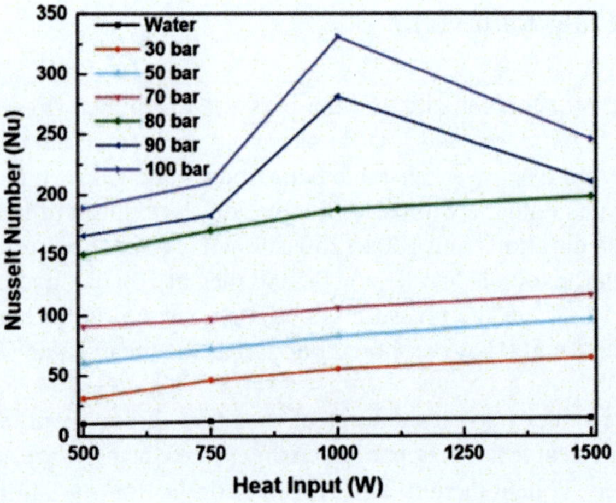

Fig. 6 Variation of Nusselt number of subcritical/supercritical CO_2 and water at different heat inputs

heat inputs and operating pressures. The following conclusions are made from the simulations:

1. Instability in a supercritical loop is maximum, whereas in a water-based loop, it is minimum.
2. Supercritical CO_2 flow exhibits repeated fluctuation or bidirectional reversal phenomenon, and the amplitude of fluctuations tends to diverge with time. Maximum fluctuations are obtained when the average fluid temperature is at pseudo-critical temperature. The flow changes from bidirectional to stable single-direction flow if the temperature is higher or lower to pseudo-critical temperature.
3. The energy transport is more stable in subcritical CO_2 compared to supercritical CO_2-based NCL.
4. For the heat input considered in the present study, the Nusselt number for supercritical is found more than subcritical CO_2 and water.

References

1. Keller, J.B.: Periodic oscillations in a model of thermal convection, J. Fluid Mech. **26**, 599–600 (1966)
2. Welander, P.: On the oscillatory instability of a differentially heated fluid loop, J. Fluid Mech. **29**, 17–30 (1967)
3. Chen, K.: On the oscillatory instability of closed-loop thermosyphons, J. Heat Transf. **107**, 826–832 (1985)
4. Vijayan, P.K.: Experimental observations on the general trends of the steady state and stability behaviour of single-phase natural circulation loops, Nucl. Eng. Des. **215,** 139–152 (2002)

5. Nayak, A.K., Vijayan, P.K., Saha, D., Venkat, R.V.: Mathematical modelling of the stability characteristics of a natural circulation loop. Mathl. Comput. Modelling. **22**, 77–87 (1995)
6. BWR linear stability analysis: Peng S.J., Podowsk M.Z.i, Lahey Jr R.T. Nucl. Eng. Des. **93**, 25–37 (1986)
7. Xiao, M., Chen, X.J., Zhang, M.Y., Veziroglu, T.N., Kakac S.: A multivariable linear investigation of two-phase flow instabilities in parallel boiling channels under high pressure, Int. J. Multiphase Flow. **19**, 65–77 (1993)
8. Chatoorgoon, V.: SPORTS-A simple non-linear thermal hydraulic stability code, Nucl. Eng. Des. **93**, 51–67 (1986)
9. Rohatgi, U.S., Neymotin, L.Y., Wul, W.: Assessment of RAMONA-3B methodology with oscillatory flow tests, Nucl. Eng. Des. **143,** 69–82 (1993)
10. Sarkar, M.K.S., Basu, D.N.: Numerical comparison of thermal hydraulic aspects of supercritical carbon dioxide and subcritical water-based natural circulation loop. Nucl. Eng. Technol. **49**, 103–112 (2017)
11. Thippeswamy, L.R., Yadav, A.K.: Heat transfer enhancement using CO_2 in a natural circulation loop, Sci. Rep. 1–10 (2020)
12. Chen, L., Zhang, X.R., Yamaguchi, H., Liu, Z.S.: Effect of heat transfer on the instabilities and transitions of supercritical CO_2 flow in a natural circulation loop. Int. J. Heat Mass Transfer. **53**, 4101–4111 (2010)
13. Yadav, A.K., Ramgopal, M., Bhattacharyya, S.: Transient analysis of subcritical/supercritical carbon dioxide based natural circulation loops with end heat exchangers: numerical studies. Int. J. Heat Mass Transf. **79**, 24–33 (2014)
14. Thimmaiah, S., Wahidi, T., Yadav, A.K., Mahalingam, A.: Comparative computational appraisal of supercritical CO_2 based natural circulation loop: effect of heat-exchanger and isothermal wall, J. Therm. Anal. Calorim. 1–11 (2020)
15. Misale, M., Frogheri, M.: Stabilization of a single-phase natural circulation loop by pressure drops, Ther. Scie. Proc. Begel House Inc, **150**, 121–126 (2000)
16. Wahidi, T., Yadav, A.K.: Instability Mitigation by Integrating Twin Tesla Type Valves in Supercritical Carbon dioxide based Natural Circulation Loop, Appl. Therm. Eng. 116087 (2020)
17. Wahidi, T., Chandavar, R.A., Yadav, A.K.: Stability enhancement of supercritical CO_2 based natural circulation loop using a modified Tesla valve, The J. Supercrit. Fluids. **166**, 105020 (2020)
18. Yadav, A.K., Ramgopal, M., Bhattacharyya, S.: Effect of Tilt Angle on Subcritical/Supercritical Carbon dioxide based Natural Circulation Loop with Isothermal Source and Sink. J. Therm. Sci. Eng. Appl. **8**, 011007–011008 (2015)
19. Chen, L., Zhang, X.R., Jiang, B.: Effects of heater orientations on the natural circulation and heat transfer in a supercritical CO_2 rectangular loop, J. Heat Transfer. **136**, 052501 (2014)
20. Yadav, A.K., Ramgopal, M., Bhattacharyya, S.: CO_2 based natural circulation loops: new correlations for friction and heat transfer. Int. J. Heat Mass Transfer. **55**, 4621–4630 (2012)
21. Vijayan, P.K., Austregesilo, H., Teschendro, V.: Simulation of the unstable oscillatory behaviour of single-phase natural circulation with repetitive flow reversals in a rectangular loop using the computer code athlet, Nucl. Eng. Des. **155**(3), 623–641 (1995)
22. NIST, Standard Reference Database-REFPROP, Version 9.1, National Institute of Standards and Technology, Gaithersburg, MD, (2013).

An Inverse Design Method for Caudal Fin of a Biomimetic Propulsion System for AUVs Using Artificial Neural Networks

K. L. Vidhu Manohar and Ranjith Maniyeri

Abstract Biomimetic propulsion systems offer higher endurance and efficiency as compared to conventional rotating propellers. In recent advances, studies have been focusing on identifying caudal fin shapes which would give optimal performance. Many studies have done comparison among existing fish tail fins using experimental and numerical models. A simple forked rigid caudal fin with four geometrical parameters such as caudal peduncle depth, leading-edge angle, trailing edge angle and chord length is studied. The present study aims to develop an inverse design procedure for predicting the caudal fin shape for the required flow conditions and performance parameters for one-meter-long autonomous underwater vehicle (AUVs) in the range of 0.15 to 0.4 of Strouhal number. The AUV is to be propelled by a body caudal fin (BCF) propulsion system with a tail fin of length 100 mm. The design factors will be average thrust and input power per cycle at a particular tail-beat frequency, amplitude and swim velocity. Eight geometries are generated by varying caudal peduncle depth or leading-edge angle separately. The geometries are analyzed numerically in ANSYS Fluent for each flow condition. These datasets are then fed to a feed-forward backpropagation neural network designed in MATLAB. The neural networks are designed to interpolate single geometric parameter. The network designed to predict caudal peduncle is provided with 72 data sets, and it is able to predict the caudal peduncle value with 94% efficiency. The neural net which is trained with same number of data sets for leading-edge angle predicted to 97% accuracy.

Keywords Caudal fin · Strouhal number · Caudal peduncle

K. L. V. Manohar · R. Maniyeri (✉)
Biophysics Laboratory, Department of Mechanical Engineering, National Institute of Technology Karnataka (NITK), Surathkal, Mangalore, Karnataka 575025, India
e-mail: mranji1@nitk.edu.in

R. P. Bharti and K. M. Gangawane (eds.), *Recent Trends in Fluid Dynamics Research*, Lecture Notes in Mechanical Engineering,
https://doi.org/10.1007/978-981-16-6928-6_23

1 Introduction

Biomimetics is one of the emerging research areas of interest. The conventional manmade designs are evolved in few decades. But on the other hand, the biological systems have been evolving for millions of years. By this very fact, the biological systems are more efficient and adapted to their environment when compared to manmade systems. The science behind the actual biological systems is much complex in nature.

The aim of the study focuses on design and selection of caudal fin for body caudal fin (BCF) propulsion system for an observation-class autonomous underwater vehicle (AUVs) using the capability of artificial neural networks (ANN) in multivariable optimization. The BCF propulsion system for observation class is selected for better efficiency, speed and endurance (for further insight on different BCF swimming modes refer to Lindsey [1], Roper et al. [2], Fish [3] and Phillips et al. [4]). Biomimetic systems are employed in limited applications. The gap for utilization can be minimized by more comprehensive studies on the factors which lead to evolution of the system to suit its environment. A wide variety of caudal fin shapes exist in nature. The studies which focus on the effect of geometric parameters are limited. Also, it is important to mention that the flexibility of the biological systems is not modeled for simplification of analysis. Hence, a reverse engineering method to predict the geometry of the rigid caudal fin which suits the requirement is proposed.

Zhang et al. [5] studied three caudal fin shapes of tuna, dolphin and whale from which it was found that with increase in Strouhal number, thrust developed is found to be increasing and efficiency to be decreasing in all three cases. The experimental studies of Feilich and Lauder [6] on various caudal fins with a fish body involved varying stiffness. Blake [7] studied pectoral fins of labriform locomotion and concluded that area has significant effect on performance of fins. Also, Geder et al. [8] concluded from their study on aerial vehicles that increase in threefold of area has increase in the thrust developed by nine times.

Arun et al. [9] conducted an exhaustive study to identify the optimal shape of caudal fin. The performance of fins in heaving and pitching motion is found by CFD implementation in ANSYS. This CFD scheme is validated with the experimental results of an oscillating foil from Anderson et al. [10]. Rai and Madavan [11] had used artificial neural networks for the prediction of geometry of turbomachinery airfoils using a triad formed by the turbine stator and rotor from Dring et al. [12]. The study uses pressure distribution as the input parameter for the neural network which predicts the airfoil geometry from the trained set of airfoil geometries. The inverse design proposed in this paper is a combination of methods mentioned in Arun et al. [9] and Rai and Madavan [11] and applied to caudal fin shapes of BCF propulsion system of AUV.

2 Methodology

The methodology followed in the study has two major phases. First phase consists of developing geometries and conducting numerical study of those geometries using an established CFD code. Second phase is developing a suitable neural network to suit the geometric parameters and data at hand for training and prediction of the geometry. While the whole procedure gets completed, the procedure and data need to be revisited if the prediction of the ANN is not up to the desired accuracy.

2.1 Generation of Data for ANN

One cannot make physical models of fin as per requirement due to different factors involved and setting up of experiments, the authors will adhere to numerical methods. With a validated CFD code, one can generate performance data of oscillating caudal fins with much higher accuracy and without much burdens of experimental setups and physical models.

Geometry model. A reference fin is selected from Arun et al. [9] for developing the fin geometries. A caudal fin has four geometric parameters as shown in Fig. 1. The 8 fins are generated for analysis by varying caudal peduncle and forking angle separately. …For each set other parameters are kept constant.

Motion model of fin. The fins are assumed to execute heaving and pitching motion about an axis in the middle of the length of the fin. The motion is as shown in Fig. 2 as described by Arun et al. [9].

Heaving and pitching motions of the fin are expressed mathematically as follows;

For heaving

$$h(t) = A\sin(2\pi f t + \psi) \tag{1}$$

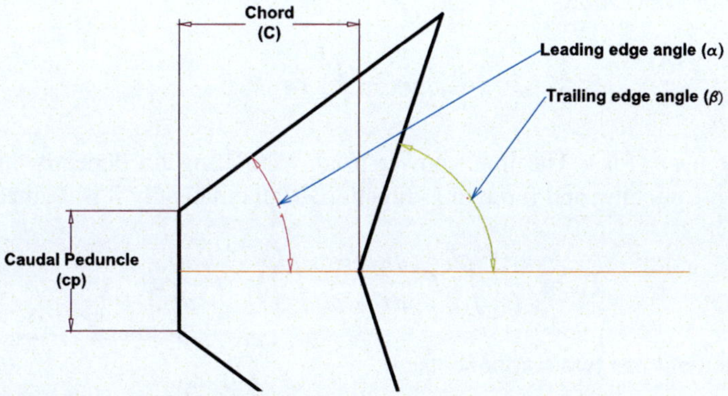

Fig. 1 Geometric parameters deciding the shape of a caudal fin

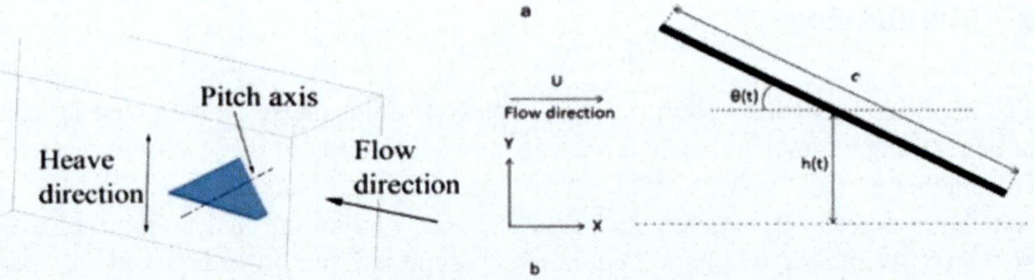

Fig. 2 Heave direction, pitch axis and flow direction of the fin (Arun et al. [9])

For pitching

$$\theta(t) = \theta_{max}\sin(2\pi f t) \tag{2}$$

Operating conditions. The flow is assumed to be laminar, incompressible in nature. Flow Reynolds number is in the vicinity of 20,000. Operating variables of an oscillating fin are given as follows,

Swim Velocity (U). The swim velocity is chosen as the velocity of incoming stream at the inlet.

Amplitude of tail flapping (A). It is the maximum movement of tail in a cycle. The value of A is selected as 0.1 m.

Frequency of tail beating (f). The frequency of tail beating is one of the important variables in deciding performance of fin. The values used for simulations are 1 Hz, 1.5 Hz and 2 Hz.

Strouhal number (St). Strouhal number is the important non-dimensional number defining performance of an oscillating fin propulsion systems. The range of Strouhal number tested is from 0.15 to 0.40.

Performance measures. Each fin is analyzed for following performance measures in each flow condition.

Average thrust force per cycle (T_{avg}). This value is equal to time-averaged force measured in x-direction.

$$T_{avg} = \frac{1}{T}\int_0^T F_x dt \tag{3}$$

Input power (Pin). The input power to an oscillating fin depends on moment provided for pitching and force in y-direction. Mathematically it is defined as,

$$P_{in} = \frac{1}{T}\left[\int_0^T F_y \frac{dh}{dt} dt + \int_0^T M_z(t)\frac{d\theta}{dt} dt\right] \tag{4}$$

The moment has two components,

$$M_z(t) = M_z^S(t) + F_x(t)[h(t) - A] \tag{5}$$

First term is the moment provided for pitching and measured by ANSYS Fluent monitor, and second term is the moment acting on the fin by the force in x-direction due to the orientation of the fin at various instances.

CFD implementation. CFD scheme is developed in ANSYS 16.2 [13] and uses first-order implicit formulation in time domain and second-order upwind scheme in space domain with SIMPLE algorithm for pressure–velocity coupling. Since it is a transient problem, one need to use moving mesh for the fin. For each Strouhal number and flapping frequency, the flow velocity is calculated and applied as the inlet velocity. For more details on the CFD scheme, refer to Arun et al. [9].

2.2 *Developing ANN Architecture for Inverse Design*

The inverse design method will be defining the geometry of the caudal fin which will give the desired performance in a given set of operating conditions. Three of the four parameters that decide the shape of fin are kept constant in our procedure, and hence, the neural net designed should predict the fourth parameter. Hence, the output from the neural network will be the fourth parameter. Since the ANN uses predetermined data for different fins, a feedforward backpropagation neural network is always suitable. This particular network utilizes user-controlled data and learning algorithm which ensures that the neural net will predict geometry parameters without ambiguity. The network is trained to do multi-dimensional interpolation for inverse design. Table 1 shows the variables selected as input for neural network. The architecture for a feedforward backpropagation network is designed which is shown in Fig. 3. The number of neurons and hidden layer is selected by trial and error for the data available and accuracy of prediction. The neural network is implemented in MATLAB.

3 Results and Discussions

Two separate designs are done for predicting the geometry of the fin with one parameter.

Table 1 Inputs selected for neural network

Type	Input
Operating conditions	Swim velocity (U)
	Frequency of tail beating (f)
Performance measures	Average thrust over a cycle (T_{avg})
	Input power (P_{in})

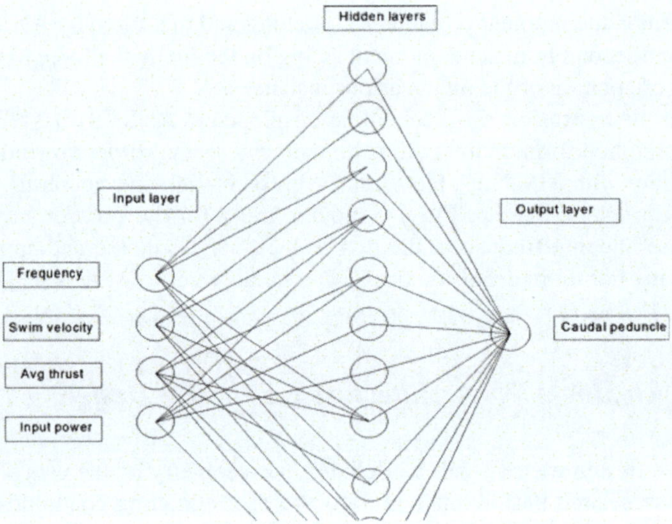

Fig. 3 Architecture for feedforward backpropagation network for prediction of caudal peduncle as geometric parameter

3.1 *Inverse Design for Caudal Peduncle*

Four fins are generated by varying caudal peduncle length from 20 to 80 mm. These are shown in Fig. 4.

Each fin is simulated for six Strouhal numbers (St), mainly 0.15, 0.2, 0.25, 0.3, 0.35 and 0.4. Three sets of frequency (f) values are used. For each value of St and f, inlet velocity is calculated and separate simulations are conducted. Thus, datasets for above geometries are of the order 72 (3 frequencies, 6 St numbers and 4 geometries). The simulated results are normalized with the help of maximum value in each range. Table 2 shows a sample of input and output data given to neural network for predicting.

The details of neural network architecture are given in Table 3.

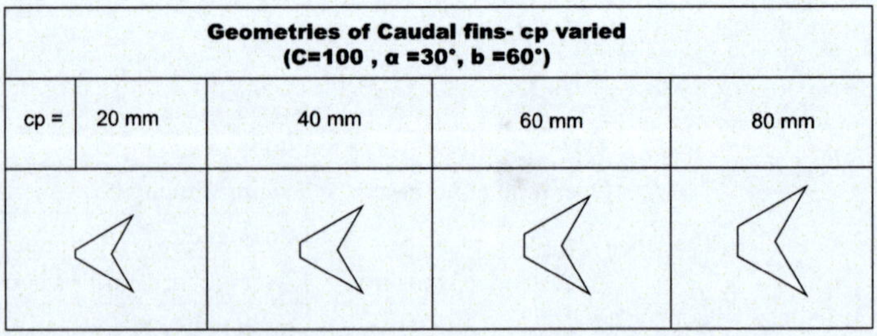

Fig. 4 Fins with varying caudal peduncle

Table 2 Data given to neural network for prediction of caudal peduncle

INPUTS				OUTPUT
Frequency (f) (Hz)	Swim Velocity (U) (m/s)	Avg. Thrust (T_{avg}) (N)	Input Power (P_{in}) (W)	Caudal Peduncle (cp) (mm)
1	0.5	0.084	0.1378	80
1.5	0.5	0.2155	0.4911	80
2	0.5	0.4103	1.1853	80

Table 3 Details of neural network architecture

Design parameter	Value
Number of inputs	(f, U, T_{avg}, P_{in})
Number of outputs	1 (cp)
Hidden layers	2
Number of neurons	20 (1st layer), 15 (2nd layer)
Training function	TRAINLM
Adaption learning function	LEARNGDM
Transfer function	TRANSIG (all layers)

The trained neural network has been provided with performance measures used for training of a fin with 40 mm caudal peduncle. The neural net predicted a value of 42.7 mm for caudal peduncle.

3.2 *Inverse Design for Leading-Edge Angle*

Four fins are generated by varying leading-edge angle from 10 to 40°. These are shown in Fig. 5

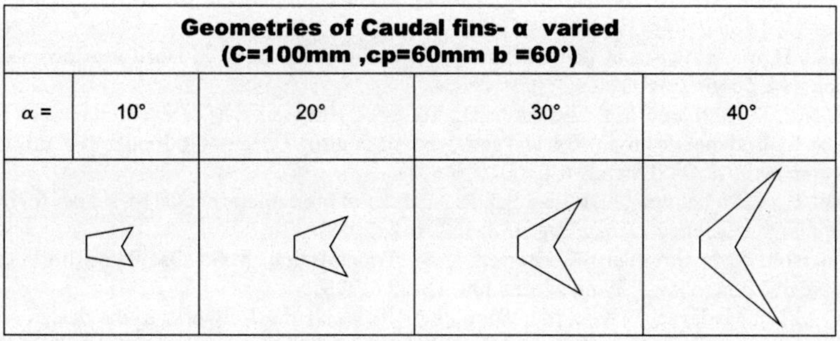

Fig. 5 Fins with varying leading-edge angle

Each fin is simulated with similar flow conditions, and data is trained with identical neural network architecture as of previous case. Operating conditions and performance measures of a fin with 20° leading-edge angle is provided to the trained neural network, which predicted fin with leading-edge angle of 19.5°.

4 Conclusion

The developed ANN model designed for predicting the caudal fin geometry with caudal fin varying is able to make prediction with 95% accuracy to the actual value and that of leading-edge angle predicted with 97% accuracy. The predictions of ANN are found to be in good interpolations in design space. One can definitely increase accuracy with more data points. We believe that with improved computing capacity, one can make the procedure easy and simple.

References

1. Lindsey C. C.: Form, function and locomotory habits in fish, Fish Physiology, Locomotion. Hoar, W.S., Randall, D.J. (eds.) vol. 7, 1st edn., pp. 1–100. New York. Academic. (1978)
2. Roper, D.T., Sharma, S., Sutton, R., Culverhouse, P.: A review of developments towards biologically inspired propulsion systems for autonomous underwater vehicles, Proceedings of the Institution of Mechanical Engineers, Part M: J. Eng. Maritime Environ. **225**, 77–96 (2011)
3. Fish, F.E.: Power output and propulsive efficiency of swimming bottlenose dolphins (Tursiopa truncatus). J. Exp. Biol. **185**, 179–193 (1993)
4. Phillips, A.B., Haroutunian, M., Murphy, A.J., Boyd, S.W., Blake, J.I.R., Griffiths, G.: Understanding the power requirements of autonomous underwater systems, Part I: an analytical model for optimum swimming speeds and cost of transport. Ocean Eng. **133**, 271–279 (2017)
5. Zhang, X., Su, Y.M., Wang, Z.L.: Numerical and experimental studies of influence of the caudal fin shape on the propulsion performance of a flapping caudal fin. J. Hydrodyn. **23**(3), 325–332 (2011)
6. Feilich, K. L., Lauder, G.V.: Passive mechanical models of fish caudal fins: effects of shape and stiffness on self-propulsion, Bioinspiration and Biomimetics, vol.10 (2015). doi.https://doi.org/10.1088/1748-3190/10/3/036002.
7. Blake, R.W.: Influence of pectoral fin shape on thrust and drag in labriform locomotion. J. Zool. **194**, 53–66 (1981)
8. Geder, J. D., Ramamurti, R., Edwards, D., Young, T., Pruessner, M.: Development of a robotic fin for hydrodynamic propulsion and aerodynamic control, Ocean—St. John's. 1–7. doi: https://doi.org/10.1109/OCEANS.2014.7003090.
9. Arun, K., Ravichandran, S., Rajagopal, P.: Analysis of biomimetic caudal fin shapes for optimal propulsive efficiency. Ocean Eng. **153**, 132–142 (2018)
10. Anderson, J.M., Streitlien, K., Barrett, D.S., Triantafyllou, M.S.: Oscillating foils of high propulsive efficiency. J. Fluid Mech. **360**, 41–72 (1998)
11. Rai, M.M., Madavan, N.K.: Application of Artificial Neural Networks to the design of turbo machinery airfoils. J. Propul. Power **17**, 176–183 (2001)
12. Dring, R.P., Joslyin, H.D., Hardin, L.W., Wagner, J.H.: Turbine rotor—stator Interaction. Journal of Engineering for power **104**, 729–742 (1982)

13. ANSYS Inc.: ANSYS Fluent User's Guide (2015).
14. MathWorks Inc.: MATLAB Documentation (2015).